ATLAS AND GLOSSARY OF PRIMARY SEDIMENTARY STRUCTURES

by

F. J. PETTIJOHN

Professor of Geology
The Johns Hopkins University

and

PAUL EDWIN POTTER

Associate Professor of Geology
Indiana University

Translations into Spanish, French, and German
by JUAN CARLOS RIGGI, MARIE-HÉLÈNE SACHET,
and HANS-ULRICH SCHMINCKE

With 117 Plates

SPRINGER-VERLAG NEW YORK, INC. 1964

Library of Congress Catalog Card Number 63-21507

Printed in Germany

Printed by Universitätsdruckerei H. Stürtz AG, Würzburg

Title No. 0769

Inadequate observation of sedimentary structures has been responsible for incorrect interpretation of the order of superposition in deformed beds and this has led, in turn, to gross errors in stratigraphy and structure. Failure to recognize and utilize those structures which indicate direction of current flow has also led to incorrect, or at least incomplete, understanding of basin development.

We believe, therefore, that there is need for a work which constitutes a *field guide* to the study of these structures — a book in which these structures, so difficult to describe or define in words, are adequately recorded in pictures. As in paleontology, where good illustrations are fundamental and without which identification of fossils is impossible, so also only from proper illustrations can the variety and variability of most sedimentary structures be comprehended.

The concept of an atlas of geological photographs is not new. There are, for example, the several atlases on microfacies, published by BRILL, of which "Fazies und Mikrofauna der Gesteine der Bayerischen Alpen" by H. HAGN was the first. These are, however, restricted to microphotographs and of interest therefore, primarily to sedimentary petrographers. The Soviets have also published several works consisting of pictures of geological subjects — outcrops, cores, and thin sections. Most of these have been restricted to particular strata from a particular region, such as "Atlas of Lithogenic Types of Coal-Bearing Deposits of the Middle Carboniferous of the Donetz Basin" by BOTVINKINA and others, or of photomicrographs similarly restricted such as KHVOROVA's "Atlas of Middle and Upper Carboniferous Carbonate Rocks of the Russian Platform" or rather specialized in content such as

TRUSHKOVA and KUKHARENKO's "Atlas of Placer Minerals." The most comprehensive atlas is the "Atlas of Textures and Structures of Sedimentary Rocks" edited by A. V. KHABAKOV (1962).

Our Atlas is an outgrowth of our work on "Paleocurrents and Basin Analysis," a book in which directional sedimentary structures are described and interpreted with special reference to the evolution of sedimentary basins. That work, however, contains minimal photographic material — just enough to give the reader some concept of the sedimentary structures described.

With two exceptions our atlas contains only photographs of subjects as they appear to the unaided eye. The pictures were selected from a large, mainly unpublished collection and, although most pictures are from North America, we have included a considerable number from other areas. There is no duplication in the photographic material in this book and our paleocurrent volume. Unless otherwise stated the pictures were taken by the authors—most of them on 35 mm film.

We have included a glossary. We found this necessary to explain many of the terms used in the plate captions. The glossary will guide the reader to papers in which the structures are further illustrated and where their origin and meaning is discussed. Moreover the glossary was enlarged to contain terms not illustrated in the atlas. The renewed interest in sedimentary "hieroglyphs" in the past ten years has led to the coining of new terms not found in any other glossary or dictionary. This proliferation of terms is perhaps deplorable, but it is a cogent reason for the assembling and publication of a glossary at this time.

Sedimentary structures are the same the world over and to make this book more generally useful the plate captions and text are also given in German, French and Spanish. This has been made possible only because of the interest and dedication of our geological colleagues, JUAN CARLOS RIGGI, of the University of Buenos Aires, who made the Spanish translation, and HANS-ULRICH SCHMINCKE, of The Johns Hopkins University, who provided the German version. The French translation was made by MARIE-HÉLÈNE SACHET of the Pacific Science Board of the National Research Council. We hope the multilingual presentation will prove of value also to geology students who know one language and are learning one or more of the others.

This book was prepared for students to whom the entire subject is new and whose field study of sedimentary structures is just beginning. For the study of sedimentary structures is primarily a field study—a microscope is not necessary. The student should learn that expensive and sophisticated instrumentation is not always needed to make significant observations and that field studies will yield important data. We think the Atlas will be of value also to teachers and to professional geologists. Many sedimentary structures are unfamiliar even to experienced geologists whose field activity has been largely in areas lacking some kinds of structures, such as the sole markings of turbidites, or who have worked in metamorphic terranes or with unconsolidated sediments in which the soles of beds are rarely, if ever, seen.

A book of this kind would be impossible or at least much less valuable were it not for the cooperation and help of many persons. We are indebted to The Johns Hopkins University and to the Guggenheim Foundation for their basic support of our enterprise. Our Atlas would indeed be far less complete if others had not generously supplied many pictures. We have received photographs from many individuals and organizations. These are acknowledged in the captions. Members of organizations who provided help in obtaining pictures from these organizations include KEITH BELL, WALTER F. FAHRIG, H. A. LEE, and B. R. PELLETIER of the Geological Survey of Canada; LLOYD G. HENBEST, PARKE D. SNAVELY, Jr., and JOHN H. STEWART of the U.S. Geological Survey; W. B. STEINRIEDE of the U.S. Corps of Engineers; C. V. CAMPBELL of the Jersey Production Research Company; J. E. LAMAR, WAYNE MEETS, J. A. SIMON and H. B. WILLMAN of the Illinois Geological Survey; C. J. STUBBLEFIELD of the Geological Survey of Great Britain; HANS-ERICH REINECK of the Institute of Marine Geology and Marine Biology of Wilhelmshaven and V. A. VANONI of the California Institute of Technology. We are also indebted to many persons for reading our glossary and for making helpful suggestions. Among these were MAHLON BALL of the Shell Development Company of Miami, Florida, W. K. HAMBLIN of the University of Georgia, E. F. McBRIDE of the University of Texas, J. C. RIGGI of the University of Buenos Aires, LOUIS RIEG and WAYNE PRYOR of the Gulf Research Corporation of Pittsburgh, RAYMOND SIEVER and ALAN JOPLING of Harvard University, and F. B. VAN HOUTEN of Princeton University. We are indebted also to BEVAN FRENCH for some help with the Russian literature. Special thanks are due H. E. CLIFTON and COLIN MCANENY, graduate students at Johns Hopkins University, for their substantial help in compiling the glossary. WILLIAM HILLER and CHARLES WEBER photographed some of our specimen material and also made most of the prints from our negatives. We wish to thank MARY GILL for her patience and skill in typing our manuscript. And lastly we are indebted to our publishers for seeing the book through the press and for their help and encouragement.

Baltimore, Maryland, January 1, 1963

F. J. PETTIJOHN
PAUL EDWIN POTTER

In Gebieten gefalteter Schichten hat ein unzureichendes Verständnis von Sedimentstrukturen oft zu falschen Schlüssen über die stratigraphische Abfolge geführt und somit schwerwiegende Folgen in Stratigraphie und Tektonik nach sich gezogen. Auch wurden Sedimentstrukturen, die Strömungsrichtungen anzeigen, oft nicht erkannt oder nur spärlich gebraucht: unrichtige oder zumindest mangelhafte Vorstellungen in der Entwicklung von Sedimentationsbecken waren die Folge.

Es scheint uns daher ein Bedürfnis zu bestehen nach einem Buch — sozusagen einem Geländeführer — in dem diese Strukturen, die so schwierig in Worten zu beschreiben und zu definieren sind, angemessen bildlich dargestellt werden. Gute Illustrationen sind entscheidend in der Paläontologie, und ohne sie können Fossilien nicht bestimmt werden. Aus dem gleichen Grunde kann man die außerordentliche Vielfalt der Sedimentstrukturen nur mit geeignetem Bildmaterial verstehen.

Die Idee, geologische Bilder in Atlasform zusammenzufassen, ist nicht neu. Es gibt zum Beispiel eine Reihe von Atlanten über Mikrofazies, die bei BRILL veröffentlicht werden, und unter ihnen war „Fazies und Mikrofauna der Gesteine der Bayerischen Alpen" von H. HAGN der erste. Diese Bände beschränken sich aber auf Mikrophotographien und interessieren deshalb in erster Linie den Sedimentpetrographen. Auch in der russischen Literatur sind Bildsammlungen geologischer Objekte (Aufschlüsse, Bohrkerne, Dünnschliffe) herausgekommen. Sie beschränken sich jedoch im allgemeinen auf bestimmte Schichten bestimmter Gegenden, wie zum Beispiel der „Atlas der lithogenetischen Typen der kohleführenden Schichten des Mittelkarbons des Donezbeckens" von BOTWINKINA et al., oder, ebenso beschränkt, auf Mikro-

photographien wie KHWOROWAS „Atlas der kohleführenden Schichten des Mittel- und Oberkarbons der russischen Plattform" oder auf hochspezialisierte Werke wie TRUSCHKOWAS und KUCHARENKOS „Atlas der seifenbildenden Minerale". Das umfassendste Werk dieser Art ist der von A. V. KHABAKOW (1962) herausgegebene „Atlas der Gefüge und Strukturen der Sedimentgesteine".

Der vorliegende Atlas ist eine Frucht unserer Arbeit über „Paleocurrents and Basin Analysis", ein Buch, in dem gerichtete Sedimentstrukturen beschrieben und mit besonderem Bezug auf die Entwicklung des Sedimentationsbeckens gedeutet werden. Jener Band enthält nur wenige Photos — gerade genug, um dem Leser eine gewisse Vorstellung über die beschriebenen Strukturen zu geben.

Mit zwei Ausnahmen zeigt dieser Atlas nur Photographien, wie sie dem bloßen Auge erscheinen. Sie wurden aus einer großen Anzahl im wesentlichen unveröffentlichter Bilder ausgewählt, und wenn auch die meisten aus Nordamerika stammen, so wurde doch eine ansehnliche Zahl von Bildern aus anderen Ländern mit aufgenommen. Sämtliche Aufnahmen dieses Bandes sind von denen in „Paleocurrents usw." verschieden. Soweit nicht anders vermerkt, stammen alle Photos, zumeist Kleinbildaufnahmen, von den Verfassern.

Wir haben ferner ein ausführliches Verzeichnis von Fachausdrücken beigegeben, um viele der in den Bildbeschreibungen gebrauchten Bezeichnungen zu erläutern. Dieses Verzeichnis verweist auf Veröffentlichungen, in denen die Strukturen ausführlicher illustriert und ihr Ursprung und ihre Bedeutung diskutiert werden. Darüber hinaus enthält das Verzeichnis Ausdrücke, die nicht abgebildet sind. Das in den vergangenen zehn Jahren aufgeblühte Interesse an sedimentären „Hieroglyphen" führte zur Bildung vieler neuer Begriffe, die man in keinem anderen Verzeichnis findet. Dieses Überwuchern von Namen ist vielleicht bedauerlich — aber ein um so zwingenderer Grund, jetzt ein Verzeichnis der Fachausdrücke zusammenzustellen und zu veröffentlichen.

Sedimentstrukturen sind in der ganzen Welt gleich. Um daher den allgemeinen Nutzen dieses Buches zu erhöhen, wurden Bilderklärungen und Text auch ins Deutsche, Französische und Spanische übersetzt. Dieses war nur möglich durch die Bereitwilligkeit und Mühe unserer Fachkollegen JUAN CARLOS RIGGI, University of Buenos Aires, der den Text ins Spanische übersetzte und HANS-ULRICH SCHMINCKE, The Johns Hopkins University, der die deutsche Fassung besorgte. Ins Französische wurde der Text von MARIE-HÉLÈNE SACHET vom „Pacific Science Board of the National Research Council" übertragen. Wir hoffen, die mehrsprachige Ausführung wird auch für solche Geologiestudenten von Wert sein, die eine Sprache kennen und die eine oder andere lernen.

Dieses Buch ist für Studenten gedacht, die sich noch nie mit Sedimentstrukturen beschäftigt haben oder erst anfangen, sie im Gelände zu beobachten; denn das Studium der Sedimentstrukturen ist in erster Linie Geländebeobachtung — ein Mikroskop ist nicht nötig. Der Student soll lernen, daß teuere, hochentwickelte Instrumente nicht immer nötig sind, um entscheidende Beobachtungen zu machen, und daß Geländearbeit noch immer wichtige Daten liefert. Wir glauben, der Atlas wird auch für den Lehrenden und Berufsgeologen von Wert sein. Viele Sedimentstrukturen sind auch dem erfahrenen Geologen unbekannt, der seine Geländearbeit weitgehend in solchen Gebieten durchführt, denen Strukturen der einen oder anderen Art fehlen, wie z.B. die Schichtflächen-Marken der Turbidite, oder der in Gebieten metamorphen oder unverfestigten Gesteins gearbeitet hat, in denen die Unterseiten der Schichten selten aufgeschlossen sind.

Ein Buch dieser Art wäre nicht möglich oder nur halb so wertvoll gewesen ohne die hilfreiche Zusammenarbeit mit vielen Personen. Wir sind der Johns Hopkins University und der Guggenheim-Stiftung für ihre grundlegende Unterstützung verpflichtet.

Der Atlas insbesondere wäre weit weniger vollständig, hätten wir nicht von vielen Seiten zahlreiches Bildmaterial erhalten. Personen und Organisationen, von denen die einzelnen Bilder stammen, sind in den Bildunterschriften angegeben. Unter Mitgliedern von Organisationen, die uns bei der Bildbeschaffung halfen, möchten wir besonders erwähnen: KEITH BELL, WALTER F. FAHRIG, H. A. LEE und B. R. PELLETIER, Geological Survey of Canada; LLOYD G. HENBEST, PARKE D. SNAVELY jr. und JOHN H. STEWART, U.S. Geological Survey; W. B. STEINRIEDE, U.S. Corps of Engineers; C. V. CAMPBELL, Jersey Research Company; J. E. DAMMAR, WAYNE MEETS, J. A. SIMON und H. B. WILLIAM, Illinois Geological Survey; C. J. STUBBLEFIELD, Geological Survey of Great Britain; HANS-ERICH REINECK, Forschungsanstalt für Meeresgeologie und Meeresbiologie „Senckenberg" in Wilhelmshaven, und V. A. VANONI, California Institute of Technology. Ferner sind wir vielen Personen verpflichtet, die das Verzeichnis der Fachausdrücke gelesen und hilfreiche Anregungen gegeben haben. Unter ihnen seien erwähnt: MAHLON BALL, Shell Development Company, Miami, Florida; W. K. HAMBLIN, University of Georgia; E. F. McBRIDE, University of Texas; J. C. RIGGI, University of Buenos Aires; LOUIS RIEG und WAYNE PRYOR, Gulf Research Corporation, Pittsburgh; RAYMOND SIEVER und ALAN JOPLING, Harvard University; und F. B. VAN HOUTEN, Princeton University. Wir danken BEVAN FRENCH für die Übersetzung russischer Arbeiten. Besonderen Dank schulden wir H. E. CLIFTON und COLIN McANENY, „graduate students" an der Johns Hopkins University, für ihre große Hilfe bei der Zusammenstellung der Fachausdrücke. WILLIAM HILLER und CHARLES A. WEBER photographierten einige unserer Handstücke und fertigten auch einen Großteil der Abzüge von unseren Negativen an. MARY GILL möchten wir für ihre Geduld und Fertigung bei der Reinschrift des Manuskriptes sehr danken. Und schließlich sind wir unseren Verlegern zu

Dank verpflichtet, den Text zum Druck gebracht zu haben und für ihre Hilfe und Ermunterung.

Baltimore, den 1. Januar 1963

F. J. Pettijohn
Paul Edwin Potter

Vorwort des Übersetzers

Primäre Sedimentstrukturen haben in jüngster Zeit die Aufmerksamkeit der Geologen vieler Länder auf sich gezogen. Einer Flut von Spezialarbeiten stand bisher keine Zusammenfassung gegenüber. Ich bin daher gerne der Anregung der Verfasser gefolgt, die vorliegende zusammenfassende Darstellung der Sedimentstrukturen ins Deutsche zu übersetzen. Die Übersetzung hat jedoch eine Reihe von Fragen aufgeworfen, deren Beantwortung nicht selbstverständlich ist. Es sei deshalb kurz auf einige Punkte eingegangen.

Aus im Vorwort (S. VI) dargelegten sprachpädagogischen Überlegungen, nämlich den Vergleich der einzelnen Sprachen zu fördern, hält sich die Übersetzung eng an die Originalfassung. In der gleichen Absicht und entgegen den Empfehlungen des Duden, jedoch im Anschluß an bei Richter (1937) und Wo. Schmidt (1959) ausführlich diskutierte Gründe, werden zusammengesetzte Wörter, wo immer es nötig scheint, durch Bindestriche voneinander getrennt.

Im Deutschen werden nach Möglichkeit die rein beschreibenden Bezeichnungen „Marke" für Unregelmäßigkeiten der Schichtfläche und „Struktur" für Inhomogenitäten der Schichtung gebraucht.

Die englischen Gesteinsnamen „silt" und „siltstone" sind mit „Silt" und „Siltstein" wiedergegeben (siehe G. Müller, 1961). Um Verwirrung zu vermeiden, bleiben die amerikanischen lithostratigraphischen Bezeichnungen „Member", „Formation" und „Group" („Series") unübersetzt.

Das „Glossary" (S. 283) will eine vollständige Übersicht der im Englischen bestehenden Namen für primäre Sedimentstrukturen geben (s. S. VI). Das Ziel der deutschen Übersetzung, indes, ist nicht ganz so weit gesteckt. Wer sich je in das Dickicht der Klassifikation und Terminologie von Sedimentstrukturen begeben hat, der wird verstehen, daß es nicht mehr als ein unzureichender Versuch sein kann, die schlecht definierten und zu definierenden Bezeichnungen der einen Sprache mit den ebenso vagen der anderen zu einer befriedigenden Übereinstimmung zu bringen (s. unten). Infolgedessen bringt die vorliegende Übersetzung eine durchaus unvollständige Bestandsaufnahme der in die deutsche Fachliteratur eingeführten Bezeichnungen primärer Sedimentstrukturen.

Da geologische Wörterbücher kaum Namen von Sedimentstrukturen führen — außer ganz alltäglichen —, wurde die wichtigste einschlägige Literatur der letzten fünfzig Jahre durchgesehen. In den zwanziger und besonders in den dreißiger Jahren schlugen sich in Deutschland zahlreiche Diskussionen um Sedimentstrukturen vor allem in den „Senckenbergiana"-Veröffentlichungen nieder, die deshalb eine ergiebige Quelle für die Terminologie darstellen.

Unter jüngeren wichtigen Arbeiten, die terminologische Fragen erörtern, seien besonders die von Illies (1949), Rabien (1956), Niehoff (1958), Seilacher (1959), Plessmann (1961) und Einsele (im Druck) genannt, die sich obendrein je durch ein ausführliches Schriftenverzeichnis auszeichnen. Im einzelnen ergeben sich für die Übersetzung der Fachausdrücke folgende Möglichkeiten:

a) Übersetzung mit im Deutschen eingebürgerten Ausdrücken

Sofern für englische Ausdrücke eindeutig sinn-entsprechende, in die deutsche Fachsprache eingeführte Bezeichnungen bekannt sind, werden diese als Übersetzung angegeben. Zweifellos der Idealfall, ist dies doch nur bei wenigen, häufigen Strukturen möglich.

Manchen Strukturen hat man viele synonyme Namen gegeben. Diese Synonyme werden teilweise in der Reihenfolge ihrer aus der Literatur bestimmten Häufigkeit angeführt. Auf der anderen Seite ist häufig ein und derselbe Begriff von verschiedenen Bearbeitern je andersartig definiert oder gebraucht worden. Solche Ausdrücke, wie

VII

etwa „Diagonalschichtung", finden sich daher im Verzeichnis an mehreren Stellen.

b) Keine Übersetzung

Der Vollständigkeit halber tauchen in der englischen Ausführung viele synonyme Bezeichnungen auf, die im Deutschen keine oder nur sehr unklare Äquivalente haben. Sie bleiben weitgehend unübersetzt.

c) Wörtliche Übersetzung

Für sehr viele englische Ausdrücke gibt es keine deutschen Entsprechungen; sei es, daß die zugehörige Struktur im deutschen Schrifttum nicht beschrieben, sei es, daß bestehende Namen bei Durchsicht der Literatur entgangen sind. In all diesen Fällen wird wörtlich übersetzt und dieses durch ein nachstehendes (w) angezeigt. Auch werden einige englische Synonyme derart behandelt.

d) Nichtwörtliche Übersetzung

Eine wörtliche Übersetzung kann im Deutschen umständlich oder bedeutungslos und deshalb für den Gebrauch ungeeignet sein. Es werden daher für die englischen Namen einiger weniger, aber häufig auftretender Strukturen einprägsame, nichtgenetische Bezeichnungen gewählt, die weder einen genau wörtlichen noch einen im deutschen Schrifttum eingeführten Ausdruck darstellen, und durch ein nachstehendes (n) gekennzeichnet. Auch unter der Gefahr einer unnötigen Beschwerung der ohnehin gewichtigen Last von Ausdrücken, schien dieses Verfahren in einigen Fällen doch die beste Lösung zu sein. „Ball-and-pillow structure", eine in gewissen Fazies häufige Struktur, wird daher mit „Ballenstruktur" übersetzt. Ebenso verfahren wurde in ganz wenigen Fällen, in denen zwar im Deutschen entsprechende Bezeichnungen vorliegen, diese jedoch mehrdeutig oder anderweitig belegt sind. „Groove cast" z.B. übersetzte man in den letzten Jahren im deutschen Schrifttum häufig mit „Schleifmarke". Es entspricht aber der Name „Schleifmarke" nur dem, was im Englischen (KUENEN, 1957) unter „drag mark" verstanden wird, und deshalb wird für das mehr umfassende „groove cast" der Name „Rinnenmarke" gewählt.

Prof. Dr. E. CLOOS danke ich sehr für Durchsicht eines Teils der Übersetzung und Dozent Dr. G. EINSELE für Einsicht in ein im Druck befindliches Manuskript. Zahlreiche Fragen zur Übersetzung wurden in Diskussionen mit Prof. Dr. F. J. PETTIJOHN geklärt, für die ich ihm sehr zu Dank verpflichtet bin.

HANS-ULRICH SCHMINCKE

VIII

Préface

Il arrive que les figures sédimentaires soient examinées de manière insuffisante, et il en résulte des erreurs dans l'interprétation de l'ordre de superposition des couches déformées, menant à leur tour à des erreurs grossières de stratigraphie et de structure.

Si l'on néglige de reconnaître et d'utiliser les structures qui indiquent les directions des courants, on arrive à des interprétations sinon incorrectes, du moins incomplètes du développement des bassins sédimentaires.

Nous avons donc été amenés à penser qu'un travail qui constituerait un guide sur le terrain pour l'étude de ces formes serait utile — un livre dans lequel ces formes, si difficiles à décrire ou à définir, seraient illustrées. Comme en paléontologie où de bonnes illustrations sont indispensables pour la détermination des fossiles, dans l'études des formes sédimentaires seules de bonnes illustrations peuvent démontrer la variété et la variabilité des formes.

L'idée d'un atlas géologique de photographies est loin d'être nouvelle. Il existe par exemple plusieurs atlas de microfaciès, édités chez BRILL, dont le premier fut l'ouvrage de H. HAGN, «Fazies und Mikrofauna der Gesteine der Bayerischen Alpen». Ces ouvrages ne contiennent que des microphotographies, et n'intéressent donc que les pétrographes. En Russie soviétique on a publié aussi plusieurs ouvrages consacrés à des photos de phénomènes géologiques: affleurements, carottes de sondage, et lames minces. La grande majorité de ces atlas sont limités à certaines couches d'une région donnée, comme par exemple «l'Atlas des types lithogéniques des dépôts houillers du Carbonifère moyen du Bassin du Donetz» de BOTVINKINA et al., ou à des microphotographies telles que celles de KHVOROVA dans son «Atlas des couches calcaires du Carbonifère moyen et supérieur de la plateforme russe». On a aussi des

ouvrages encore plus spécialisés, comme «l'Atlas des minéraux de placer» de TRUSKHOVA et KUKHARENKO. L'Atlas dont le sujet est le plus étendu est l'Atlas des Textures et Structures des Roches Sédimentaires, A. V. KHABAKOV, éditeur, 1962.

Notre Atlas représente une ramification de notre travail «Paleocurrents and Basin Analysis», dans lequel nous décrivons les structures sédimentaires orientées, et les interprétons dans le cadre de l'évolution des bassins sédimentaires. Mais ce livre ne contient qu'un minimum de photographies, juste assez pour donner au lecteur une idée des formes sédimentaires étudiées.

A deux exceptions près, l'Atlas ne contient que des photographies de sujets vus à l'oeil nu. On les a choisies parmi un très grand nombre de clichés, dont la plupart sont inédits; les exemples illustrés sont surtout pris en Amérique du Nord, mais il y en a beaucoup aussi d'autres régions. Aucune photo ne se trouve à la fois dans l'Atlas et le livre sur les paléocourants. Sauf indication contraire, tous les clichés ont été pris par les auteurs, la grande majorité sur film de 35 mm.

Nous avons joint à l'Atlas un glossaire, car il s'est révélé nécessaire de définir bien des termes utilisés dans les légendes des planches. Le glossaire aidera le lecteur à trouver des ouvrages qui offrent d'autres illustrations des formes en question et où leur origine ou leur signification sont discutées. Le glossaire s'étend de plus à des termes qui ne sont pas illustrés dans l'atlas. Le renouveau d'intérêt qui s'est manifesté pendant ces 10 dernières années à l'égard des «hiéroglyphes» sédimentaires a amené la création d'un grand nombre de termes nouveaux qui ne se rencontrent dans aucun autre glossaire ou dictionnaire. Ceci est peut-être regrettable, mais constitue une raison valable pour la préparation et la publication d'un glossaire à l'heure qu'il est.

Les formes sédimentaires sont les mêmes dans le monde entier, et pour rendre cet ouvrage plus utile, les légendes et le texte sont donnés en quatre langues, anglais, français, allemand et espagnol. Seul l'intérêt qu'ont pris nos collègues à cette entreprise, et leur dévouement, ont permis cette réalisation: nous voulons parler de JUAN CARLOS RIGGI, de l'Université de Buenos Aires, qui a préparé le texte espagnol, et de HANS-ULRICH SCHMINCKE de l'Université de Johns Hopkins à qui nous devons le texte allemand. La traduction française est de MARIE-HÉLÈNE SACHET, du Pacific Science Board, National Research Council. Nous espérons que cette présentation multilingue se révèlera utile pour les étudiants en géologie qui connaissent une de ces langues et étudient les autres.

Cet ouvrage a été préparé à l'intention des étudiants pour qui le sujet est entièrement nouveau, et qui débutent dans l'étude des formes sédimentaires sur le terrain. Car cette étude est fondamentalement une étude sur le terrain, il n'est pas besoin de microscope. L'étudiant apprendra que des appareils coûteux et compliqués ne sont pas indispensables pour faire des observations de valeur et que le travail sur le terrain reste une source fondamentale de données. Nous croyons que notre Atlas sera utile aussi aux professeurs et géologues professionnels. Bien des structures sédimentaires restent mal connues des géologues dont l'expérience est plutôt confinée à des régions où ces formes se rencontrent rarement, par exemple les accidents des faces inférieures des turbidites, ou des spécialistes de terrains métamorphiques ou de sédiments meubles dans lesquels on ne voit pratiquement jamais la face inférieure des couches.

Un ouvrage de ce genre serait impossible à réaliser, ou aurait une bien moindre valeur, sans l'aide et la coopération de nombreuses personnes. Nous devons beaucoup à l'Université de Johns Hopkins et à la Fondation Guggenheim sous les auspices de qui s'est réalisé ce projet. Notre atlas aurait été bien moins complet n'était le fait que beaucoup de photographies nous ont été communiquées par de nombreuses personnes et organisations. Les sources des clichés sont mentionnées dans les légendes. Les personnes dont les noms suivent nous ont procuré des photographies appartenant à leurs organisations: KEITH BELL, WALTER F. FAHRIG, H. A. LEE et B. R. PELLETIER, du Service Géologique du Canada; LLOYD G. HENBEST, Parke D. SNAVELY, Jr., et JOHN H. STEWART du U.S. Geological Survey; W. B. STEINRIEDE du U.S. Corps of Engineers, C. V. CAMPBELL, de la Jersey Production Research Company; J. E. LAMAR, WAYNE MEETES, J. A. SIMON et H. B. WILLMAN du Illinois Geological Survey; C. J. STUBBLEFIELD du Geological Survey de Grande-Bretagne; et enfin HANS-ERICH REINECK de l'Institut de Géologie et de Biologie Marines de Wilhelmshaven. Nous devons aussi beaucoup aux personnes qui ont lu notre glossaire et dont les commentaires nous ont été fort utiles. Tels sont MAHLON BALL de la Shell Development Company de Miami, Floride, W. K. HAMBLIN de l'Université de Georgia, E. F. McBRIDE de l'Université du Texas, J. C. RIGGI de l'Université de Buenos Aires, LOUIS RIEG et WAYNE PRYOR de la Gulf Research Corporation de Pittsburgh, RAYMOND SIEVER et ALAN JOPLING de l'Université de Harvard, et F. B. VON HOUTEN de l'Université de Princeton. Nous remercions particulièrement H. E. CLIFTON et COLIN McANENY, étudiants de l'Université de Johns Hopkins qui nous ont beaucoup aidés dans la préparation du glossaire, et WILLIAM HILLER et CHARLES WEBER qui ont pris certains clichés de nos échantillons et ont préparé presque tous les agrandissements de nos négatifs. Nous sommes reconnaissants aussi à MARY GILL de la patience et du soin avec lesquels elle a tapé le manuscrit. Enfin, notre gratitude va à nos éditeurs, qui nous ont prodigué leur aide et leurs encouragements, et ont assuré la belle présentation de notre ouvrage.

Baltimore, Maryland, 1er Janvier 1963

F. J. PETTIJOHN
PAUL EDWIN POTTER

X

Notes sur le texte français

La multiplicité de termes anglais que déplorent les auteurs du présent volume ne se retrouve heureusement pas dans le vocabulaire sédimentologique français. De nombreux géologues et géographes français, belges, et suisses ont étudié des formes sédimentaires comme celles qui sont illustrées dans cet atlas, en particulier dans les bassins houillers, le Sahara et les régions alpines du Flysch. Mais les auteurs de langue française tendent, soit à utiliser les termes anglais, soit à employer des tournures descriptives. Ils ont plus de scrupules à utiliser des noms qui impliquent la genèse supposée de l'accident considéré. Certains facilitent le travail de la traductrice en indiquant entre parenthèses l'expression anglaise qu'ils essaient de reproduire dans notre langue. D'une manière générale, il ressort de leurs travaux une tendance conservatrice, un désir d'éviter de baptiser à tort et à travers toutes les variantes possibles et imaginables, de créer un vocabulaire très étendu de termes qui ne seront peut-être jamais repris parce qu'ils s'appliquent à des formes rarement rencontrées, ou difficiles à reconnaître comme ayant déjà été nommées. Dans un domaine qui s'est récemment et très rapidement développé, cette attitude paraît raisonnable, elle évitera bien des confusions, des applications érronées, et des listes de synonymes pour l'avenir. Toutefois, elle complique la tâche de la traductrice, car la grande majorité des termes définis ici dans le glossaire, et utilisés dans les légendes des planches n'ont pas d'équivalents français. Laisser les termes anglais serait une solution de facilité, qui ne serait pas très juste à l'égard des lecteurs qui ignorent cette langue. Dans l'absence de termes équivalents français j'ai donc procédé comme suit: lorsque les terms anglais sont formés de mots courants, je les ai traduits littéralement. Voir par exemple Structure en boules-et-coussins. Ces traductions sont précédées de l'abréviation *Lit.* Dans certains cas, il me semble qu'elles pourraient bien être utilisées dans le vocabulaire sédimentologique français (et elles le sont dans les légendes des planches). Dans d'autres elles serviront seulement à indiquer au lecteur l'origine, le sens, de l'expression anglaise. Dans le cas où la définition anglaise, et sa traduction, sont très courtes, cette dernière doit pouvoir suffire, si l'expression anglaise est intraduisible littéralement. Voir par exemple la définition de «Choppy» cross-lamination.

Dans la traduction scientifique, on ne se sert peu ou pas de dictionnaires. Les termes scientifiques ne s'y trouvent pas et les mots courants sont bien connus. Parfois le traducteur a la chance de disposer de glossaires ou de vocabulaires spécialisés. C'est ainsi que j'ai utilisé les ouvrages préparés par BAULIG (1956) et SCHIEFER-DECKER (1959) ainsi que l'Essai de nomenclature des roches sédimentaires (Chambre syndicale ... 1961). Mais rien ne peut remplacer la connaissance du sujet et si le traducteur n'est pas spécialiste en la matière, il lui faut le devenir un peu. J'ai donc lu, d'une part un grand nombre des livres et articles en anglais dont sont tirés les termes définis et illustrés ici, et d'autre part tous les ouvrages en français que j'ai pu me procurer. Le plus important, de loin, est le monumental travail de LOMBARD (1956). Je m'en suis servie constamment, bien qu'il ne soit pas d'un maniement facile, car le plan en est compliqué et l'Index très incomplet; il contient certainement des termes utiles qui m'ont échappé. Les livres de GUILCHER, de CAROZZI, et des TERMIER, les articles de BERSIER, DANGEARD *et al.*, GRANGEON, KUENEN *et al.*, MACAR, et STROKHOV qui m'ont été le plus utiles sont inclus dans la bibliographie générale de l'Atlas. En outre j'ai consulté un grand nombre d'articles qui sont cités dans les ouvrages précédents, surtout dans Les Séries Marines de Lombard (travaux de CAROZZI, GULINCK, MACAR, PRUVOST, et bien d'autres), ou qui sont rassemblés dans les Comptes Rendus du Ve Congrès International de Sédimentologie (Eclogae Geologicae Helvetiae, vol. 51, no. 3, 1959). Il y a certainement des termes français disponibles que je n'ai pas utilisés, soit faute de lire les ouvrages nécessaires, soit parce qu'il est parfois très difficile de s'assurer que deux termes, français et anglais, s'appliquent bien à la même chose. On trouvera certainement des erreurs d'interprétation, et je m'en excuse d'avance, en espérant qu'elles ne seront pas trop nombreuses.

La traduction française n'est pas indépendante du texte anglais. C'est ainsi que dans les légendes des planches, on n'a pas répété les coordonnées des sites photographiés, les noms de lieu ni l'origine des photographies. De même, dans le glossaire, les rubriques qui ne sont que des renvois ne sont pas traduites. Que le lecteur veuille bien garder en mémoire que «see» veut dire Voir, et «see also» Voir aussi, et se référer aux termes anglais indiqués de la sorte, et en faire de même pour les renvois aux planches. On n'a pas non plus répété les références bibliographiques qui suivent chaque en-tête anglais. Un index des termes français permettra de trouver le terme anglais à chercher dans le glossaire pour obtenir une définition ou une explication des expressions françaises utilisées dans les légendes des planches ou ailleurs dans le glossaire.

MARIE-HÉLÈNE SACHET

Prefacio

Como resultado de observaciones incompletas efectuadas en el estudio de estructuras sedimentarias, se ha llegado a interpretaciones incorrectas en lo concerniente al orden de superposición de estratos deformados. Es obvio que ello ha incidido desfavorablemente en la apreciación de las relaciones estructurales y estratigráficas.

La omisión o descuido en el reconocimiento y utilización de estructuras que pueden indicar una dirección y sentido en el movimiento de las corrientes, ha conducido también a equívocos o al menos a interpretaciones parciales en la evolución de cuencas.

Por ello creemos necesario publicar la presente obra, como guía de campo para el estudio de estructuras sedimentarias. Además, así como es imprescindible el empleo de ilustraciones en paleontología, sin las cuales la identificación de fósiles sería prácticamente imposible, también el reconocimiento de las estructuras sedimentarias se ve facilitado mediante el auxilio de ilustraciones, cuya objetivización supera a las descripciones.

La idea de difundir un atlas de fotografías sobre temas geológicos no constituye una novedad. Existen, por ejemplo, diversos atlas de microfacies editados por BRILL, de los cuales el de HAGN, titulado "Fazies und Mikrofauna der Gesteine der Bayerischen Alpen", es el primero. Sin embargo estos trabajos están restringidos a microfotografías y por lo tanto de interés primordial para petrógrafos de rocas sedimentarias. Los investigadores soviéticos también han publicado diversas obras que incluyen fotografías de temas geológicos, sobre afloramientos, núcleos mineralizados y corte delgados. La mayoría se refieren a problemas particulares de una región, como el "Atlas of Lithogenic Types of Coal-Bearing Deposits of the Middle Carbonif-

erous of the Donetz Basin" de BOTVINKINA y otros, el "Atlas of Middle and Upper Carboniferous Carbonate Rocks of the Russian Platform", o aún el más especializado "Atlas of Placer Minerals" de TRUSHKOVA y KUKHARENKO. El atlas más completo es el "Atlas of Textures and Structures of Sedimentary Rocks" publicado por A. V. KHABOKOW (1962).

El presente atlas constituye en realidad una ampliación de nuestro trabajo titulado "Paleocurrents and Basin Analysis", en el cual las estructuras sedimentarias direccionales son descriptas e interpretadas con relación a la evolución de las cuencas sedimentarias. El mismo contiene un material fotográfico reducido, pero suficiente como para aclarar al lector las estructuras analizadas.

Las fotografías han sido seleccionadas de colecciones e informes inéditos y aunque la mayoría corresponden a Norteamérica, se incluyen un número considerable pertenecientes a otros países. Con dos excepciones este atlas presenta sólo fotografías de estructuras observables a simple vista y no figuran ilustraciones incluidas en nuestro volúmen sobre paleocorrientes. Salvo que no se especifique especialmente, ellas fueron obtenidas por los autores, la mayoría en película de 35 mm.

Hemos agregado un glosario que consideramos necesario para explicar los términos empleados en las descripciones de las fotografías; al mismo tiempo servirá para indicar las fuentes bibliográficas donde las estructuras son analizadas con mayor detalle. Debe advertirse que en este glosario figuran muchos términos no ilustrados.

El interés suscitado en los últimos diez años en los problemas vinculados a los *hieroglifos* sedimentarios, ha motivado la aparición de nuevas acepciones. Tan abundante terminología justifica la compilación y publicación de un glosario.

Las estructuras sedimentarias, por ser el resultado de procesos naturales, son comunes en todas la regiones del orbe. Por ello y con la finalidad de conferirle a esta obra un alcance de mayor utilidad, se la ha traducido al alemán, francés y castellano. Su realización ha sido posible merced al interés y dedicación de nuestros colegas, JUAN CARLOS RIGGI de la Dirección Nacional de Geología y Minería, y de la Universidad de Buenos Aires, quien tuvo a su cargo la traducción al castellano; HANS-ULRICH SCHMINCKE de la Universidad de Johns Hopkins, quien realizó la traducción al alemán; y MARIE-HÉLÈNE SACHET del Pacific Science Board of the National Research Council, al francés. Es nuestro deseo que el presente atlas, publicado conjuntamente en cuatro idiomas, sea de provecho para aquellos estudiantes que se inician en la investigación de estos problemas, como así también para profesores y geólogos.

Debe recalcarse que el estudio de las estructuras sedimentarias tiene lugar, primordialmente, en el campo; el uso del microscópio no es necesario. Los estudiantes deben tener en cuenta que para realizar observaciones de valor, no siempre es indispensable contar con instrumentos especiales y costosos.

Muchas estructuras sedimentarias son desconocidas aún por geólogos experimentados, cuya actividad de campo se ha desarrollado principalmente en áreas carentes de ciertos tipos de estructuras, tales como las marcas de base en las turbiditas; del mismo modo no resultan familiares para aquéllos que han trabajado en ambientes metamórficos o en regiones con sedimentos no consolidados, donde las marcas de base aparecen por excepción.

La realización de esta obra hubiera resultado imposible, de no contar con la cooperación de muchos colaboradores y patrocinadores de la "The Johns Hopkins University" y de la "Guggenheim Foundation", quienes apoyaron nuestra tarea. Además, hemos contado con la colaboración individual y de instituciones en la obtención del material fotográfico; la cita figura en las descrip-

ciones de las láminas. Entre los miembros de instituciones que facilitaron fotografías figuran: KEITH BELL, WALTER F. FAHRIG, H. A. LEE y B. R. PELLETIER del "Geological Survey of Canada"; LLOYD G. HENBEST, PARKE D. SNAVELY, Jr., y JOHN H. STEWART del "U.S. Geological Survey"; C. V. CAMPBELL de la "Jersey Production Research Company"; J. E. LAMAR, WAYNE MEETS, J. A. SIMON y H. B. WILLMAN del "Illinois Geological Survey"; C. J. STUBBLEFIELD del "Geological Survey of Great Britain"; y HANS-ERICH REINECK del "Institute of Marine Geology and Marine Biology of Wilhelmshaven". Dejamos constancia de nuestro reconocimiento a quienes contribuyeron con su crítica y sus sugerencias; entre ellos figuran: MAHLON BALL de la "Shell Development Company of Miami, Florida"; W. K. HAMBLIN de la "University of Georgia"; E. F. MCBRIDE de la "University of Texas"; LOUIS RIEG y WAYNE Pryor de la "Gulf Research Corporation of Pittsburgh"; RAYMOND SIEVER y ALAN JOPLING de la "Harvard University"; y F. B. VAN HOUTEN de la "Princeton University". Agradecemos especialmente a H. E. CLIFTON y COLIN MCANENY, estudiantes graduados de la "Johns Hopkins University", por la colaboración en la compilación del glosario; a WILLIAM HILLER y CHARLES WEBER quienes fotografiaron algunos ejemplares y realizaron las copias fotográficas; y a MARY GILL por la realización a máquina del manuscrito. Por último, agradecemos a los editores el interés y apoyo prestado en la realización del presente atlas.

Baltimore, Maryland, Enero 1, 1963

F. J. PETTIJOHN
PAUL EDWIN POTTER

Prólogo del traductor

La importancia y desarrollo que han alcanzado últimamente los estudios sobre estructuras sedimentarias en algunos países del Hemisferio Norte, determinó la formación de una rica terminología. Se comprende por lo tanto la necesidad de contar con términos equivalentes en castellano, para facilitar así a los estudiosos de esta disciplina las descripciones y por ende, la expresión de conceptos y su difusión en los países de habla hispana.

La realización en castellano del presente glosario no ha sido empresa sencilla. He tropezado frecuentemente con la imposibilidad de traducir vocablos literalmente. Como consecuencia se proponen nuevas acepciones; algunas son expresiones libres del lenguaje corriente, otras pocas han sido formadas con raíces griegas o latinas. Por otro lado, muchas son conocidas y empleadas en la literatura geológica de nuestro idioma; ellas han sido de utilidad en la formación de términos compuestos.

En casos frecuentes, por la multiplicidad de términos referidos a una estructura, unos descriptivos y otros genéticos, he creído conveniente seleccionar un vocablo general que los represente; la finalidad es obvia. Sin embargo se han mantenido aquéllos creados por VASSOEVICH, porque su ensayo de clasificación tiene carácter general y un patrón que responde a raíces griegas y en consecuencia de mayor comprensión.

Para aclarar la terminología propuesta, es conveniente hacer las siguientes consideraciones. Los vocablos en desuso correspondientes a otras lenguas, tienen sus equivalentes en castellano y están referidos a aquellos sinónimos actualmente utilizados. Con respecto a la expresión calco, su empleo en singular puede significar el calco de varias formas en conjunto o una forma en particular de ese mismo conjunto; el plural se utiliza cuando se desean indicar dos o más formas comprendidas en un calco, para diferenciarlas de otras marcas o para indicar ciertas características de un número determinado de formas.

Algunos términos han adquirido categoría internacional, como por ejemplo *flaser*, que tipifica cierta textura originada por procesos dínamo-metamórficos. En sedimentos, particularmente pelíticos, suele desarrollarse una estructura semejante, en sus formas, a la mencionada; por ello resulta apropiado el empleo de la expresión *estratificacion flaser*.

Para finalizar, deseo manifestar que la necesidad de enriquecer la nomenclatura sedimentológica de nuestra lengua, ha sido claramente comprendida por los doctores F. J. PETTIJOHN y P. E. POTTER. Si el objetivo ha sido logrado, a ellos deben y me es grato recordarlo, su agradecimiento los países de habla castellana.

JUAN CARLOS RIGGI

Table of Contents · Inhaltsverzeichnis · Table des Matières · Indice

Introduction

The study of sedimentary structures is as old as the study of geology itself. Bedding is the most characteristic structure displayed by sediments and, although there are exceptions, it is almost a sure criterion of a sedimentary origin. Like texture and composition, structure is an inherent property of a rock and a guide to its origin. Whereas texture deals with the grain to grain relations in a rock, structure has to do with discontinuities and major inhomogeneities. Structure is concerned with the organization of the deposit—the way in which it is put together. Hence structures are the larger features that, in general, are best studied in the outcrop rather than in the hand specimen or thin section. Structure is best defined by example. Included here are various aspects of bedding, such as cross-bedding and graded bedding, ripple marks and mud cracks.

We found it a little difficult at times to determine what should be considered a texture and what is more properly called a structure. Is the imbricate arrangement of pebbles in a conglomerate a texture or structure? It is perhaps an expression of gravel fabric and not strictly a structure, although we have included several illustrations of this feature. Are the grooves and striations found on some bedding planes actually structures? They can hardly be called textures.

Some structures, however, are texture dependent. Ripple marks and cross-bedding for example, characterize only those sediments which have a grain size in the sand range. These structures are, however, independent of composition as both appear in limestones, provided that these rocks are formed from carbonate sands. Ripple marks are even reported from rock salt (KAUFMAN and SLAWSON, 1950). Other structures, such as parting lineation (plate 76B) are also texture dependent. It is not yet clear what causes parting lineation but it is known to be closely correlated with the grain fabric of the rock.

Other examples of texture-dependent structures include mud cracks, a structure capable of formation only in cohesive muds, and the ball-and-pillow structure seen in some sandstones. This structure is independent of composition as it also appears in some mechanically-deposited fine lime sands and silts.

Are there structures peculiar to non-clastic sediments or structures which are composition-dependent? Some of the rhythmic bedding displayed by salt and anhydrite and some of the nodular bedding seen in some limestones has no obvious analogue in the clastic sediments. Likewise stromatolitic bedding seems restricted to limestones or dolomites. For the most part, however, the sedimentary structures depicted in this atlas are those produced by current action or by soft-sediment deformation and can occur in rocks of any composition provided that such materials were current transported and deposited and have the proper texture.

It is worthy of note that some of the structures which characterize granular sediments also appear in some of the layered basic igneous rocks. Graded bedding and cross-bedding are examples (GATES, 1961, figs. 15 and 18). A photographic record of igneous bodies with primary sedimentary structures is a worthy project which, however, the authors have not undertaken. Obviously air- and water-borne pyroclastic materials show most of the structures displayed by nonvolcanic sediments.

A structure cannot be defined in the same precise manner as can a geometrical figure or solid such as a cube or sphere. As is the case with fossils and organic forms in general, a picture is required to define the object. Our inability to analytically define these objects, however, does not mean that we cannot make meaningful measurements of them. Because sedimentary structures, like organisms, display a great variety of forms, a single picture is incapable of conveying this variety and variability. An atlas is required to do the job.

We do not attempt to discuss the origin of these structures nor to elaborate on their use in determining stratigraphic order nor their value in mapping paleocurrent systems. Their use to determine sequence of beds has been set forth by SHROCK (1948) and the significance of the directional structures in paleocurrent and basin analysis has been treated elsewhere by us (1963). An atlas of photographs of sedimentary structures to be useful however, needs to be organized in a logical manner. Such organization presupposes some kind of classification. We shall, therefore, after defining our subject further, discuss the classification and nomenclature which may be applied to sedimentary structures.

Definition and scope

We are concerned here with *primary sedimentary structures*. By "primary" we mean formed at the time of deposition or shortly thereafter and before consolidation of the rock in which they are found.

The primary structures dealt with include bedding or stratification, especially the external form of the bed and its continuity and uniformity of thickness, the internal structure and organization of the bed, the interfacial or bedding plane hieroglyphs or markings, both on top and on the bottom of the bed, and the structures produced by soft-sediment deformation. In short, this Atlas deals largely with bedding or some aspect of it.

We have excluded all secondary structures formed by tectonic movements. On the other hand, we do include those faults, folds and related structures produced by pre-consolidation movement due to unequal loading or to downslope slide or slump. We have omitted also any consideration of the

manner of breaking of the rocks such as fissility and jointing. Such structures, although controlled perhaps by primary features, presuppose a consolidated state. We exclude also all concretionary or accretionary structures such as nodules, concretions, septaria and the like even though it is claimed that some of these structures may be formed during sedimentation or before consolidation. After some hesitation we excluded those structures due to frost action or cryoturbation including involutions and ice wedges as well as those structures related to the formation of soils. Likewise we have passed over lightly any structures produced by organisms. Organic trails, tracks and burrows are given some consideration but a full treatment of ichnofossils, as these features are called, seems to us too specialized for most users of the Atlas. The interested reader is referred to the publications of ABEL (1935), SEILACHER (1953), and CASTER (1957). We have included, however, a few other structures of organic origin such as stromatolites inasmuch as they are a species of bedding—a type of growth bedding or at least bedding modified by organic action.

Classification of primary sedimentary structures

The authors do not know of any serious attempt to classify all sedimentary structures. Several partial classifications, however, have been proposed. Most important are those which deal primarily with bedding. The most comprehensive of these are those of ANDRÉE (1915), HOPPE (1930), ANDERSEN (1931), ZHENCHUZHNIKOV (1940), BRUNS (1954), BIRKENMAJER (1959) and BOTVINKINA (1960, 1962). These classifications deal not only with external form but also with internal structure and some of them include the sequential arrangements, in part cyclical, of differing lithologies. Other partial classifications of sedimentary structures include those dealing with hieroglyphs of KREJCI-GRAF (1932) and VASSOEVICH (1953). Recent classifications of sole marks, another group of structures, are those of

CROWELL (1955) and KUENEN (1957). Even more restricted in scope are the various classifications of cross-bedding and ripple marks.

It is not our intent to review these various classifications nor to attempt to frame a comprehensive system of our own. Instead we propose a simple scheme so that the atlas materials can be arranged in a logical order. We have, therefore, divided the sedimentary structures into four major groups:

I. Bedding, external form
II. Bedding, internal organization and structure
III. Bedding plane features
IV. Bedding, disturbed and deformed

A few structures fail to fall readily into any of these groups or could logically be put in more than one class. Worm borings, for example, could be considered (in some cases) as a bedding plane feature, but they also may be internal or may even be regarded as a kind of disturbance of the bedding. Mud balls, beach cusps, and sandstone dikes are examples of structures difficult to classify.

Bedding, external form:

Although bedding has been recognized and described for many years, there is as yet no wholly satisfactory or agreed upon classification of this structure. We present here only a minimal treatment of this problem, leading to a system that perhaps would not be practical in everyday use but which permits us to classify or arrange our photographs of bedded sequences.

The basic entity is a single bed or *sedimentation unit* which has been defined as "that thickness of sediment which was deposited under essentially constant physical conditions" (OTTO, 1938). The qualification "essentially", however, allows for short-term fluctuations in current direction or velocity which may produce internal laminations or for progressively changing conditions, without major interruptions, that result in a graded structure. The bedding planes which separate the various strata are made evident because of the unlike texture and composition of the several beds.

The basic properties of a bed or sedimentation unit are (1) its thickness or scale, (2) its lateral dimensions, and (3) its internal structure. Bedding sequences can be described independently of their internal organization or structure. We shall defer our discussion of the latter to a separate section.

We have to first consider the matter of thickness. Strata are commonly described as thick-bedded, thin-bedded or laminated. More precise definitions of these terms have been proposed. PAYNE (1942), for example, used the term *stratum* for beds greater than 1 cm in thickness which are clearly separated from over- and underlying like units. He restricted the term *lamination* to layers less than 1 cm in thickness. McKEE and WEIR (1953) likewise restrict lamination to layers under 1 cm in thickness. They define *very thin* as those beds 1—5 cm in thickness, *thin* as 5—60 cm in thickness, *thick* as 60—120 cm and apply the term *thick-bedded* to those over 120 cm. These are attempts to quantify our observations of thickness. Others who have written on this subject include INGRAM (1954), GRAY (1955) and BOKMAN (1956).

If many beds are present, then we need also to consider both the *mean* thickness and the variations in thickness expressed as the *standard deviation*, or better perhaps as the *coefficient of variation*. Some sequences have a remarkably low coefficient of variation (pls. 2 and 8B); others do not. Some writers have attempted to analyze the variations in thickness in terms of cycles, such as varves (pl. 1 A), whereas others have noted that the variations of thickness are in many cases log normal (SIMONEN and KOUVO, 1951, fig. 4 and ATKINSON, 1962, p. 357). Clearly the field geologist should be aware of possible repetitive or cyclical patterns (*cf.* pls. 4 and 9B) in bedding thickness. Such variations may be correlated with corresponding variations in lithology.

In addition to absolute thickness, a bed is characterized by its variability in thickness in a lateral direction. Some beds are remarkably uniform throughout their ex-

4

posed length. Others pinch and swell or display wavy bedding (*cf.* pls. 17 A and 11 B). The extreme case is represented by beds that pinch out completely. Such lateral discontinuities may be due to local unconformities, scour-and-fill or wash-outs (pls. 19 A to 21 A), or to a type of soft-sediment boudinage (pl. 15 A). Some lithologies, such as limestone, contain beds that, if traced laterally, break up into a series of nodules. In some cases the whole formation displays nodular bedding (pl. 15 B).

We can classify a sequence of beds into several groups which form a series from those exhibiting the most regular or ordered character to those which present a state of maximum irregularity or disorder. One can, for example, imagine a sequence in which the beds are all of equal or near equal thickness, which are uniform in thickness laterally and apparently of unlimited extent. Such a sequence would represent the highest degree of order (pl. 2). On the other hand one can picture a group of beds of unlike thicknesses, each of variable thickness, and each discontinuous within the limits of the outcrop (pl. 21 B). Such would constitute a disordered arrangement. Various intermediate states of disorder are also possible.

To illustrate these concepts we can define four classes:

Class A: 1) Beds equal or subequal in thickness
2) Beds laterally uniform in thickness
3) Beds continuous

Class B: 1) Beds *unequal* in thickness
2) Beds laterally uniform in thickness
3) Beds continuous

Class C: 1) Beds *unequal* in thickness
2) Beds laterally *variable* in thickness
3) Beds continuous

Class D: 1) Beds *unequal* in thickness
2) Beds of *variable* lateral thickness
3) Beds *discontinuous*

Examples of Class A are shown by plates 1 A, 2, and 8 B, of Class B by plates 1 B, 3, 4, 6 A, and 7, of Class C by plates 11, 12, and 13, and of Class D by plates 15 A, 17, and 21 B.

Bedding, internal organization and structure:

The internal organization of a bed is as important as its external form. Various patterns of internal arrangement are possible. Important types of internal structures are:

Class A: Massive

Class B: Laminated
1) Horizontal laminations
2) Cross-laminations (simple and multiple)

Class C: Graded

Class D: Imbricated (gravels) and other oriented internal fabrics.

Class E: Growth structures (such as exhibited by stromatolitic limestones, some reef-rock, travertine, etc.)

A bed may be *massive*, that is seemingly without internal structure (pls. 25 and 26). Recent work (HAMBLIN, 1962) suggests that the massive appearance is deceiving as many "massive" beds display laminations or other structures when *x*-rayed as shown by plate 23.

A bed may also be *laminated*. Such laminations record momentary and minor changes in physical conditions during the accumulation of the bed. Laminations may be strictly parallel to one another and parallel to the bounding planes of the bed itself (pls. 8 A and 20 B, lower part). Beds of sandstone displaying these characteristics commonly split with ease along the laminations and form excellent flagstones. The surfaces of separation commonly show parting lineation (pl. 76 A and B).

In granular materials of sand size, the laminations are commonly oblique to the bounding surfaces of the bed. Such diagonal laminations or *cross-bedding* is very common in some sand facies. These cross-bedded strata, both sandstones and limestones, are abundantly illustrated (pls. 29 B to 37). The cross-bedding may be on a very small scale, or it may be on a very large scale (cf. pls. 36 and 38 B). In many cases the whole bed is involved and the foresets extend from the base to the top of the bed (pl. 34 A); in other cases the bed may be composite, showing several smaller cross-laminated zones (pl. 37). The foresets may be straight in section, or curved and tangential to the base of the bed (pls. 32 A and 33 B) In some cases the foresets are oversteepened, even overturned or crumpled (pl. 110). In complex cases, the bed may be massive in its lower part, display a horizontally laminated central part and a cross-laminated top.

Another feature of some beds is grading. *Graded bedding* is in actuality a regular textural variation. It is an important feature, generally readily seen by the unaided eye, and hence we have included several photographs which illustrate this feature (pls. 1 B and 42 A and B).

Graded bedding may occur in gravel, sand or silt. It is most characteristic in the repetitive bedding of flysch sequences—in the so-called turbidite sands. The grading is evident even from a distance by the difference in sharpness of the lower and upper contacts of the sandstone beds (pl. 1 B). On close inspection the change in grain size, correlated with a change in color from light (sand) to dark (silt or clay), can be seen with the naked eye (pls. 8 B and 42 B). In metamorphic terranes the slaty cleavage, formed at an angle to the bedding, changes direction or is refracted in passing from the coarser, lighter-colored sand fraction to the finer darker slate (pl. 6 A).

Although graded bedding and cross-bedding are more or less mutually exclusive, there are beds which show both, generally showing grading in the lower part and very small scale cross-bedding near the top. The latter may be closely associated with convolute bedding.

Another property of a bed is its *fabric*. Like gradation in grain size, this is a textural rather than a structural attribute. Normally, microscopic examination is needed to discern the fabric pattern of sands and finer materials. The fabric of gravels, however, is easily seen by the field geologist. Normally the longer dimensions of the pebbles are parallel to the surface of deposition. In many fluvial gravels, however, the flatter pebbles or cobbles display an imbrication in which the pebbles overlap one another in shingle fashion. They dip upstream (pl. 43 A). This arrangement is most evident in vertical sections parallel to the stream flow. A few intraformational conglomerates show pebbles with an "edgewise" arrangement and these are termed edgewise conglomerates.

The various types bedding and internal structures thus far described occur in clastic sediments and are largely the product of sedimentation from moving fluids. We should also consider briefly the bedding of nonclastic deposits. Many limestones are current deposited and hence they too exhibit all the features described above including cross- and graded bedding. Other limestones, however, are formed *in situ* — in some cases deposited by organisms. Such rocks may display bedding quite unlike that produced by currents. Notable is *stromatolitic bedding*, a growth bedding, presumed to be related to the formation of mats of blue-green algae. Stromatolitic bedding varies widely but is generally characterized by laminations. Unlike current laminations, the growth laminations show irregularities, such as bifurcation, crinkling, arrangement in varied convex-upward patterns. It is not our task to display the manifold forms assumed by stromatolites. The interested reader is referred to the specialized literature in this field (see, *e.g.* CLOUD, 1942, and REZAK, 1957). It is probable that much "algal bedding", not organized into well-defined structures, is overlooked by the field geologist. We have shown several examples of stromatolitic bedding, both structured and nonstructured (pls. 48, 49 A and B, and 50 B).

Other carbonates may display an internal growth structure related to organisms, such as corals, stromatoporoids and the like, and in still other cases, the structures may be growth structures formed inorganically. In the latter category are the banded structures of some travertines and onyx.

Bedding plane marking and irregularities:

Many bedding planes, if examined closely, show a variety of structures. These may be divided into those which are found on the *bottom* or "sole" of the bed, on the *top* of the bed and on planes *within* the bed. Each of these classes consists of a variety of structures as outlined below:

Class A: On base of bed

    1) Load structures (load casts)
    2) Current structures (scour marks and tool marks)
    3) Organic markings (trail and burrow casts; "fucoids")

Class B: Within the bed (parting lineation)

Class C: On top of bed

    1) Ripple marks
    2) Erosional marks (rill marks, current crescents)
    3) Pits and small impressions (bubble and rain prints, etc.)
    4) Mud cracks, mud crack casts, ice-crystal casts, salt crystal casts, etc.
    5) Organic markings (tracks, trails, footprints, etc.)

Although sole markings—hieroglyphs of various kinds—have been known for many years, only within the last decade have they been studied intensively. As a result, many new structures have been described and to them has been applied a large and confusing nomenclature. This nomenclature is a mixed one embodying genetic as well as descriptive aspects.

Sole marks generally are found on sandstone or limestone beds overlying shales. They are described as "casts" inasmuch as they are fillings of depressions formed on the surface of the underlying mud. These structures originate by (1) unequal loading of the soft hydroplastic mud, (2) by the action of currents on the mud surface, or by (3) the activity of organisms on this surface.

The *load structures* are most commonly called "load casts", though they form as the result of downsinking of the overlying sand into the mud and not by the filling of a depression. Load casts are varied in size, shape, and abundance (pls. 52A to 54A). Some of these varieties have been given special names such as torose load casts (pl. 54B) and squamiform load casts. Normally these structures are shown as bulbous irregularities on the underside of the bed. In cross-section they appear as sinuosities, such as the so-called "flame structure" (pl. 53B).

The *current marks* belong to two groups, namely, those formed by the current itself and those produced by "tools", such as shell and wood fragments, shale chips and sand grains, swept along by the current. The first are called *scour marks*; the most important of these being flutes which appear as "flute casts" on the soles of many sandstone beds and also on the underside of some limestones. They vary widely in size, shape, and pattern of arrangement on the sole (pls. 55A to 59B). Strong current action may also produce small channels, wash-outs and the like (pls. 20A and 20B).

The *tool marks* are a greatly varied group of structures. Some are produced by movement of an engraving tool, in continuous contact with the bottom. These include striations and grooves which appear as striation and groove casts (pls. 61 to 67). These markings are analogous to those made by rock fragments embedded in moving glacial ice—the striations and grooves of glacial pavements (pls. 73A to 74). Other tool marks are the product of intermittent contact of various objects with the bottom. The corresponding structures on the sole of the overlying sandstone are variously termed skip casts (pl. 68), bounce or brush casts (pls. 68 and 69A), and prod casts (pls. 61, 66, and 68). If the object is propelled by rolling, then various types of roll marks are produced. In some cases the object responsible for the roll mark can be identified by the signature left on the muddy bottom (see marks produced by the fish vertebrae, pl. 68). If the engraving tool has a vibratory motion, vibration marks and chevron marks pl. 62 and 63A) or ruffled groove casts are produced.

A somewhat different structure is a slide mark (pl. 65B) produced by the sliding of masses of plant debris, mud or like material. These appear as multiple, parallel striations, obviously all made at the same time by one object and not by individual particles moving independently.

The distinction between load and current marks is not always obvious. Many load structures were initiated as current marks which as a result of load deformation became enlarged, swollen and misshapen (cf. pls. 63 A and 63 B).

Bedding soles also display a variety of irregular markings left by organisms (ichnofossils). These include various kinds of tracks, trails, and burrows. Most of these found on soles of sandstone beds were produced on the underlying mud surface and are now preserved as "casts" (pls. 70 A and 70 B). Some, however, post-date the deposition of the sand and extend both upward and downward across the bedding (pl. 116 B).

Some sandstone beds part readily along internal bedding planes. Such surface may display parting lineation or "current lineation" (cf. 76 A and 76 B) and also may show the bedding plane pattern of micro-crossbedding which STOKES (1953, p. 17 to 21) called the rib-and-furrow structure (pl. 38 A).

The upper surfaces of beds also display a variety of structures. Included here are the many and varied patterns of sand waves, large and small. The smallest include ripple marks such as the regular oscillation or wave ripple patterns (pls. 85 B and 87 B), interference ripple patterns or "tadpole nests" (pl. 87 A), and the less regular linguoid or cuspate ripple patterns produced by currents (pls. 84 A and 84 B). Of larger size are the giant ripples of many fluvial, tidal and marine environments (pls. 27, 89 A and 89 B). Of greatest magnitude are the sand-wave patterns of subaerial and subaqueous dune fields (pl. 28). The large sand-waves are rarely seen as surface features on a bed. They more commonly are expressed as cross-bedding within the sand bed.

Another top surface feature, seldom if ever seen in older rocks, are beach cusps (pl. 98 B).

The top surface of sand beds also is subject to current action. Such structures as rill marks, with both tributary and distributary patterns (pl. 92 A), current crescents (pls. 91 B, 93 A and B), and swash marks, belong here. The top surface of many fine-grained beds are various small pits and depressions such as rain prints (pl. 94) and related spray and hail prints. These closely resemble gas bubble impressions. Larger structures of this type are springs pits. Perhaps related to these are the rare but curious sandstone cylinders (pl. 115 A). If the sediment is mud, subject to shrinkage on drying, the surface may be broken into polygonal blocks separated by a network of "mud-cracks" or desiccation cracks (pl. 94). A somewhat similar crack pattern is formed by sublimation of ice-crystals (pl. 96 A). Related to these are mud-crack casts (pl. 96 B) and ice-crystal casts — the result of filling of the openings by sand or similar material. Salt crystal casts (pl. 98 A) are also known and owe their origin to filling of cavities left by the solution of halite or other soluble crystals which were embedded in the mud.

The top surface of fine grained sediments is also marked by various organic trails and tracks (pl. 99).

Bedding deformed by penecontemporaneous processes:

Sediments exhibit various structures which have been attributed to deformation occurring during or shortly after sedimentation and before consolidation of the rock. Such deformation is ascribed to non-tectonic processes including foundering and gravity slump or downslope sliding. The effects of these processes vary from slight modification of bedding and other primary structures to profound folding, brecciation or even liquification and injection with complete loss of original structure. Deformation structures can be grouped into several classes:

Class A: Founder and load structures (load casts, ball-and-pillow structures, *etc.*)

Class B: Convolute bedding

Class C: Slump structures (fold, faults, breccias)

Class D: Injection structures (sandstone dikes, sill, *etc.*)

Class E: Organic structures (burrows, "churned" beds, *etc.*)

The deformation varies from purely vertical displacements to large lateral displacement. In the first category is the structure here termed "ball-and-pillow structure", also known as "flow rolls", "pseudonodules" and the like. These occur both in fine-grained sandstones (pls. 100 B to 101 B and 103 B) and in mechanically-deposited limestones of similar texture (pls. 102 A and 102 B). Although commonly attributed to slump, it is more likely a mechanical re-organization as a result of foundering.

A structure of somewhat uncertain origin is convolute bedding or convolution lamination, also called "curly bedding." This structure is characteristic of some coarse silt or fine sand beds and involves only the internal laminations of the bed itself and not the bed which remains undeformed. Convolute bedding is characterized by extreme disorganization of the folds, lack of faulting, and confinement to a single, relatively thin undeformed layer. We present the structure as it appears in plan view (pl. 106 A) and in cross-section (pl. 106 B).

Larger scale deformation is generally attributed to slump, especially large-scale subaqueous gliding. Such action involves many beds, resulting in piling up of deformed strata in some places with lacunae or gaps in the stratigraphy at other places. Such structure in the Late Glacial deposits has been thought by some to be due to overriding glacial ice or to grounded icebergs. In some cases the folds are of the décollement type (pls. 107 B and 108 A and B). Brecciation (pl. 105 A), faulting (pl. 111 B), and flowage (pl. 112) are accompaniments of subaqueous slump.

Under certain conditions water-saturated sands and silts can be transformed into a state in which they behave as a liquid and can move and be injected into adjacent sediments. Structures produced by such movements include sandstone dikes and sills and sand volcanoes. Sandstone dikes may be sharp-walled and tabular, as in plate 113 B, in some cases with conglomeratic materials in them, or they may be highly irregular in form as in plate 113 A. Sand volcanoes are a relatively rare feature that have been described as structures superimposed on a slump sheet (pl. 114).

Other problematica of uncertain origin include whirl balls and sandstone balls (pl. 112 A). Armored mud balls (pl. 117) are another primary sedimentary structure difficult to classify.

A resumé of deformational structures would not be complete without mention of the role of organisms in destroying primary sedimentary lamination resulting in homogenization of the rock. Mud-eating and burrowing organisms may be numerous enough to riddle the sediment with burrows and destroy much or all of the primary bedding producing a "churned" deposit almost devoid of original structure (see pls. 116 A and 116 B).

Einleitung

Seit den Anfängen der Geologie haben Sedimentstrukturen besondere Aufmerksamkeit genossen: ist doch Schichtung, hervorragend unter den Strukturen, ein fast untrügliches Kennzeichen für sedimentären Ursprung überhaupt. Ähnlich Gefüge und Komposition ist die Struktur eine primäre Eigenschaft des Gesteins, die mancherlei Auskunft über dessen Entstehung geben kann. Gefüge betrifft die Korn-zu-Korn-Verhältnisse; Struktur dagegen umgreift größere Inhomogenitäten und Unterbrechungen des Gesteins. Struktur meint vor allem: Organisation einer Ablagerung; die Art und Weise, in der eine Schichtenfolge zusammengesetzt ist.

Aus diesem Grunde sind Strukturen Großmerkmale eines Gesteins, die man besser im Aufschluß als im Handstück oder gar Dünnschliff studiert. Sie werden am besten durch Beispiele definiert. Wir schließen hier nun verschiedene Arten und Merkmale der Schichtung ein, z.B. Schrägschichtung, gradierte Schichtung, Rippelmarken und Trockenrisse.

Es war nicht immer leicht, zwischen Gefüge und Struktur im engeren Sinne zu unterscheiden. Ist z.B. Dachziegel-Lagerung von Geröllen in einem Konglomerat ein Merkmal des Gefüges oder der Struktur? Vielleicht ist sie Ausdruck des Schottergefüges und weniger eine Struktur im engen Sinne, auch wenn sie mit einigen Bildern vertreten ist. Sind die Rinnen und Riefen auf einigen Schichtflächen in Wirklichkeit Strukturen? Kaum können sie Gefügemerkmale genannt werden.

Einige Strukturen hängen jedoch vom Gefüge ab. So kommen Rippeln und Schrägschichtung nur in solchen Gesteinen vor, deren Korngröße im Sandbereich liegt. Freilich sind diese Strukturen unabhängig von der Zusammensetzung; denn man findet

sie auch in Kalkstein, wenn dieser aus Kalksand besteht. Rippelmarken sind sogar aus Salzablagerungen bekannt (KAUFMAN und SLAWSON, 1950). Andere Strukturen, wie Strömungs-Streifung (Tafel 76B), hängen ebenfalls vom Gefüge ab. Wenn man auch nicht weiß, wie Strömungs-Streifung entsteht, so steht jedenfalls ein enger Zusammenhang mit dem Korngefüge fest.

Unter anderen Beispielen gefügeabhängiger Strukturen seien Trockenrisse erwähnt, die sich nur in kohäsivem Schlamm formen können, und die in gewissen Sandsteinen auftretende Ballenstruktur. Diese Struktur ist unabhängig von der Zusammensetzung, denn man findet sie auch in mechanisch abgelagerten Kalksanden und -silten.

Gibt es nun Strukturen, die nichtklastischen Sedimenten eigen sind? Oder solche, die von der Komposition abhängen? Rhythmische Schichtung in Salz- und Anhydritablagerungen und die knollige Schichtung mancher Kalksteine haben keine augenfällige Entsprechung in klastischen Sedimenten. In ähnlicher Weise ist stromatolithische Schichtung auf Kalksteine und Dolomite beschränkt. Zum größten Teil sind jedoch die hier abgebildeten Sedimentstrukturen von der Strömung oder durch Verformung des weichen Sedimentes hervorgerufen. Sie können in Gesteinen jeder Zusammensetzung vorkommen, vorausgesetzt, daß solches Material von Strömungen transportiert und abgelagert wurde und ein geeignetes Gefüge zeigt.

Gewisse Strukturen, die für körnige Sedimente charakteristisch sind, treten sogar in einigen geschichteten basischen Erstarrungsgesteinen auf. Gradierte Schichtung und Schrägschichtung mögen als Beispiele dienen (GATES, 1961, Abb. 15 und 18). Eine Zusammenstellung von Bildern sedimentärer Strukturen in Erstarrungsgesteinen

10

wäre eine vielversprechende Aufgabe, derer sich jedoch die Verfasser nicht unterzogen haben. Es ist einleuchtend, daß wind- oder wassertransportiertes pyroklastisches Material fast alle Strukturen aufweist, die man in den nichtpyroklastischen Sedimenten findet.

Strukturen kann man nicht so genau definieren wie geometrische Figuren oder Körper, wie z.B. einen Würfel oder eine Kugel. Wie bei den Fossilien oder organischen Formen im allgemeinen ist eine Abbildung unbedingt erforderlich, um das Objekt zu definieren. Weil Sedimentstrukturen, ähnlich Organismen, eine große Formfülle aufweisen, ist ein einzelnes Bild gänzlich ungeeignet, um diese Mannigfaltigkeit und Verschiedenheit aufzuzeigen. Nur ein Atlas kann diese Aufgabe erfüllen.

Wir werden nicht versuchen, den Ursprung dieser Strukturen zu diskutieren, noch uns über ihren Gebrauch zur Bestimmung des stratigraphischen Oben und Unten oder ihren Wert im Kartieren fossiler Strömungen verbreiten. Ihren Gebrauch beim Bestimmen der stratigraphischen Abfolge hat SHROCK (1948) dargestellt, und die Bedeutung der gerichteten Strukturen in der Analyse fossiler Strömungen und Sedimentationsräume ist an anderer Stelle von uns behandelt worden (1963). Ein Bilderatlas über Sedimentstrukturen ist jedoch nur dann nützlich, wenn er auf logische Weise aufgebaut ist. Solch eine Organisation setzt irgendeine Art von Klassifikation voraus. Wir werden also zunächst abgrenzen, was wir unter primären Sedimentstrukturen verstehen, und uns dann mit der Klassifikation und Nomenklatur der Sedimentstrukturen etwas eingehender befassen.

Definitionsbereich

Wir befassen uns hier mit *primären Sedimentstrukturen*. Mit „primär" meinen wir: sie sind während oder kurz nach der Ablagerung entstanden, jedenfalls vor der Verfestigung des Gesteins, in dem sie sich finden.

Zu den hier behandelten primären Strukturen gehören Schichtung, insbesondere die äußere Form der Schichten, ihre Beständigkeit und Mächtigkeit, innere Struktur und Aufbau der Schichtung, Hieroglyphen oder Marken, die auf Schichtflächen auftreten, sowohl an Schicht-Ober- wie -Unterseiten und schließlich durch synsedimentäre Deformation entstandene Strukturen. Kurzum, der vorliegende Atlas beschäftigt sich mit der Schichtung im weiten Sinne oder mit gewissen Merkmalen der Schichtung.

Die sekundären, von *tektonischer* Deformation geschaffenen Strukturen haben wir nicht berücksichtigt. Wir haben aber die Verwerfungen, Falten und verwandte Strukturen aufgenommen, welche auf Grund ungleicher Belastung, Hangrutsch oder Gleitung, durch Bewegungen vor der Verfestigung entstanden sind. Ferner haben wir primäre Schieferung und Klüftung ausgeschlossen, die die Spaltbarkeit eines Gesteins bestimmen. Vielleicht sind solche Strukturen von primären Ursachen bedingt; sie setzen jedoch ein halbwegs verfestigtes Gestein voraus. Alle konkretionären und akkretionären Strukturen, wie Knollen, Konkretionen, Septarien und dergleichen, sind ebenfalls nicht aufgenommen, obwohl einige dieser Strukturen während der Sedimentation oder doch vor der Verfestigung entstanden sein sollen. Nach einigem Zögern schieden wir solche Strukturen aus, die auf Frostwirkung oder Kryoturbation beruhen, einschließlich Einwickelstrukturen und Eiskeile, als auch bei der Bodenbildung entstandene Strukturen. In ähnlicher Weise haben wir uns nur flüchtig mit organisch entstandenen Strukturen befaßt. Organische Spuren, Fährten und Bauten werden zwar angeführt; indessen scheint uns eine eingehendere Darstellung der Ichnofossilien, wie man diese Hieroglyphen nennt, zu speziell für diesen Atlas. Der interessierte Leser sei auf die Veröffentlichungen von ABEL (1935), SEILACHER (1953) und CASTER (1957) verwiesen. Wir haben andere organische Strukturen wie Stromatolithen aufgenommen insofern, als sie einen Typus Schichtung darstellen — eine Art Wachstumsschichtung oder zumindest Schichtung, die durch organische Tätigkeit beeinflußt ist.

Einteilung der primären Sedimentstrukturen

Es ist uns kein ernsthafter Versuch bekannt, alle Sedimentstrukturen zu klassifizieren. Teil-Klassifikationen sind jedoch vorgeschlagen worden, und die wichtigsten unter ihnen behandeln Schichtung. Als umfassendste Arbeiten seien die von ANDRÉE (1915), HOPPE (1930), BRUNS (1954), BIRKENMAJER (1959) und BOTWINKINA (1960) genannt. Diese Klassifikationen befassen sich jedoch nicht nur mit der äußeren Form, sondern beziehen auch das Gefüge und z.T. rhythmische und sogar zyklische Folgen ein, die in verschiedenen Gesteinen auftreten. Teil-Klassifikationen anderer Art behandeln z.B. Hieroglyphen, wie die von KREJCI-GRAF (1932) und WASSOEWITSCH (1935). Moderne Einteilungen von Schichtflächen-Marken — einer anderen Art von Strukturen — sind die von CROWELL (1955) und KUENEN (1957). Viel spezialisierter noch sind die verschiedenen Einteilungen von Rippelmarken und Schrägschichtung.

Es ist weder unsere Absicht, diese verschiedenen Einteilungen zu referieren, noch eine umfassende eigene Klassifikation vorzuschlagen. Statt dessen wollen wir eine einfache Einteilung benutzen, um das Bildmaterial im Atlas in logischer Weise anzuordnen. Wir haben deshalb die Sedimentstrukturen in vier größere Gruppen eingeteilt:

I. Schichtung, äußere Form
II. Schichtung, Gefüge und Struktur
III. Schichtflächen-Merkmale
IV. Schichtung, gestört und verformt

Einige Strukturen sind leicht in irgendeine dieser Gruppen einzuordnen oder können mehr als einer Klasse angehören. Wurmbauten z.B. können in einigen Fällen als Schichtflächen-Merkmale angesehen werden, in anderen als Gefüge oder als Störung der Schichtung. Tonkugeln, Strandhörner und Sandstein-Gänge sind Beispiele von Strukturen, die schwierig zu klassifizieren sind.

Die äußere Form der Schichtung

Obwohl Schichtung seit recht langer Zeit bekannt und beschrieben ist, gibt es dennoch bis heute keine befriedigende oder allgemein anerkannte Analyse und Klassifizierung dieser Struktur. Bei minimaler Behandlung des Problems legen wir hier ein System vor, das für den täglichen Gebrauch vielleicht nicht praktisch ist, uns aber gestattet, unsere Bilder von Schichtgesteinen zu klassifizieren und anzuordnen.

Die Grundeinheit ist die Einzelschicht oder *Sedimentations-Einheit*, von OTTO (1938) definiert als „die Sedimentdicke, die unter im wesentlichen konstanten physikalischen Bedingungen abgelagert wurde". Die Qualifikation „wesentlich" erlaubt jedoch für kurzzeitige Schwankungen in Strömungsrichtung oder -geschwindigkeit, die innere Feinschichtungen hervorrufen kann, oder fortlaufend wechselnde Bedingungen, ohne größere Unterbrechung, die gradierte Schichtung erzeugen. Schichtfugen treten durch Unterschiede in Gefüge und Zusammensetzung der Einzelschichten hervor. Die grundlegenden Eigenschaften einer Schicht oder Sedimentations-Einheit sind 1. Mächtigkeit oder Dicke, 2. seitliche Erstreckung und 3. Gefüge. Man kann Schichtfolgen unabhängig von ihrer inneren Organisation oder Struktur beschreiben. Letzteres werden wir in einem späteren Abschnitt besprechen.

Zunächst die Mächtigkeit. Schichten werden gemeinhin als bankig, dünnschichtig oder feinschichtig beschrieben. Genauere Definitionen dieser Begriffe sind vorgeschlagen worden.

PAYNE (1942), z.B. schlägt die Bezeichnung „Stratum" (Schicht) für mehr als 1 cm dicke Schichten vor, die klar von über- und unterliegenden ähnlichen Einheiten abgetrennt sind. Den Begriff „Lamination" (Feinschicht) beschränkt er auf Schichten von weniger als 1 cm Dicke. In ähnlicher Weise beziehen sich die Feinschichten von MCKEE und WEIR (1953) auf weniger als 1 cm dicke Schichten. Bei ihnen bezeichnet „sehr dünnschichtig" 1—5 cm dicke Schichten, „dünnschichtig" reicht von 5—60 cm, „bankig" von 60—120 cm, während alle mächtigeren Schichten „grobbankig" heißen. Dies sind Beispiele von Versuchen, über die reine Beschreibung hinaus quantitative Aussagen über Schichtung zu machen. Zum gleichen Thema liegen u.a.

Arbeiten von INGRAM (1954), GRAY (1955) und BOKMAN (1956) vor.

Wenn man mit vielen Schichten arbeitet, muß man auch mittlere Mächtigkeiten und Wechsel in der Mächtigkeit in Betracht ziehen. Dies läßt sich am besten als Standardabweichung oder vielleicht noch besser als Variationskoeffizient ausdrücken. Einige Schichtfolgen weisen erstaunlich geringe Variationskoeffizienten auf (Tafeln 2, 8 B). Während einige Autoren Mächtigkeitswechsel unter Zuhilfenahme von Zyklen — z. B. Varven — analysieren, weisen andere auf die Häufigkeit von log-normalen Verteilungen in Schichtmächtigkeiten hin (SIMONEN und KOUVO, 1951, Abb. 4 und ATKINSON, 1962, S. 357). Jedenfalls sollte der Geländegeologe sich der Möglichkeit von sich wiederholenden oder zyklischen Anordnungen (Tafeln 4, 9 B) der Schichtmächtigkeiten bewußt sein. Derartige Wechsel sind oft mit entsprechendem Gesteinswechsel verbunden.

Neben der absoluten Mächtigkeit zeichnet sich eine Schicht durch seitliche Veränderlichkeit der Dicke aus. Während einige Schichten eine außerordentliche konstante Mächtigkeit in ihrer ganzen aufgeschlossenen Länge aufweisen, keilen andere öfters aus oder zeigen wellenartige Schichtung (s. Tafeln 17 A, 11 B). Den Extremfall stellen Schichten dar, die vollständig auskeilen. Solche lateralen Diskontinuitäten mögen von örtlichen Diskordanzen, kleinen Erosionsrinnen oder Prielen (Tafeln 19 A bis 21 A), oder von einer Art synsedimentärer Boudinage verursacht sein (Tafel 15 A). Manche Gesteine wie Kalkstein enthalten Schichten, die seitlich in eine Reihe von Knollen übergehen. In anderen Fällen zeigt das ganze Gestein knollige Schichtung (Tafel 15 B).

Wir können Schichtung in verschiedene Gruppen einteilen, die alle Stufen von höchster Ordnung bis zu größter Unordnung und Unregelmäßigkeit darstellen. Man kann sich z. B. eine Abfolge vorstellen, in der die Schichten alle von gleicher oder nahezu gleicher Mächtigkeit sind, von seitlich konstanter Mächtigkeit und scheinbar unbegrenzter Ausdehnung. Eine solche Abfolge würde den höchsten Grad von Ordnung darstellen. Auf der anderen Seite kann man sich Gruppen von Schichten vorstellen, die untereinander verschieden mächtig sind, deren Mächtigkeit im einzelnen schwankt und die innerhalb der Aufschlußgrenzen auskeilen (Tafel 21 B). Dieses wäre ein Fall großer Unordnung. Darüber hinaus sind verschiedene Zwischenstufen im Ordnungsgrad möglich.

Zur Illustration dieser Konzepte können wir vier Klassen definieren:

Klasse A: 1. Schichten gleicher oder fast gleicher Mächtigkeit
2. Schichten seitlich konstanter Mächtigkeit
3. Kontinuierliche Schichten

Klasse B: 1. Schichten *ungleicher* Mächtigkeit
2. Schichten seitlich konstanter Mächtigkeit
3. Kontinuierliche Schichten

Klasse C: 1. Schichten *ungleicher* Mächtigkeit
2. Schichten seitlich *wechselnder* Mächtigkeit
3. Kontinuierliche Schichten

Klasse D: 1. Schichten *ungleicher* Mächtigkeit
2. Schichten seitlich *wechselnder* Mächtigkeit
3. Schichten *diskontinuierlicher* Mächtigkeit

Beispiele für Klasse A finden sich in Tafeln 1 A, 2 und 8 B; für Klasse B: 1 B, 3, 4, 6 A und 7, für Klasse C: 11, 12, 13 und für Klasse D: 15 A, 17, 21 B.

Innere Organisation und Struktur der Schichtung

Die innere Organisation einer Schicht ist so wichtig wie ihre äußere Form. Verschiedene Modelle innerer Ordnung sind möglich. Wichtige Typen innerer Struktur sind:

Klasse A: Massig

Klasse B: Feinschichtig
1. horizontale Feinschichten
2. Schräg-Feinschichtung (einfach und vielfach)

Klasse C: Gradiert

Klasse D: Dachziegel-Lagerung (Schotter) und andere orientierte innere Gefüge

Klasse E: Wachstums-Strukturen (stromatolithischer Kalkstein, einige Riffkalke, Travertin usw.)

Eine Schicht kann *massig* genannt werden, wenn sie allem Anschein nach kein inneres Gefüge aufweist (Tafeln 25 und 26). Neuere Untersuchungen (HAMBLIN, 1962) zeigen jedoch, daß der Ausdruck „massig" irreführend sein kann; denn Röntgenaufnahmen scheinbar massiger Gesteine zeigen Feinschichtung und andere Strukturen (s. Tafel 23).

Eine Schichtlage kann *feinschichtig* sein. Solche Feinschichten spiegeln kurzfristige Änderungen in den physikalischen Bedingungen während der Ablagerung der Schicht wieder. Feinschichten mögen streng parallel zueinander und zu den Begrenzungsflächen der Schicht sein (Tafeln 8A und 20B, untere Hälfte). Sandstein-Schichten solcher Struktur sind gewöhnlich parallel zur Feinschichtung leicht spaltbar und stellen ausgezeichnete Platten-Sandsteine dar. Die Schichtfugen solcher Sandsteine weisen häufig Strömungs-Streifung auf (Tafeln 76A und B).

In körnigem Material von Sandgröße liegen die Feinschichten gewöhnlich schräg zur Begrenzungsfläche der Bank. Solche diagonalen Feinschichten oder *Schrägschichten* sind häufig in gewissen Sandfazies. Diese schräggeschichteten Sand- oder Kalksteine sind reichlich illustriert (Tafeln 29B bis 37). Schrägschichtung kommt in kleinen bis zu sehr großen Dimensionen vor (Tafeln 36, 38B). In vielen Fällen wird die ganze Schicht einbezogen, und die Leeblätter reichen von der oberen bis zur unteren Begrenzungsfläche einer Schicht (Tafel 34A); in anderen Fällen kann die Schicht zusammengesetzt sein und somit mehrere schräggeschichtete Zonen zeigen (Tafel 37). Die Leeblätter mögen gerade oder bogig und tangential zur Schichtbasis sein (Tafeln 32A und 33B). In einigen Fällen sind die Leeblätter übersteilt oder gar überkippt und gerunzelt (Tafel 110). In verwickelten Fällen kann eine Schicht massig im Unterteil, horizontal feingeschichtet in der Mitte und schräggeschichtet im Oberteil sein.

Gradierte Struktur ist ein anderes Merkmal der Schichtung. Eigentlich ist *gradierte Schichtung* eine gewöhnliche Abart des Gefüges. Sie ist ein besonderes Kennzeichen und gemeinhin mit bloßem Auge leicht erkennbar. Wir zeigen deshalb verschiedene Abbildungen gradierter Schichtung (Tafeln 1B und 42A, B).

Gradierte Schichtung kommt in Schottern, Sanden und Silten vor. Sie ist ganz besonders charakteristisch für Repetitions-Schichtung in Flyschabfolgen — in den sog. Turbidit-Sanden. Gradierte Schichtung kann man auch von weitem an den unterschiedlich scharfen Unter- und Oberseiten einer Sandlage erkennen (Tafel 1B). Bei näherem Hinsehen wird einem ein Wechsel in der Korngröße und gleichzeitig ein Übergang von hell (Sand) nach dunkel (Silt oder Ton) auffallen (Tafeln 8B und 42B). Wenn in metamorphen Gebieten Transversalschieferung von hellen kompetenten Sandlagen in dunkle inkompetente Tonlagen übergeht, wird sie die Richtung ändern oder gebrochen werden (Tafel 6A).

Obwohl gradierte Schichtung und Schrägschichtung sich im allgemeinen ausschließen, gibt es doch Schichten, in denen man beide findet: In den meisten Fällen findet sich dann gradierte Schichtung im Unterteil der Bänke, während kleindimensionale Schrägschichtung (u. U. zusammen mit Wulstschichtung) im Oberteil auftritt.

Ein anderes Kennzeichen der Schichtung ist ihr *Gefüge.* Ähnlich Gradierung in der Korngröße ist das Gefüge mehr ein mikroskopisches denn ein strukturelles Merkmal. Normalerweise braucht man nämlich ein Mikroskop, um in Sanden und feinerem Material ein Gefüge zu erkennen. Nur Schottergefüge kann man auch mit bloßem Auge wahrnehmen. In der Regel liegen die Gerölle mit ihrer Längsachse parallel zur Ablagerungsfläche. In vielen Flußschottern zeigen jedoch die flachen Gerölle Dachziegel-Lagerung, in der die einzelnen geneigten Gerölle wie Dachziegel gestapelt sind und nach stromaufwärts einfallen (Tafel 43A). Augenfällig wird diese An-

ordnung in Profilen parallel zum Strom-
verlauf. Einige wenige synsedimentäre
Konglomerate zeigen Gerölle die „auf der
Kante" stehen, weshalb sie „kantengestell-
te" Konglomerate genannt werden.

Die bis jetzt beschriebenen verschiedenen
Arten von Schichtung und innerer Struktur
kommen in klastischen Sedimenten vor und
sind weitgehend das Ergebnis von Sedimentation in bewegtem Medium. Wir soll-
ten auch einen kurzen Blick auf die
Schichtung nichtklastischer Ablagerungen
werfen.

Viele Kalke sind von Strömungen abgelagert, und so zeigen auch sie alle oben ge-
nannte Merkmale, wie Schrägschichtung,
gradierte Schichtung usw. Andere Kalksteine bilden sich jedoch *an Ort und Stelle*,
in einigen Fällen aus Organismen. Die
Schichtung solcher Gesteine ist oft der von
Strömungen erzeugten sehr unähnlich. Dies
gilt besonders für *stromatolithische Schichtung*, eine Wachstums-Schichtung, die man
im allgemeinen auf die Wirkung von rasen-
bildenden blaugrünen Algen zurückführt.
Stromatolithische Schichtung ist außerordentlich unterschiedlich, zeichnet sich jedoch im allgemeinen durch Feinschichtung
aus. Unähnlich strömungsbedingter Feinschichtung zeigen die Wachstums-Lagen
Unregelmäßigkeiten, wie Gabelung, Runzelung und Reihen verschiedenartiger nach
oben gewölbter Gebilde. Es ist nicht
unsere Absicht, die mannigfaltigen stromato-
lithischen Erscheinungsformen darzustellen.
Der interessierte Leser sei auf diesbezügliche
Spezialliteratur verwiesen (s. z.B. CLOUD,
1942 und REZAK, 1957). Sehr wahrscheinlich wird „Algenschichtung", die keine auf-
fälligen Formen zeigt, oft vom Gelände-
geologen übersehen. Wir haben etliche Beispiele strukturierter wie auch formloser
stromatolithischer Schichtung abgebildet
(Tafeln 48, 49A, B und 50B).

Andere Karbonatgesteine mögen innere
Wachstums-Strukturen zeigen, die sich
von Organismen herleiten, wie Korallen,
Stromatoporen u. a., und in wiederum
anderen Fällen werden scheinbare Wachstums-Strukturen auf anorganische Weise
geformt. Zur letzten Art gehören die ge-
bänderten Strukturen in einigen Traver-
tinen und in Onyx.

Schichtflächen-Marken und -Unregelmäßigkeiten

Auf vielen Schichtflächen findet sich bei
näherem Zusehen eine Fülle von Strukturen.
Man kann diese unterteilen in solche, die
auf Schicht-*Unterseiten*, auf Schicht-*Oberseiten* oder *innerhalb* einer Schicht auftreten. Jede dieser Klassen besteht aus
einer Reihe von Strukturen, die wir wie folgt
eingeteilt haben:

Klasse A: Schicht-Unterfläche

1. Belastungsstrukturen
   (Belastungsmarken)
2. Strömungsstrukturen (Ero-
   sions- und Gegenstands-
   marken)
3. Organische Marken (Spuren
   und Bauten; „Fukoiden")

Klasse B: Schicht-Innenfläche
(Strömungs-Streifung)

Klasse C: Schicht-Oberfläche

1. Rippelmarken
2. Erosionsmarken (Rieselmarken, Strömungskämme)
3. Trichter und Eindrücke
   (Blasen- und Regentropfen-
   Eindrücke usw.)
4. Trockenrisse, Netzleisten,
   Eiskristall-Abdrücke,
   Salzpseudomorphosen usw.
5. Organische Marken (Spuren,
   Fährten usw.)

Sind Schichtflächen-Marken auch seit vielen
Jahren bekannt — die verschiedenen
Hieroglyphen —, so wurden sie doch erst
in der letzten Dekade intensiv studiert. Infolgedessen wurde eine große Zahl neuer
Strukturen beschrieben, und eine umfang-
reiche, verwirrende Nomenklatur häufte
sich an. Es entstand eine gemischte
Nomenklatur, die sowohl genetische als
auch beschreibende Bezeichnungen enthält.

Schichtflächen-Marken finden sich im allgemeinen in Sandstein oder Kalkstein, die
über Schieferton liegen. Man kann sie
durchweg als Ausgüsse beschreiben insofern, als sie Mulden ausfüllen, die in der
Oberfläche des darunterliegenden Schlammes geformt wurden. Diese Strukturen

entstehen 1. bei ungleicher Belastung eines weichen hydroplastischen Schlammes, 2. durch Einwirken von Strömungen auf die Schlamm-Oberfläche, oder 3. durch die Tätigkeit von Organismen auf jener Fläche. Die Belastungsstrukturen werden meist Belastungsmarken genannt, auch wenn sie sich durch Einsacken von Sand in Schlamm formen und nicht durch Ausfüllen von Mulden, wie die meisten anderen Marken. Belastungsmarken unterscheiden sich ziemlich in Größe, Form und Häufigkeit (Tafeln 52 A bis 54 A). Einigen dieser Abarten hat man Spezialnamen gegeben, wie „wulstige Belastungsmarken" (Tafel 54 B) und „schuppenförmige Belastungsmarken". Normalerweise erscheinen diese Strukturen als unregelmäßige Wülste an Schicht-Unterflächen. Im Profil sind sie gekrümmt wie die sog. Flammenstrukturen (Tafel 53 B).

Die *Strömungsmarken* gehören zu zwei Gruppen: solche, die von der Strömung selbst geformt sind, und solche, die von Gegenständen hervorgerufen werden, wie z. B. von Schalen, Holz-Bruchstücken, Schieferfetzen und Sandkörnern, die von der Strömung über den Grund getrieben werden. Die ersteren nennt man *Erosionsmarken*. Die wichtigsten unter ihnen sind die Mulden, die als „Strömungswülste" an der Unterseite vieler Sandsteine und Kalksteine auftreten. Sie wechseln erheblich in Größe, Form und allgemeiner Anordnung auf den Schicht-Unterflächen (Tafeln 55 A bis 59 B). Starke Strömungen können kleine Priele und dergleichen mehr hervorrufen (Tafeln 20 A, B).

Die *Gegenstands-Marken* sind eine sehr unterschiedliche Gruppe. Einige entstehen durch Eindrücken eines sich kontinuierlich am Boden fortbewegenden Gegenstandes. Diese schließen Riefen und Rillen ein, und ihre Ausgüsse sollen als Riefen- und Rillenmarken bezeichnet werden (Tafeln 61 bis 67). Ähnliche Marken werden von Gesteinsbruchstücken hervorgerufen, die in sich bewegendem Eis eingeschlossen sind (Tafeln 73 A bis 74). Andere Gegenstands-Marken entstehen bei unterbrochenen Berührungen verschiedener Objekte mit dem Boden. Die korrespondierenden Strukturen auf der Unterseite des hangenden Sandsteins werden wechselweise Hüpfmarken (Tafel 68),

Aufprall- oder Quastenmarken (Tafeln 68, 69 A) und Aufstoßmarken (Tafeln 61, 66 und 68) genannt. Wenn der Gegenstand auf dem Boden entlangrollt, dann entstehen die verschiedenen Arten der Rollmarken. In einigen Fällen kann das Objekt, das für die Rollmarken verantwortlich ist, an den Eindrücken erkannt werden, die es im Schlamm hinterläßt (s. von Fischwirbeln hervorgerufene Marken, Tafel 68). Bei der Vibration eines sich eindrückenden Gegenstandes entstehen Schwingungs- und Fiedermarken (Tafeln 62, 63 A) oder gefiederte Rillenmarken.

Wiederum andersartig sind die Gleitmarken (Tafel 65 B), die beim Abgleiten von angehäuften Massen von Pflanzenhäcksel, Schlamm oder dergleichen mehr entstehen. Sie treten als vielfache, parallele, Riefen auf, offensichtlich alle zum gleichen Zeitpunkt vom selben Objekt und nicht von individuellen Teilchen in unabhängiger Bewegung hervorgerufen.

Der Unterschied zwischen Belastungs- und Strömungsmarken ist nicht immer eindeutig. Viele Belastungsmarken begannen als Strömungsmarken, die dann durch Belastung vergrößert, vergröbert und umgestaltet wurden (s. Tafeln 63 A, B).

Schicht-Unterflächen weisen eine Vielfalt unregelmäßiger von Organismen hinterlassener Spuren auf (Ichnofossilien). Sie umfassen verschiedene Arten von Fährten, Spuren und Bauten. In der Mehrzahl entstanden diese Spuren auf Schlammflächen und finden sich jetzt als Ausgüsse auf den Schicht-Unterseiten darüberliegender Sandsteine (Tafeln 70 A, B). Einige formen sich jedoch nach der Ablagerung des Sandsteins und erstrecken sich sowohl in unter- wie überliegende Schichten (Tafel 116 B).

Manche Sandsteinlagen sind leicht nach inneren Schichtfugen spaltbar. Solche Flächen zeigen häufig Strömungs-Streifung (s. Tafeln 76 A, B) und mögen auch Schichtflächen-Muster kleindimensionaler Schrägschichtung aufweisen, das STOKES (1953, S. 17—21) „Kamm- und Furchenstruktur" nannte (Tafel 38 A).

Auch Schicht-Oberflächen zeigen eine Reihe von Strukturen, zu denen mannigfaltige Muster und große und kleine Sandwellen gehören. Die kleinsten unter ihnen umfassen

Rippelmarken, wie die regulären Oszillationsrippeln oder Wellenrippel-Muster (Tafeln 85 B, 87 B), Interferenzrippel-Muster oder „Kaulquappen-Nester" (Tafel 87 A) und die strömungsbedingten, weniger regelmäßigen zungen- oder bogenförmigen Rippelmuster (Tafeln 84 A, B). Von beträchtlichem Ausmaße sind die Großrippeln vieler fluviatiler, Watten- oder mariner Milieus (Tafeln 27, 89 A, B). Am gewaltigsten sind die Sandwellen-Muster subaerialer und subaquatischer Dünenfelder (Tafel 28). Selten finden sich die großen Sandrücken als Oberflächen-Formen auf Schichtflächen. Öfters zeigen sie sich als Schrägschichtung innerhalb einer Sandschicht.

Eine andere Oberflächenform, selten — wenn überhaupt — fossil vorhanden, sind Strandhörner (Tafel 98 B).

Auch die Oberfläche von Sandsteinen ist der Tätigkeit von Strömungen ausgesetzt. Hierher gehören Strukturen wie Rieselmarken, mit Zufluß- als auch Abflußmustern (Tafel 92 A), Strömungskämme (Tafeln 91 B, 93 A, B) und Spülmarken. Die Oberfläche vieler freinköniger Schichten ist von Trichtern und Einsenkungen mancherlei Art besetzt, wie Regentropfen-Eindrücke (Tafel 94) und ähnlich Spritz- und Hageleindrücke. Diese sehen Gasblasen-Trichtern sehr ähnlich. Größere Strukturen dieser Art sind Quelltrichter. Vielleicht sind die merkwürdigen, jedoch seltenen Sandstein-Zylinder (Tafel 115 A) ihnen verwandt. Die Oberfläche eines schlammigen Sedimentes, der Eintrocknung oder Schrumpfung ausgesetzt, kann in polygonale Blöcke zerbrechen, die durch ein Netzwerk von Trockenrissen voneinander getrennt sind (Tafel 94). Ein etwas ähnliches Muster von Rissen entsteht bei Sublimation von Eiskristallen (Tafel 96 A). Diesen ähnlich sind Netzleisten (Tafel 96 B) und Ausgüsse von Eiskristall-Eindrücken — das Ergebnis einer Füllung von Mulden mit Sand oder ähnlichem Material. Salzkristall-Pseudomorphosen (Tafel 98 A) sind ebenfalls bekannt und verdanken ihre Entstehung der Ausfüllung von Hohlräumen, die nach der Lösung von Halit oder anderen löslichen Kristallen entstanden, die im Schlamm eingebettet waren.

Die Oberfläche feinkörniger Sedimente ist ebenfalls von allerlei organischen Spuren und Fährten bedeckt (Tafel 99).

Durch synsedimentäre Prozesse verformte Schichtung

Sedimente weisen verschiedene Strukturen auf, welche der Deformation zugeschrieben werden, die während oder kurz nach der Ablagerung und vor der Verfestigung des Gesteins eintritt. Für solch eine Deformation werden atektonische Prozesse verantwortlich gemacht, einschließlich Sakkung und der Schwerkraft zugeschriebene Rutschung und Hanggleitung. Die Auswirkungen dieser Prozesse reichen von leichter Verformung der Schichtung und anderer primärer Strukturen bis zu ausgesprochener Faltung, Brecciierung oder gar Verflüssigung und Injektion unter vollständigem Verlust der Originalstruktur. Verformungsstrukturen können in verschiedene Klassen eingeteilt werden:

Klasse A: Sackungs- und Belastungsstrukturen (Belastungsmarken, Ballenstrukturen, usw.)

Klasse B: Wulstschichtung

Klasse C: Rutschstrukturen (Falten, Verwerfungen, Breccien)

Klasse D: Injektionsstrukturen (Sandstein-Gänge, -Lagergänge usw.)

Klasse E: Organische Strukturen (Bauten, Wühlstrukturen usw.)

Verformung reicht von rein vertikaler Versetzung bis zu ausgedehnten seitlichen Verschiebungen. Zur ersten Klasse gehört die Struktur, die hier „Ballenstruktur" genannt wird, auch bekannt unter dem Namen Sedimentrolle, „Pseudoknolle" u.ä.m. Man findet sie in feinkörnigen Sandsteinen (Tafeln 100 B bis 101 B und 103 B) und auch in mechanisch abgelagerten Kalksteinen ähnlicher Korngröße (Tafeln 102 A, B). Wenn man sie auch im allgemeinen Rutschungen zuschreibt, so ist doch mechanische Reorganisation als Folge von Sackung viel wahrscheinlicher.

Eine Struktur etwas ungewissen Ursprungs ist Wulstschichtung oder Wulst-Feinschichtung, auch „Kräuselschichtung" genannt.

Diese Struktur ist charakteristisch für einige grobkörnige Silte oder feinkörnige Sandsteine. Es sind nur die einzelnen Feinschichten innerhalb einer Schicht verformt, während die Schicht als solche ungestört bleibt. Wulstschichtung zeichnet sich durch extreme Regellosigkeit der Kleinfalten, Abwesenheit von Verwerfungen und strenge Beschränkung auf eine einzelne, relativ dünne, als solche nicht gefaltete Schicht aus. Unsere Abbildungen zeigen die Struktur von oben (Tafel 106 A) und von der Seite (Tafel 106 B).

Deformation größeren Ausmaßes wird im allgemeinen Rutschungen, besonders großräumigen subaquatischen Gleitungen, zugeschrieben. Solch ein Ereignis betrifft viele Schichten. Als Folge von Rutschungen häufen sich deformierte Schichten an einigen Stellen auf, während sie an anderen Stellen fehlen und Lücken in der stratigraphischen Abfolge hinterlassen. Ähnliche Strukturen, die sich in Ablagerungen der letzten Eiszeit finden, sind von einigen Bearbeitern der Wirkung darüberfahrenden Eises oder grundgehender Eisberge zugeschrieben worden. In manchen Fällen sind die Falten vom „Abscherungstyp" (Tafeln 107 B und 108 A, B). Brecciierung (Tafel 105 A), Verwerfungen (Tafel 111 B) und Fließfältelung (Tafel 112) sind Begleiterscheinungen von subaquatischen Rutschungen.

Unter gewissen Bedingungen können wassergesättigte Sande und Silte in einen Zustand versetzt werden, in dem sie sich als Flüssigkeiten benehmen und in umgebende Sedimente injiziert werden können. Unter den solcherart entstandenen Strukturen sind Sandstein-Gänge, -Lagergänge und Sandvulkane die bekanntesten. Sandstein-Gänge können tafelförmig sein, mit glatten und scharfen Wänden (s. Tafel 113 B), konglomeratisches Material einschließen, oder sie können außerordentlich unregelmäßig sein wie in Tafel 113 A. Die sehr seltenen Sandvulkane sind als Oberflächen-Kegel von Rutschungsmassen beschrieben worden (Tafel 114).

Zu anderen Problematika ungewissen Ursprungs gehören Wirbelbälle und Sandsteinballen (Tafel 112 A). Gespickte Tongerölle sind eine andere primäre Sedimentstruktur, die schwierig einzuordnen ist.

Eine Zusammenfassung von Verformungsstrukturen wäre nicht vollständig ohne einen Hinweis auf die Rolle, die Organismen bei der Zerstörung primärer sedimentärer Feinschichtung spielen. Durch ihre Tätigkeit entsteht häufig ein vollkommen homogenisiertes Gestein. Schlammfressende und bohrende Organismen mögen zahlreich genug sein, um das Sediment mit Bauten zu durchlöchern und die primäre Schichtung nahezu oder vollständig zu zerstören. Sie hinterlassen dann eine zerwühlte Ablagerung, die fast keine ursprüngliche Struktur mehr zeigt (s. Tafeln 116 A und 116 B).

18

L'étude des formes sédimentaires est aussi ancienne que celle de la géologie elle-même. La stratification est un des traits les plus caractéristiques des sédiments, et, à quelques exceptions près, est un critère presque infaillible d'une origine sédimentaire.Comme la texture et la composition, la structure* est une propriété inhérente d'une roche et une indication de son origine. Alors que la texture caractérise les relations de grain à grain dans une roche, la structure décrit les discontinuités et les changements majeurs d'homogénéité. La structure se rapporte à l'organisation du dépôt, la façon dont il est agencé. Il s'en suit que les structures sont des formes plus importantes, et en général mieux visibles sur une coupe de terrain que sur un échantillon ou une lame mince. C'est par un exemple que la structure est le mieux définie. On a inclu ici divers types de stratification, tels que la stratification croisée et le granoclassement, les rides et les fentes de dessiccation. Il s'est parfois révélé difficile de distinguer ce qu'on doit considérer comme une texture, et ce qui est plus exactement appelé une structure. L'imbrication des galets dans un conglomérat est-elle une texture ou une structure ? On pourrait la considérer comme une caractéristique de l'assemblage des grains (de la taille de galets) du sédiment, plutôt qu'une structure ; malgré cela, nous avons inclus plusieurs illustrations de ce phénomène. Les rainures et stries que l'on rencontre sur certains plans de stratification sont-elles vraiment des structures ? On ne peut guère les qualifier de textures.

* Les termes anglais «texture» et «structure» ne sont pas exactement équivalents des mêmes mots en français (qui sont d'ailleurs assez souvent confondus), mais en ce qui concerne les roches sédimentaires du moins, on les utilise à peu près de la même manière dans les deux langues, et, pour faciliter les choses, ils seront presque toujours rendus par les analogues français.

Certaines structures, toutefois, sont sous la dépendance de la texture. Les rides et la stratification croisée, par exemple, ne se rencontrent que dans les sédiments dont la granulométrie tombe dans l'échelle des sables. Mais ces structures sont indépendantes de la composition, car elles se rencontrent aussi bien dans les roches calcaires, à condition que ces dernières se soient formées à partir de sable. On a même décrit des rides dans du sel (KAUFMAN et SLAWSON, 1950). D'autres structures, telles que les linéations de délit sont aussi sous la dépendance de la texture. La cause des linéations n'a pas encore été élucidée, mais on sait qu'elle est étroitement liée au grain de la roche.

D'autres exemples de structures liées à la texture sont les fentes de dessiccation, qui ne peuvent se former que dans les argiles cohésives, et la structure «en boules-et-coussins» que l'on rencontre dans certains grès. Cette structure est indépendante de la composition, puisqu'elle se montre aussi bien dans des sables calcaires fins ou pulvérulents résultant d'un dépôt mécanique.

Existe-t-il des structures caractéristiques des sédiments non clastiques, ou qui dépendent de la composition ? Certaines formations rythmiques de sel et d'anhydrite, certaines couches à nodules dans des calcaires sont sans analogues dans les sédiments clastiques. De même la stratification stromatolithique parait limitée aux calcaires et dolomies. Toutefois la plupart des formes sédimentaires illustrées dans cet atlas sont produites par l'action d'un courant, ou par la déformation de sédiments mous, et peuvent se rencontrer dans des roches de composition quelconque à condition que ces matériaux aient été transportés et déposés par un courant et aient la texture nécessaire.

Il vaut la peine de noter que certaines des structures caractéristiques des sédiments granuleux se rencontrent aussi dans les roches volcaniques basiques litées. Le granoclassement et la stratification croisée en sont des exemples (GATES 1961). Un catalogue photographique de formations volcaniques montrant des structures sédimentaires primaires serait un projet du plus haut intérêt mais que nous n'avons pas cru devoir entreprendre. Bien évidemment, les matériaux pyroclastiques transportés par l'air ou l'eau montrent la plupart des structures qu'on rencontre dans les sédiments non-clastiques. On ne peut pas définir une structure avec la précision qu'on emploierait dans le cas d'une figure géométrique, d'un cube ou d'une sphère. Comme dans le cas des fossiles et des formes organiques en général, il faut une image pour définir l'objet en question. Bien que nous ne puissions pas définir analytiquement ces objets, nous pouvons toutefois les mesurer. Les structures sédimentaires, de même que les organismes, offrent une grande variété de formes, une seule image ne suffit donc pas à donner une idée de leur variété et variabilité, et il faut un atlas pour le faire.

Nous n'avons pas essayé de discuter de l'origine de ces structures, ni de nous étendre sur leur utilité pour déterminer l'ordre stratigraphique, ou pour cartographier les paléosystèmes de courants. Leur utilisation dans la détermination des séquences de roches a été décrite par SHROCK, et nous avons traité ailleurs de l'importance des structures orientées dans l'analyse des courants et des bassins anciens. Pour être utile, un atlas photographique de structures sédimentaires doit avoir une organisation logique, laquelle présuppose une classification. Nous allons donc définir plus avant notre sujet, et présenter la classification et la nomenclature qui peuvent s'appliquer aux formes sédimentaires.

Définition et étendue du sujet

Ce qui nous concerne ici est l'étude des *formes sédimentaires primaires*. «Primaire» indique des formes de genèse contemporaine du dépôt, ou légèrement postérieure, mais précédant toujours la consolidation de la roche où on les rencontre.

Les formes primaires traitées ici comprennent la stratification, en particulier la forme extérieure de la couche, la continuité et l'uniformité de son épaisseur, sa structure et son organisation internes, les hiéroglyphes ou accidents des plans de séparation ou de stratification, et les figures produites par la déformation des sédiments mous. En bref, cet atlas traite de divers types de dépôt des sédiments.

Nous en avons exclu toutes les structures secondaires formées par des mouvements tectoniques. Mais nous y laissons les failles, plis, et structures apparentées, qui se sont formés avant la consolidation par suite de mouvements dûs à une charge inégale ou à un éboulement ou glissement sur une pente. Nous avons omis toute considération des cassures telles que la fissilité et les joints. Ces structures sont peut-être déterminées par des formes primaires, mais elles présupposent un état consolidé. Nous avons exclu aussi les concrétions ou formes d'accrétion, telles que nodules, concrétions, septaria, etc., bien qu'on puisse prétendre qu'elles soient parfois formées pendant la sédimentation ou avant la consolidation. Après quelques hésitations, nous avons fini par exclure les structures dues au gel et à la cryoturbation, y compris les involutions et coins de glace, de même que les formations pédologiques. De même, nous ne faisons qu'effleurer le sujet des structures produites par les organismes vivants. Les pistes, empreintes et tubulures d'origine organique sont mentionnées en passant, mais une discussion approfondie de ces ichnofossiles, comme on les appelle, serait sans doute trop spécialisée pour la plupart des lecteurs de l'Atlas. Le lecteur que ce sujet intéresse peut consulter les ouvrages d'ABEL, SEILACHER et CASTER. Nous avons traité de certaines autres structures d'origine organique, telles que les stromatolithes, parce qu'elles constituent un genre de stratification — un type de stratification dû à la croissance organique, ou du moins modifié par l'activité organique.

20

Classification des structures sédimentaires primaires

Nous n'avons pas connaissance d'un essai conséquent de classification de toutes les formes sédimentaires. On a toutefois proposé plusieurs classifications partielles. Les plus importantes sont celles qui concernent plus spécialement la stratification. Les plus complètes sont celles de Andrée (1915), Hoppe (1930), Andersen (1931), Zhenchuzhnikov (1940), Bruns (1954), Birkenmajer (1959), et Botvinkina (1960). Ces classifications ne concernent pas seulement les formes externes mais aussi la structure interne, et certaines comprennent même l'arrangement en séquences, souvent cyclique, de différentes roches. D'autres classifications partielles de structures sédimentaires sont celles des hiéroglyphes de Krejci-Graf (1932) et Vassoevich (1953). Parmi les classifications récentes d'empreintes de mur on peut citer celles de Crowell (1955) et Kuenen (1957). Encore plus restreintes comme sujet sont les diverses classifications de stratification croisée ou d'empreintes de rides.

Nous n'avons pas l'intention de passer en revue ces diverses classifications, ni d'essayer de créer tout un système de notre crû. Nous offrons au contraire un système très simple qui permette d'organiser le contenu de l'atlas d'une manière logique. Nous avons donc divisé les formes sédimentaires en quatre groupes principaux:

I. Stratification externe
II. Stratification interne
III. Accidents des surfaces de stratification
IV. Stratification dérangée ou déformée.

Il existe certaines structures qui ne se placent exactement dans aucune de ces catégories, ou que l'on pourrait placer aussi bien dans l'une que dans l'autre. Les perforations dues aux vers marins, par exemple, peuvent dans certains cas être considérées comme des accidents de la surface de stratification mais peuvent être aussi des caractéristiques internes de la stratification, ou même peuvent représenter un certain type de stratification dérangée.

Stratification externe:

Il y a fort longtemps qu'on a reconnu et décrit la stratification, mais il n'en existe pas encore de classification tout à fait satisfaisante, et sur quoi l'accord soit fait. Nous offrons ici seulement un traitement minimum de cette question, un système qui ne serait peut-être pas commode dans la pratique, mais qui nous permet de classer ou d'ordonner nos photographies de séquences lithologiques.

L'unité de base est une couche unique, ou *unité de sédimentation*, qu'on a définie comme «l'épaisseur de sédiment déposée dans des conditions physiques essentiellement inchangées» (Otto 1938). L'adverbe «essentiellement» permet toutefois des fluctuations dans la direction ou la vitesse du courant qui peuvent produire des laminations internes, ou des changements progressifs, sans interruptions brusques, et qui produisent une structure granoclassée. Les surfaces de stratification qui séparent les diverses strates sont révélées par le fait que les lits successifs ont une texture ou une composition différente.

Les caractères essentiels d'une couche ou unité sédimentaire, sont 1. son épaisseur, ou son échelle, 2. son étendue latérale et 3. sa structure interne. Les séquences sédimentaires peuvent être décrites indépendamment de leur organisation ou structure interne. La discussion de cette dernière sera reportée à une autre section.

Il nous faut d'abord considérer l'épaisseur. Les couches sont souvent décrites comme étant épaisses, minces ou très fines (feuilletées). On a proposé des définitions plus précises de tels termes. Payne (1942) par exemple utilise le terme «strate» pour des lits de plus d'1 cm d'épaisseur, nettement distincts des unités semblables qui les encadrent. Il réserve le terme «lamination» pour des couches de moins d'1 cm d'épaisseur. McKee et Weir (1953) limitent de même «lamination» à des couches de moins d'1 cm d'épaisseur. Ils appellent «très minces» des couches de 1 à 5 cm d'épaisseur, minces de 5 à 60 cm, épaisses de 60 à 120 cm, et utilisent l'expression «en lit épais» pour celles de plus de 120 cm. Ces essais de classification sont basés sur des observations quantitatives. Ce même sujet a été traité aussi par des auteurs tels que INGRAM (1954), GRAY (1955) et BOKMAN (1956).

Si les couches sont nombreuses, il faut alors considérer leur épaisseur *moyenne*, ainsi que les variations que l'on peut exprimer en termes d'*écart type* ou mieux de *coefficient de variation*. Certaines séquences ont un coefficient de variation extrêmement peu élevé (pl. 2 et 8 B), d'autres pas. Certains auteurs ont essayé d'analyser les variations d'épaisseur en termes de cycles, tels que les varves, alors que d'autres ont remarqué que les variations suivent une distribution logarithmique gaussienne (SIMONEN et KOUVO 1951, fig. 4 et ATKINSON 1962, p. 357). Il est bien évident que le géologue sur le terrain devrait être conscient de l'existence possible de séquences répétitives ou cycliques (cf. pl. 4 et 9 B). De telles variations peuvent être mises en corrélation avec des variations lithologiques correspondantes.

Outre son épaisseur absolue, une couche est caractérisée par sa variation latérale d'épaisseur. Certains lits sont remarquablement uniformes sur toute la coupe. D'autres s'amincissent et s'élargissent, ou forment une stratification ondulée (cf. pl. 17 A et 11 B). Le cas extrême est représenté par des couches qui s'amenuisent jusqu'à disparaître. Ces discontinuités latérales peuvent être le résultat de discordances locales, de phénomènes de creusement et remplissage, ou de dichotomies (wash-out) (pl. 19 A à 21 A), ou encore d'un type de boudinage de sédiments mous (pl. 15 A). Certains types lithologiques, tels que les calcaires, renferment des couches qui, si on les suit latéralement, passent à une série de nodules. Dans certains cas, toute la formation montre une stratification en nodules (pl. 15 B).

Nous pouvons classer les séquences de couches entre plusieurs groupes formant une série depuis ceux qui montrent le caractère le plus régulier, ou ordonné, jusqu'à ceux qui atteignent le maximum d'irrégularité, ou de désordre. Par exemple, on peut imaginer une séquence dans laquelle les couches sont toute d'épaisseur égale ou presque, d'épaisseur continue latéralement

et apparemment de très grande étendue. On aurait là le plus haut degré d'ordonnance (pl. 2). D'autre part, on peut se représenter un groupe de couches d'épaisseur différente, chaque couche d'épaisseur variable, et chacune discontinue au sein de la coupe (pl. 21 B). On aurait alors un arrangement désordonné. Il existe divers types intermédiaires.

Pour illustrer ces concepts, on peut reconnaître 4 classes:

Classe A: 1. couches d'épaisseur égale ou presque
2. couches d'épaisseur uniforme latéralement
3. couches continues

Classe B: 1. couches d'épaisseur *inégale*
2. couches d'épaisseur uniforme latéralement
3. couches continues

Classe C: 1. couches d'épaisseur *inégale*
2. couches d'épaisseur *variable* latéralement
3. couches continues

Classe D: 1. couches d'épaisseur *inégale*
2. couches d'épaisseur *variable* latéralement
3. couches *discontinues*

La classe A est illustrée dans les planches 1 A, 2 et 8 B, la classe B dans les planches 1 B, 3, 4, 6 A et 7, la classe C dans les planches 11, 12 et 13, et la classe D dans les planches 15 A, 17 et 21 B.

Stratification interne:

L'organisation interne d'une couche est aussi importante que son aspect externe. Divers types d'arrangement interne sont possibles. Certains types de structure interne sont par exemple:

Classe A: Massive
Classe B: Litée
1. Litage horizontal
2. Stratification croisée (simple ou multiple)
Classe C: Granoclassée
Classe D: Imbriquée (galets), et autres textures internes orientées

22

Classe E : Structures de croissance (telles qu'on en trouve dans les calcaires stromatolithiques, certains calcaires récifaux, travertins, etc. en un mot, les calcaires construits)

Les couches peuvent être *massives*, c'est-à-dire sans structure interne apparente (pl. 25 et 26). Des travaux récents (HAMBLIN 1962) semblent indiquer que leur aspect massif est illusoire, car beaucoup de couches «massives» montrent une structure feuilletée ou d'autres accidents quand on les examine aux rayons X (*cf.* pl. 23).

Les couches peuvent aussi être *litées*. Le litage est la marque de changements mineurs et de courte durée dans les conditions de milieu pendant le dépôt de la couche. Le litage peut être en feuillets parallèles les uns aux autres et parallèles aux surfaces limites de la couche (pl. 8 A, 20 B, partie inférieure). Les grès ainsi lités se divisent aisément le long des feuillets et forment d'excellentes dalles. Les surfaces de délit montrent fréquemment des linéations de délit (pl. 76 A, 76 B).

Dans les matériaux granuleux de la taille de sable, les feuillets sont souvent obliques par rapport aux surfaces limites de la couche. Ces feuillets en diagonale ou en *stratification oblique* sont très fréquents dans certains faciès sableux. De telles couches obliques, soit dans des grès soit dans des calcaires, sont abondamment illustrées (pl. 29 B, 37). La stratification oblique peut être à très petite échelle, ou à très grande échelle (cf. pl. 36, 38 B). Dans bien des cas, toute la couche est en cause, et les couches frontales vont de la base au sommet de la couche (pl. 34 A) ; dans d'autres cas, la couche est composée de plusieurs zones à stratification oblique (pl. 37). Les couches frontales peuvent être rectilignes en coupe, ou arquées et tangentielles à la base de la couche (pl. 32 A, 32 B). Dans certains cas, les couches frontales sont redressées, ou même renversées ou plissotées (pl. 110). Dans les situations complexes, la couche peut être massive à la base, litée horizontalement au centre, et à lits obliques au sommet.

Un autre caractère des couches est le classement des sédiments. Le *granoclasse-ment*, en fait, est une variation régulière de texture. C'est un caractère important, en général bien visible à l'œil nu, et de ce fait on a choisi plusieurs photographies qui le montrent (pl. 1 B, 42 A, 42 B).

Le granoclassement peut se rencontrer dans des graviers, sables ou limons (poudres). Il est particulièrement caractéristique des dépôts à alternance répétée des flyschs — des turbidites sableuses. Le classement est visible même à distance grâce à la différence de netteté des contacts inférieur et supérieur avec les couches encaissantes des grès (pl. 1 B). De plus près, on peut voir à l'œil nu le changement dans la taille des grains, qui est lié à un changement de couleur de claire (sable) à foncée (matériau fin ou argile) (pl. 8 B, 42 B). Dans les terrains métamorphiques, le clivage des ardoises se produit obliquement par rapport à la direction des lits, et peut se réfracter (changer d'angle) en passant de la partie sableuse claire à l'ardoise fine plus foncée (pl. 6 A).

Bien qu'en général le granoclassement et la stratification croisée s'excluent mutuellement, il existe des couches qui montrent ces deux phénomènes, le plus souvent granoclassement à la base et stratification croisée à petite échelle au sommet. Cette dernière peut-être associée de très près avec une stratification ondulée.

L'*arrangement des matériaux* est une autre propriété fondamentale des couches. Il relève, comme le grain, de la texture plutôt que de la structure. En général, il faut un microscope pour reconnaître la trame des sables ou des matériaux plus fins. Mais dans le cas des graviers, elle se reconnait facilement sur le terrain. D'une manière générale, la plus grande dimension des cailloux est disposée parallèlement à la surface de dépôt. Toutefois, dans beaucoup de graviers fluviaux, les cailloux plats et les gros galets sont imbriqués de telle sorte que les cailloux se recouvrent comme les tuiles d'un toit, et sont inclinés vers l'amont (pl. 43 A). Cette disposition se voit le mieux sur les coupes verticales parallèles à la direction du courant. Dans certains conglomérats intraformationnels, les galets sont de champ et on a un conglomérat de champ.

Les types de stratification et de structure interne décrits jusqu'ici se rencontrent dans les sédiments clastiques et résultent généralement d'un dépôt à partir de liquides en mouvement. Il nous faut aussi considérer rapidement la stratification de dépôts non clastiques. Bien des calcaires sont déposés par un courant et montrent les caractères décrits ci-dessus, y compris la stratification oblique et le granoclassement. D'autres se forment *in situ*, par exemple par l'action d'organismes vivants. Le litage de ces calcaires construits est très différent de celui qui est produit par les courants. La *stratification stromatolithique*, par exemple, est une stratification de croissance, qu'on croit liée à la formation de feutrages d'algues cyanophycées. La stratification stromatolithique est très variable mais est généralement caractérisée par un feuilletage. Contrairement aux feuillets déposés par les courants, ceux qui sont dûs à la croissance organique montrent des irrégularités telles que des bifurcations, replis, et dispositions en formes bombées diverses. Ce n'est pas ici le lieu d'illustrer les nombreux aspects des stromatolithes. Le lecteur que ceci intéresse est invité à consulter les ouvrages spécialisés consacrés à ce sujet (voir par exemple CLOUD 1942, et REZAK 1957). Sur le terrain, il est probable que les « stratifications d'algues » qui ne sont pas nettement organisées en structures bien définies échappent souvent à l'œil du géologue. Nous avons illustré plusieurs types de stratification stromatolithique, avec ou sans structure nette (pl. 48, 49 A, 49 B, 50 B).

D'autres calcaires construits peuvent avoir des caractères internes liés à la croissance d'organismes vivants tels que coraux, stromatopores, et autres; dans d'autres cas, les structures peuvent être dues à une croissance inorganique; c'est le cas des structures rubannées des travertins ou de l'onyx.

Figures et accidents des plans de stratification:

Lorsqu'on les examine de près, bien des plans de stratification montrent des figures diverses. On peut les classer suivant qu'elles se trouvent à la *face inférieure* de la couche, à son *sommet*, ou sur des *plans internes*. Dans chacune de ces classes on a des structures diverses comme suit:

Classe A: A la face inférieure des couches:
    1. Figures de charge (moulages de charge)
    2. Structures dues au courant (empreintes de creusement et d'outils)
    3. Empreintes organiques (moulages de pistes et de tubulures; fucoïdes)

Classe B: Au sein de la couche (linéation de délit)

Classe C: A la face supérieure de la couche:
    1. Empreintes de rides
    2. Empreintes d'érosion (rigoles de plage, moulages de cupules en croissant)
    3. Fossettes et petites dépressions (empreintes bullaires, gouttes de pluie, etc.)
    4. Fentes de dessiccation et leurs moulages, moulages de cristaux de glace ou de sel, etc.
    5. Empreintes d'organismes (pistes, empreintes de pas, etc.)

Les empreintes des faces inférieures, ou hiéroglyphes de types divers, sont connues depuis longtemps, mais ce n'est que depuis une dizaine d'années qu'on les étudie de près. Il en résulte que de nombreuses structures nouvelles ont été récemment décrites, et ont reçu des noms qui forment une nomenclature étendue et confuse, comprenant à la fois des termes génétiques et descriptifs.

Les empreintes des faces inférieures de couches se rencontrent sur des grès ou calcaires surmontant des schistes argileux. On les considère comme des moulages parce qu'elles sont les remplissages de dépressions formées à la surface de la boue sous-jacente. Elles prennent naissance 1. par suite d'une charge inégale sur la boue molle hydroplastique, 2. par l'action des courants à la surface de la boue, ou encore 3. à la suite de l'activité d'organismes vivants sur cette surface.

Les *empreintes de charge* sont souvent appelées moulages de charge bien qu'elles soient formées par l'enfoncement de sable sus-jacent dans la boue, et non pas par remplissage de dépressions. Les empreintes de charge varient quant à leur taille, leur forme et leur abondance (pl. 52A à 54A). Certaines de ces variations ont reçu des noms spéciaux, par exemple les empreintes de charge cordées (pl. 54B) et les empreintes en écaille. Généralement ces structures se montrent comme des irrégularités bulbeuses à la face inférieure des couches. En coupe, elles apparaissent comme des sinuosités, telle la soit-disant «flame structure» (pl. 53B).

Les *figures de courant* se répartissent en deux groupes, celles qui sont formées par le courant lui-même, et celles qui sont formées par des «outils», tels que des fragments de coquilles ou de bois, des écailles de schiste ou des grains de sable, emportés par le courant. Les empreintes du premier type sont des *empreintes de creusement*: les plus importantes sont les sillons (flutes) qui se montrent comme des «flute casts» à la face inférieure de bien des grès, et aussi de certains calcaires.

Les *marques d'outils* forment un groupe de structures très variées. Certaines sont produites par le mouvement d'un outil graveur, en contact continu avec le fond. Telles sont les stries et cannelures qui se voient sous forme de moulages (pl. 61 à 67). Ces empreintes sont analogues à celles que laissent des fragments rocheux inclus dans la glace en mouvement — les stries et cannelures des dallages glaciaires (pl. 73A à 74). D'autres empreintes d'outils résultent du contact intermittent d'objets divers avec le fond. Les structures correspondantes, à la face inférieure des grès sus-jacents, sont appelées empreintes de «saut» ou de rebondissement (pl. 68), éraflures (pl. 68, 69A), ou encore «prod casts» (pl. 61, 66, 68). Si l'objet se déplace en roulant, on a divers types d'empreintes de roulement. Dans certains cas les objets qui roulent peuvent être reconnus par leur «signature» sur le fond mou (*cf.* empreintes laissées par des vertèbres de poisson, pl. 68). Si l'outil graveur est animé d'un mouvement vibratoire, on obtient des empreintes de vibration et des empreintes en chevrons (pl. 62, 63A), ou encore des cannelures tuyautées.

Les empreintes de glissement (pl. 65B) sont légèrement différentes, et produites par le glissement de masses de débris végétaux, de boue, ou d'autres matériaux. Elles se montrent comme des stries multiples, parallèles, et visiblement toutes produites en même temps par le même objet et non par des fragments séparés qui se meuvent indépendamment.

La distinction entre les empreintes de charge et de courant n'est pas toujours évidente. Bien des figures de charge ont débuté comme marques de courant, qui se sont agrandies, gonflées et déformées à la suite de la pression (*cf.* pl. 63A, 63B).

Les faces inférieures des couches montrent aussi divers types d'empreintes irrégulières laissées par des organismes vivants (ichnofossiles). Elle comprennent des *empreintes de pas*, des *pistes* et des *terriers*. Celles qu'on rencontre sur les grès ont en général été formées à la surface de la boue sous-jacente, puis moulées et conservées ainsi (pl. 70A, 70B). Mais certaines sont subséquentes au dépôt et s'étendent vers le haut et le bas en travers des strates (pl. 116B).

Certains grès se délitent facilement le long des plans de litage internes. Ces surfaces peuvent montrer des linéations de délit, ou «de courant» (cf. pl. 76A, 76B) ou aussi les figures de micro-stratification croisée appelées structures «en côtes et sillons» (rib and furrow) par STOKES (1953, pp. 17—21).

Les faces supérieures des couches peuvent aussi montrer divers types de figures, y compris les nombreux réseaux de vagues de sable petites ou grandes. Les plus petites comprennent les rides de plage ou réseaux de rides d'oscillation ou de rides de vagues (pl. 85B, 86), les réseaux d'interférence de rides, rides en fossettes ou «nids de têtards» (pl. 87A), et les arrangements moins réguliers de rides linguiformes ou arquées en croissant formés par les courants (pl. 84A, 84B). Les rides géantes de régions fluviatiles, ou marines, ou de zones de balancement des marées, sont de taille plus importante (pl. 27, 89A, 89B). Les plus grandes de toutes sont les vagues de sable des champs de dunes subaériens ou sous-aquatiques (pl. 28). Les plus grandes vagues de sable se rencontrent rarement comme accidents de surface des couches sédimentaires. Elles sont indiquées le plus souvent par la stratification croisée au sein du sable.

D'autres accidents de surface rarement (peut-être jamais) présents dans les roches sont les croissants de plage (pl. 98 B).

La face supérieure des couches sableuses est soumise aussi à l'action des courants. Ici se placent des figures telles que rigoles de plage, divergentes ou convergentes (pl. 92 A), cupules en croissant (pl. 91 B, 93 A, 93 B) et marques du jet de rive (swash-marks). Les faces supérieures de couches à grain fin sont souvent marquées de petites fossettes et dépressions telles que des empreintes de gouttes de pluie (pl. 94) et empreintes d'embruns ou de grêle. Celles-ci ressemblent beaucoup à des cratères de bulles de gaz. Il existe aussi d'autres accidents du même type mais de plus grande taille (*cf.* spring pits). Les rares et curieux filons cylindriques du grès s'apparentent peut-être à ces structures (pl. 115 A). Si le sédiment est une vase, susceptible de retrait à la dessiccation, sa surface peut se diviser en blocs polygonaux séparés par un réseau de fentes de dessiccation (pl. 94). Un réseau de fentes quelque peu comparable se forme par sublimation de cristaux de glace (pl. 96 A). A ces formes s'apparentent naturellement les moulages de fentes de dessiccation (pl. 96 B) et les moulages de cristaux de glace — qui proviennent du remplissage des fentes ou des ouvertures par du sable ou autres matériaux. On rencontre aussi des moulages de cristaux de sel (pl. 98 A) qui doivent leur origine au remplissage de cavités formées par dissolution d'une halite ou d'autres cristaux solubles qui se trouvaient inclus dans la boue.

Les surfaces supérieures de sédiments à grain fin sont marquées aussi de diverses pistes et empreintes d'origine organique (pl. 99).

Stratification déformée par des processus contemporains de la fin de la sédimentation:

Les sédiments montrent des structures diverses qu'on a attribuées à des dé-formations produites pendant, ou peu après, la sédimentation, et avant la consolidation de la roche. On attribue ce genre de déformation à des processus non tectoniques, tels qu'effondrements, glissements par gravité, ou écoulements sur une pente. Les effets de ces processus varient depuis une modification légère de la stratification et autres structures primaires, jusqu'à des plissements majeurs, formation de brèches intraformationnelles ou même liquéfaction et injection suivies de la disparition totale de la structure originelle. Les structures de déformation se rangent en plusieurs classes:

Classe A: Structures d'effondrement et de charge (empreintes de charge, structures en boules-et-coussins, etc.)

Classe B: Couches contournées (convolutions)

Classe C: Structures de glissement (plis, failles, brèches)

Classe D: Structures d'injection (filons clastiques, sills clastiques, etc.)

Classe E: Structures d'origine organique (terriers, «couches brassées», etc.)

La déformation va de déplacements purement verticaux à de grands déplacements latéraux. Dans la première catégorie se rangent les structures appelées ici «en boules-et-coussins» ou encore «flow rolls», pseudonodules et autres figures du même genre. On les rencontre dans les grès à grain fin (pl. 110 b à 101 B) aussi bien que dans les calcaires de texture comparable déposés mécaniquement (pl. 102 A, 102 B). On les attribue généralement à un glissement, mais elles résultent plus probablement d'un réajustement mécanique à la suite d'un affaissement.

D'origine incertaine est la structure appelée stratification convolute ou ondulée, appelée aussi stratification «en boucles». Elle est caractéristique de sables fins et ne concerne que les feuillets internes du lit, et non le reste de la couche qui demeure intacte. Les convolutions sont caractérisées par l'extrême désorganisation des plissotements, l'absence de failles, et par le fait qu'elles sont confinées au sein d'une couche unique relativement mince et autrement intacte. Elles sont illustrées en plan (pl. 106 A) et en coupe (pl. 106 B).

La déformation à plus grande échelle est généralement attribuée à l'écoulement, en particulier aux glissements sous-aquatiques importants. Cet effet concerne de nombreuses couches, et a pour résultat ici l'empilement de strates déformées, ailleurs des lacunes ou interruptions de la stratification. Dans les dépôts de la fin de l'époque glaciaire, on a attribué des structures de ce genre au chevauchement par des glaciers ou à des icebergs échoués. Dans certains cas il y a décollement des plis (pl. 107 B, 108 A, 108 B). La formation de brèches (pl. 105 A), de failles (pl. 111 B) et de structures fluidales (pl. 112) accompagne le glissement sous-aquatique (subsolifluxion).

Dans certaines conditions, des sables ou limons gorgés d'eau peuvent atteindre un état où ils se comportent comme un liquide qui coule et peut être injecté dans les sédiments voisins. Ces mouvements produisent des structures telles que filons clastiques et sills clastiques, et volcans de sable. Les filons clastiques du grès peuvent avoir des parois nettes et rectilignes, comme dans la planche 113 B, et dans certains cas contenir des matériaux conglomératiques, ou bien il peuvent être de forme très irrégulière comme dans la planche 113 A. Les volcans de sable sont rares, et on les a décrits comme des structures reposant sur une nappe de glissement (pl. 114).

D'autres structures problématiques, d'origine inconnue, sont les «whirl balls» et boules de grès (pl. 112 A).

Cette revue des structures de déformation serait incomplète sans une mention du rôle que jouent les organismes vivants dans la destruction du litage primaire, et l'homogénéisation de la roche qui en résulte. Les organismes mangeurs de boue et ceux qui creusent des trous ou terriers peuvent être si nombreux que le sédiment est criblé de perforations, la stratification primaire disparaît, et il en résulte un dépôt «brassé», privé de structure originelle.

Introducción

El estudio de las estructuras sedimentarias es tan antiguo como el de la geología misma. La *estratificatión* constituye el rasgo más característico y aunque con excepciones, resulta un criterio seguro para corroborar su origen sedimentario. Al igual que la textura y composición, la estructura es una propiedad inherente de las rocas. Mientras la textura se refiere a la relación entre los componentes de una roca, la estructura está vinculada al ordenamiento y discontinuidades entre las partes de la misma. Por lo tanto comprende a caracteres amplios, mejor estudiados en afloramientos que en muestras o cortes delgados. Ella se define más claramente con ejemplos, entre los que se incluyen diversos aspectos de la estratificación, tales como *estratificación entrecruzada, gradada, óndulas* y *grietas de desecación*.

En muchos casos suelen presentarse dificultades para poder discernir entre textura y estructura. ¿La disposición imbricada de los rodados en un conglomerado, es un atributo de la textura o estructura? En este caso se trata en realidad de la orientación de los clastos (fábrica) y por ello es un rasgo textural. Sin embargo creemos conveniente incluir algunas ilustraciones de este tipo. ¿Los *surcos* y *estrías* halladas en algunos planos de estraficación, son realmente estructuras? En realidad es impropio considerarlas texturas.

Es necesario destacar que algunas estructuras dependen en realidad de la textura y no están relacionadas con la composición de la roca. Las *óndulas* y la *estratificación entrecruzada*, por ejemplo, se forman en sedimentos granulares del orden de las arenas. Es común la formación de *óndulas* en arenas silíceas y calcáreas, citándose algunos casos menos frecuentes de are-

nas constituidas por halita (KAUFMAN y SLAWSON, 1950). La estructura *lineación por corriente* (lám. 76 B) también resulta dependiente de la textura. Tal estructura se manifiesta por la preferencia en la separación de ciertas areniscas a lo largo de planos, lo cual parece estar vinculado a la fábrica de los granos.

Entre otros ejemplos de dependencia textural, pueden mencionarse las *grietas de desecación* y *estructuras almohadilladas*, que se originan en ciertos fangos y arenas (silíceas o calcáreas) respectivamente.

También se conocen estructuras dependientes de la composición de los sedimentos, como en el caso de ciertas estratificaciones rítmicas, referidas a depósitos de sal, anhidrita y aquéllas nodulares en calizas. Del mismo modo la estratificación estromatolítica está restringida a calizas y dolomías. Sin embargo, la mayoría de las estructuras sedimentarias que figuran en este atlas, se han originado por la acción de corrientes y deformación de sedimentos aún no consolidados, de composición variada.

Es de destacar que ciertas estructuras caracteristicas en sedimentos clásticos, también se encuentran en algunas rocas ígneas básicas que por su fluidalidad se disponen en capas. Como ejemplos puede citarse la *estratificación gradada* y *entrecruzada* (GATES, 1961, figs. 15 y 18). Un registro fotográfico de cuerpos ígneos con estructuras sedimentarias primarias es un proyecto cuya realización sería valiosa, pero no ha sido emprendida por los autores en el presente trabajo. Por otro lado, los materiales piroclásticos transportados por aire y agua pueden desarrollar estructuras características en sedimentos no volcánicos.

No se puede definir una estructura con la misma precisión de detalle que una figura geométrica o cuerpo sólido. Como en el caso de los fósiles y formas orgánicas en general, para su comprensión, se requiere el auxilio de material fotográfico. La dificultad en definir analíticamente a los mismos, no significa imposibilidad de efectuar observaciones detalladas de ellas. Además, las estructuras sedimentarias presentan una gran variedad de rasgos, por lo cual una única fotografía no resulta representativa, de ahí la conveniencia de publicar este atlas.

No es nuestra intención tratar el origen de estas estructuras, su valor estratigráfico o su importancia en el mapeo de sistemas de paleocorrientes. El primer tópico ha sido ampliamente analizado por SHROCK (1948) y el segundo por los autores de este atlas (1963). Como toda obra similar deberá ordenarse en base a una clasificación orgánica, analizada a continuación conjuntamente con la nomenclatura empleada.

Definición y campo

El presente atlas trata exclusivamente a las *estructuras sedimentarias primarias*. Por "primaria" queremos significar que ellas se han originado en el momento de la depositación o muy poco tiemp antes o después de la consolidación del sedimento.

Las estructuras primarias comprenden a todo lo relacionado con la estratificación, especialmente la forma externa del estrato, su continuidad y uniformidad del espesor, la estructura interna y ordenamiento del mismo, los hieroglifos o marcas existentes en los planos interfaciales o de estratificación y a aquellas estructuras producidas por deformación de sedimentos aún en estado plástico.

Se han omitido las estructuras secundarias originadas por movimientos tectónicos, pero se incluyen aquellos pliegues y fallas que han tenido lugar anteriormente a la consolidación de los sedimentos por carga diferencial y deslizamientos. Además, hemos omitido toda consideración acerca de la manera de partirse las rocas, tales como la

fisilidad y el *diaclasamiento*. Dichas estructuras, aunque tal vez controladas por rasgos primarios, presuponen un estado de consolidación. También se han excluido las estructuras *concrecionales* y *acrecionales*, entre ellas *nódulos, concreciones, septarias* y otras similares, aunque se afirma que algunas pueden formarse durante la sedimentación o con anterioridad a la consolidación. Ante ciertas dudas hemos excluido las estructuras debidas a fenómenos de congelamiento o crioturbación, incluyendo involuciones y grietas producidas por el hielo y aquéllas relacionadas con la formación de suelos. Las estructuras originadas por organismos son tratadas someramente; figuran *huellas, rastros* y *cavidades tubulares*. Para una información más detallada se recomiendan los trabajos de ABEL (1935), SEILACHER (1953) y CASTER (1957). Hemos incluido otras pocas estructuras de origen orgánico, tal como *estromatolitas*, puesto que son formas de estratificación originada por crecimiento o al menos estratificación modificada por actividad biológica.

Clasificación de las estructuras sedimentarias primarias

Los autores no tienen conocimiento de la existencia de una clasificación apropiada, que comprenda a todas las estructuras sedimentarias. Existen sí, trabajos que tratan la estratificación primaria donde se proponen clasificaciones parciales. Entre ellos se citan los de: ANDRÉE (1915), HOPPE (1930), ANDERSEN (1931), ZHEN-CHUZHNIKOV (1940), BRUNS (1954), BIR-KENMAYER (1959) y BOTVINKINA (1960). Estas clasificaciones versan no sólo sobre la forma externa, sino también sobre la estructura interna y algunas incluyen secuencias, en parte cíclicas, de distinta litología. Otras clasificaciones parciales vinculadas a los hieroglifos figuran en los trabajos realizados por KREJCI-GRAF (1932) y VASSOEVICH (1953). Sobre *marcas de base* se conocen clasificaciones recientes propuestas por CROWELL (1955) y KUENEN (1957). Se han publicado también clasificaciones más restringidas que tratan sobre la *estratificación entrecruzada y óndulas*.

No es nuestra finalidad analizar los ensayos de clasificación conocidos hasta el presente. Para el propósito que nos guía proponemos un esquema simple, con el objeto de lograr un ordenamiento lógico del material ilustrativo; por ello hemos dividido a las estructuras sedimentarias en cuatro grupos principales:

I. Formas externas de la estratificación
II. Ordenamiento interno y estructura de la estratificación
III. Marcas e irregularidades en los planos de estratificación
IV. Deformaciones y perturbaciones de la estratificación

Unas pocas estructuras no se ubican fácilmente en alguno de estos grupos o podrían ser incluidas en más de una clase. Las cavidades originadas por gusanos, por ejemplo, pueden ser consideradas en algunos casos, como un rasgo del plano de estratificación, pero también estas cavidades suelen desarrollarse en el seno de un estrato o aún perturbar la estratificación. Los *rodados de arcilla*, los *cuspilitos* y los *diques clásticos* son ejemplos de estructuras de difícil clasificación.

Formas externas de la estratificación

Aunque el concepto de estratificación es conocido desde hace mucho tiempo, no se ha llegado a un criterio uniforme para su clasificación. El sistema aquí adoptado tal vez no sea útil o conveniente en muchos casos, pero nos ha permitido clasificar y ordenar todo el material fotográfico existente.
La entidad básica es un estrato o *unidad sedimentaria*, definida como "aquel espesor de sedimentos depositados en condiciones físicas esencialmente constantes" (OTTO, 1938). La calificación de "esencialmente" se refiere a fluctuaciones de corto período en la dirección de la corriente o velocidad, que pueden originar laminación interna o bien cambios progresivos que conducen a estructuras gradadas. Los planos de estrati-

ficación que separan distintos estratos se manifiestan por diferencias en la textura y composición del conjunto sedimentario.
Las propiedades fundamentales de una unidad sedimentaria comprenden: 1. espesores, 2. dimensiones laterales y 3. estructura interna. Las secuencias estratificadas pueden ser descriptas independientemente de su ordenamiento interno o estructura. Diferiremos la discusión de la última para tratarla en sección aparte.
Con respecto al espesor, a los estratos se los considera de estratificación gruesa, fina o laminada. Sin embargo, se han propuesto definiciones más precisas. PAYNE (1942) utiliza el vocablo *estrato* para capas de espesor mayor de un centímetro, que se hallan nítidamente diferenciadas de las infra y suprayacentes. Este autor, al igual que McKEE y WEIR (1953), restringe el empleo del vocablo *laminación* para capas de espesor menor de un centímetro. Estos últimos adoptan la siguiente clasificación:

Estratificación muy fina:
1 a 5 cm de espesor
Estratificación fina:
5 a 60 cm de espesor
Estratificación gruesa:
60 a 120 cm de espesor
Estratificación muy gruesa:
mayor de 120 cm de espesor

Entre otros autores que han tratado el problema del espesor, figuran INGRAM (1954), GRAY (1955) y BOKMAN (1956).
Cuando se desea caracterizar una formación, es necesario tener en cuenta el *espesor medio* y las variaciones de espesor espresadas como *desviación standart* o *coeficiente de variación*. Algunas secuencias tienen un coeficiente de variación bajo (láms. 2 y 8 B). Ciertos autores han analizado las variaciones de espesor en función de ciclos, como en el caso de los varves (lám. 1 A); otros han determinado que las variaciones en el espesor es en muchos casos *log normal* (SIMONEN y KOUVO, 1951, fig. 4 y ATKINSON, 1962, p. 357). Los geólogos deben tener en cuenta las posibles repeticiones cíclicas en el espesor de los estratos (compárese láms. 4 y 9 B). Tales variaciones están vinculadas a los correspondientes cambios litológicos.

Además de su espesor absoluto, un estrato está caracterizado por sus variaciones de espesor en el sentido lateral. Algunos estratos presentan uniformidad en el sentido lateral, otros son irregulares o desarrollan estratificación ondulada (compárese láms. 17B y 11B). El caso extremo está representado por estratos que se acuñan, variaciones que se deben a discordancias locales como *relleno de cauces* (láms. 19A a 21A) o a cierto tipo deformación por fuerzas tensionales (lám. 15A). En ciertas calizas estratificadas se observa un pasaje lateral con desarrollo nodular; en otras la estructura comprende a toda la formación (lám. 15B).

Es factible clasificar una secuencia de estratos en varios grupos que varían desde aquellos de carácter regular y ordenado, hasta aquellos de máxima irregularidad y desorden. Se puede imaginar, por ejemplo, una secuencia formada por estratos de igual o casi igual espesor, uniformes lateralmente y de amplia extensión. Tal secuencia representaría el mayor grado de orden que regían en los procesos de depositación (lám. 2). En contraposición, se puede imaginar un grupo de estratos de espesores distintos y discontinuos dentro de los límites de un afloramiento (lám. 21B). Entre ambas posibilidades pueden existir varios estados intermedios de ordenamiento.

Para ilustrar estos conceptos podemos definir cuatro clases:

Clase A: 1) Estratos de espesores constantes o aproximadamente constantes

2. Estratos de espesores lateralmente uniformes

3. Estratos continuos

Clase B: 1) Estratos de espesores *desiguales*

2. Estratos de espesores lateralmente uniformes

3. Estratos continuos

Clase C: 1. Estratos de espesores *desiguales*

2. Estratos de espesores lateralmente *variables*

3. Estratos continuos

Clase D: 1. Estratos de espesores *desiguales*

2. Estratos de espesores lateralmente *variables*

3. Estratos *discontinuos*

Ejemplos de la *Clase A:* láms. 1A, 2, 8B; *Clase B:* 1B, 3, 4, 6A, 7; *Clase C:* 11, 12, 13; *Clase D:* 15A, 17, 21B.

Ordenamiento interno y estructura de la estratificación

El ordenamiento interno de un estrato es tan importante como su forma externa. Entre los tipos principales de este grupo figuran:

Clase A: Masiva

Clase B: Laminada

1. Laminación horizontal

2. Laminación entrecruzada (simple y múltiple)

Clase C: Gradada

Clase D: Imbricada (gravas) y otros tipos de fábrica

Clase E: Estructuras de crecimiento (calizas estromatolíticas, arrecifes, travertinos, etc.)

Un estrato puede ser *masivo*, es decir sin estructura interna (láms. 25 y 26). En estudios recientes (HAMBLIN, 1962) se señala que el aspecto masivo resulta engañoso, pues muchos estratos "masivos" evidencian desarrollo laminado con el auxilio de rayos x (lám. 23).

Los estratos que presentan *laminación* reflejan las fluctuaciones de las condiciones físicas reinantes durante la depositación. La laminación puede ser primordialmente paralela entre sí y con relación a los planos que limitan al estrato (láms 8A y 20B, parte inferior). Estratos areniscosos laminados, comunmente se hienden con facilidad a lo largo de planos y constituyen las areniscas lajosas. Estas superficies de separación, por regla general, se deben a una orientación granular (láms. 76A y B).

31

En materiales del tipo de las arenas, la laminación aparece inclinada con respecto a las superficies limitantes de los estratos, siendo más común en facies de areniscas y calcarenitas; tal disposición se define como *estratificación entrecruzada* (láms. 29 B y 37). Su desarrollo tiene lugar en pequeña o gran escala (compárese láms. 36 y 38 B). En muchos casos abarca a todo el estrato (lám. 34 A); en otros el estrato puede ser compuesto, mostrando distintas zonas de laminación entrecruzada (lám. 37). Las capas frontales pueden ser rectas en perfil, o bien curvas o tangentes a la base del estrato (láms. 32 A a 33 B). En algunos casos las capas frontales aparecen casi verticales, volcadas o deformadas (lám. 110); en otros — combinados — el estrato puede ser masivo en su base, horizontalmente laminado en su parte media y con laminación entrecruzada en su techo.

Otra característica de ciertos sedimentos es la *gradación*, que consiste en una variación regular de la textura (láms. 1 B, 42 A, B). La misma tiene lugar en gravas, arenas y limos, siendo muy común en las llamadas turbiditas arenosas. La gradación resulta visible aún a cierta distancia porque los límites de los estratos resaltan nítidamente (lám. 1 B). El cambio en el tamaño del grano está vinculado al cambio del color, desde claro (arena) a oscuro (limo o arcilla) y puede observarse a simple vista (láms. 8 B y 42 B). En ambientes metamórficos el clivaje secundario forma ángulo con los cambios en la dirección de la estratificación o es refractado al pasar de niveles de grano grueso a niveles de grano fino (lám. 6 A).

Aunque la estratificación gradada y la entrecruzada se forman separadamente, existen estratos en los que ambas aparecen juntas, de tal forma que la gradada se observa en la parte inferior y la entrecruzada en el techo. Esta última puede estar vinculada con la estratificación intraplegada.

Otra propiedad es la *fábrica*, que constituye un atributo textural más que estructural. En la mayoría de los casos se requiere un exámen microscópico para la determinación exacta del tipo de fábrica, sin embargo en ciertas gravas puede realizarse directamente en el terreno. Generalmente las elongaciones de los rodados se orienta paralelamente a la superficie de depositación, pero en algunas gravas fluviales las formas tabulares adoptan una disposición imbricada con inclinación en sentido opuesto a la corriente (lám. 43 A). Ciertos conglomerados intraformacionales — sesgoconglomerados — también presentan un ordenamiento de sus elementos.

Los diferentes tipos de estratificación y estructuras internas consideradas, tienen lugar en sedimentos clásticos como resultado de la depositación producida por fluídos en movimiento. Pero aún falta referirse a la estratificación de depósitos no clásticos. Muchas calizas depositadas por corrientes exhiben las características antes descriptas, incluyendo la estratificación entrecruzada y gradada. Otras, las originadas "in situ", desarrollan otros tipos de estratificación. A este grupo pertenece la *estratificación estromatolítica*, formada por crecimiento laminado atribuido a la actividad biológica de algas. Sus variedades son múltiples pero su rasgo primordial es la laminación, comunmente con bifurcaciones, rizaduras y de trazado convexo hacia arriba. El lector que desee profundizar más este tema puede recurrir a los trabajos de CLOUD (1942) y REZAK (1957). Muchos biohermas no presentan estratificación definida y en consecuencia pasan inadvertidos para el geólogo. Ejemplos de estratificación estromatolítica bien o mal definidas pueden verse en las láminas 48, 49 A, B y 50 B.

Ciertas calizas desarrollan estructuras internas de crecimiento, originadas por organismos tales como corales y estromatoporoides. En otros casos, la estructura de crecimiento es de origen inorgánico como en algunos travertinos y ónices bandeados.

Marcas e irregularidades en los planos de estratificación

Muchos planos de estratificación, si se los examina minuciosamente, muestran una gran variedad de estructuras que pueden clasificarse de acuerdo a su ubicación en el techo, base o interior del estrato.

Clase A: *En la base del estrato*
1. Estructuras de carga (calco de carga)
2. Estructuras de corrientes (marcas de desbaste y marcas labradas)
3. Marcas producidas por organismos (calcos tubulares, calcos de huellas; fucoides)

Clase B: *Dentro del estrato* (lineación por corriente)

Clase C: *En el techo del estrato*
1. Ondulas u ondulitas
2. Marcas de erosión (marcas de escurrimiento, calco en herradura)
3. Hoyos y pequeñas marcas (hoyos de burbujéo, marcas de lluvia, etc.)
4. Grietas de desecación, calco de grietas de desecación, calco de cristales de hielo, calco de cristales de sal, ect.
5. Marcas producidas por organismos (rastros, pisadas, etc.)

Si bien las *marcas de base* (hieroglifos de varios tipos) son de antiguo conocidas, recién en la última década se ha intensificado su estudio, pero debido a que la terminología empleada comprende aspectos descriptivos y genéticos, la nomenclatura resultante es confusa.

Las marcas de base se encuentran generalmente en aquellas areniscas y calizas que cubren a lutitas. Estas estructuras se originan: 1. por resistencia diferencial a la compresión por parte de sedimentos hidroplásticos, 2. por la acción de corrientes sobre la superficie de sedimentos, 3. por la actividad biológica de organismos.

Las *estructuras de carga*, comunmente llamadas *calco de carga*, se forman como resultado del asentamiento de arenas suprayacentes a pelitas plásticas y no por relleno de depresiones. Se presentan en varias escalas de tamaño, forma y abundancia (láms. 52A a 54A). Algunas de sus variedades reciben nombres particulares, tales como *calcos de carga polilobulados* (lám. 54B) y *calcos de carga escamiformes*.

Normalmente estas estructuras aparecen como irregularidades bulbosas en la base del estrato, pero en perfil se manifiestan como sinuosidades y crestas, recibiendo la denominación de *estructura flamiforme* (lám. 53B).

Las marcas originadas por corrientes se dividen en dos grupos: las debidas a la corriente misma y las formadas por cuerpos transportados, como ser fragmentos de madera, de lutitas y granos de arena. Las primeras se denominan *marcas de desbaste*. A este grupo pertenecen los *turboglifos* hallados en la base de ciertas areniscas y calizas, de tamaños y formas muy variadas (láms. 55A a 59B). Las aguas turbulentas producen cauces y estructuras similares (láms. 20A y 20B).

Las *marcas labradas* también se presentan en formas muy variadas, como consecuencia de los múltiples cuerpos que pueden originarlas, por el contacto continuo sobre el sedimento sobre el cual se desplazan. Incluyen a *estriaciones* y *surcos* que aparecen como calcos (láms. 61 a 67). Estas marcas son análogas a las originadas por fragmentos incluidos en un glaciar en movimiento (láms. 73A a 74). Otras son el resultado del contacto intermitente de diferentes cuerpos sobre un fondo y reciben distintas denominaciones: *marcas de saltación regular* (lám. 68), *calco de roce* o *calco de empuje* (láms. 68 y 69A), *calco de punzamiento* (láms. 61, 66 y 68). Si el cuerpo es transportado por rolido se producen diferentes tipos de marcas. En algunos casos puede identificarse al cuerpo que las produce por la impresión que deja en el sedimento fangoso (véase marcas originadas por vértebras de peces en lámina 68). Cuando el cuerpo se desplaza con movimiento vibratorio, origina las *marcas de vibración, marcas espigadas* (láms. 62 y 63A) o *calco de surco espigado*.

Una estructura algo diferente es la denominada *calco de deslizamiento* (lám. 65B), formada por deslizamiento de masas compuestas por restos de plantas, fango o materiales semejantes. Aparecen como estriaciones múltiples y paralelas, originadas al mismo tiempo por un cuerpo en conjunto y no por movimiento individual de partículas.

La distinción entre marcas debidas a corrientes y carga no siempre resulta fácil. Muchas estructuras de carga fueron inicialmente marcas de corrientes, posteriormente rellenadas y deformadas (compárese láms. 63 A y 63 B).

Entre otras marcas presentes en los planos de estratificación, figuran una gran variedad de formas dejadas por organismos (icnofósiles). La mayoría de ellas aparecen en la base de areniscas como calco de marcas originadas en las capas subyacentes (láms. 70 A y 70 B). Algunas sin embargo, a posteriori de la deposición del estrato, se desarrollan en todo el espesor del mismo por actividad biológica (lám. 116 B).

Ciertas areniscas se hienden fácilmente a lo largo de planos de estratificación secundarios. En tales planos suelen aparecer estructuras — *lineación por corriente* — debidas a discontinuidades en la deposición orientada del material granular (compárese láms. 76 A y 76 B) y también la *estructura de costillas y surcos* (STOKES, 1953, p. 17—21) que refleja el aspecto en planta de la *laminación entrecruzada* (lám. 38 A).

En las superficies superiores de los estratos se encuentran una gran variedad de estructuras, entre ellas diversos diseños de ondulaciones pequeñas y grandes. Los primeros incluyen aquéllas de *oscilación* (láms. 85 B y 87 B), de *interferencia* (lám. 87 A) y *linguoides* (menos regulares) originados por corrientes (láms. 84 A y B). De mayor tamaño son las ondulaciones gigantes de muchos ambientes fluviales, intercotidales y marinos (láms. 27, 89 A y B). De mayor magnitud son las *ondas de arena* subaéreas y subácueas (lám. 28), de corta duración y cuyo único rasgo conservado es la estratificación entrecruzada (en perfil).

Deben citarse también entre las estructuras de superficie a los *cuspilitos* (lám. 98 B), de existencia efímera; las *marcas de escurrimiento* (lám. 92 A) bifurcadas en sentido opuesto a la corriente y viceversa; las *marcas y calcos en herradura* (láms. 91 B, 93 A y 93 B); y las *marcas de resaca*.

En la superficie de muchas capas de grano fino se encuentran pequeños hoyos y depresiones que determinan estructuras tales como *marcas de lluvia* (lám. 94) y *marcas de granizo*, que se asemejan a aquéllas producidas por burbujas gaseosas. Como ejemplo de este último tipo puede citarse a la estructura *cratercillos de playa*, que presenta las marcas de mayor dimensión, quizá relacionada con la *estructura cilíndrica* (lám. 115 A). Fenómenos de desecación o encogimiento sobre sedimentos fangosos, originan bloques poligonales separados por una red de *grietas de descación* (lám. 94). Un diseño similar se forma por sublimación de cristales de hielo (lám. 96 A); el relleno de estas marcas por arena o material similar determina la formación de las estructuras *calco de grietas de desecación* (lám. 96 B) y *calco de cristales de hielo*. (También son conocidos los *calcos de cristales de sal* (lám. 98 A) originados por solución de cristales de halita u otras sales solubles y posterior relleno de sus moldes dejados en sedimentos fangosos. En el techo de sedimentos finos suelen hallarse diversos rastros y huellas producidas por organismos (lám. 99).

Deformación y perturbación de la estratificación

Los sedimentos suelen mostrar diversas estructuras atribuibles a deformación penecontemporánea, vinculada a procesos no tectónicos tales como asentamiento, descenso gravitacional y deslizamiento. Los efectos producidos por estos mecanismos, incluye desde la estratificación levemente deformada hasta repliegues, brechas y aún licuefacción e inyección, con pérdida completa de la estructura original.

Las estructuras de deformación pueden agruparse en las siguientes clases:

Clase A: Estructuras de carga (calcos de carga, estructura almohadillada, etc.)

Clase B: Estratificación intraplegada

Clase C: Estructuras de deslizamiento (pliegues, fallas y brechas)

Clase D: Estructuras de inyección (diques sedimentarios, filones capas, etc.)

Clase E: Estructuras por actividad biológica (estructura tubular, capas removidas, etc.)

Las deformaciones pueden ser originadas por desplazamientos verticales o laterales.

En la primer categoría figura la *estructura almohadillada*; esta tiene lugar tanto en arenas finas (láms. 100 B a 101 B; 103 B), como en depósitos calcáreos de textura similar (láms. 102 A a 102 B). Aunque comunmente dichas estructuras se atribuyen a deslizamientos, al parecer se deben a un reordenamiento mecánico como resultado de asentamientos.

Una estructura de origen incierto es la *estratificación intraplegada*, que se desarrolla en ciertas capas limosas o arenosas finas, en forma de pliegues sin fallas y limitada a un sólo estrato. Su aspecto en planta y en perfil puede observarse en las láminas 106 A y 106 B.

Deformaciones de mayor envergadura se atribuyen a deslizamientos, especialmente subácueos. Este proceso comprende a muchas capas que resultan del apilamiento de estratos deformados en algunos lugares, con interrupción en la secuencia de otros. Tal estructura hallada en los depósitos glaciares modernos se formaría, a criterio de algunos autores, por empuje de glaciares o a icebergs; es el caso de la *estructura de despegue* (láms. 107 B, 108 A y 108 B). Fenómenos de formación de brechas (lám. 105 A), fallas (lám. 111 B) y flujo viscoso (lám. 112), acompañan a deslizamientos subácueos.

Bajo ciertas condiciones las arenas y limos saturados de agua adquieren un estado por el cual se comportan como un líquido viscoso, pudiendo así ser inyectados en sedimentos vecinos. Las estructuras originadas por estos desplazamientos constituyen los *diques clásticos, filones capas y volcanes de arena*. Los primeros pueden ser de paredes regulares y tabulares, como aparece en la lámina 113 B, o bien puede presentarse con formas muy irregulares (lám. 113 A). Los *volcanes de arena* componen un tipo de estructura relativamente rara, que ha sido descripta como formas superpuestas a mantos deslizados (lám. 114). Entre otras estructuras de origen incierto se incluyen las areniscas *vorticeladas* y *almohadilladas* (lám. 112 A). Los *rodados de arcilla acorazados* (lám. 117) es otra estructura sedimentaria primaria de difícil clasificación. Para completar el resumen de estructuras originadas por deformación, debe mencionarse el rol de aquellos organismos en la destrucción de la laminación primaria con la consiguiente homogeinización de la roca. Organismos cavadores y limnófagos pueden ser los suficientemente numerosos como para cribar los sedimentos, pudiendo llegar a destruir la estructura original (láms. 116 A y B).

The plates

Die Tafeln

Les planches

Láminas

Plates

The plates are arranged according to the classification outlined and summarized in the Introduction. However, any particular plate may show several features and hence an alternate position for such a plate is possible. The reader who wishes to find a picture of a particular structure should consult the Glossary where, if the structure is illustrated, he will find the required plate numbers under the proper entry in the Glossary. Not all structures are illustrated, although all the common structures are shown, and most, but not all, the rare structures are depicted.

Les planches

Les planches sont arrangées suivant la classification esquissée dans l'Introduction, mais une planche peut illustrer plusieurs structures, et peut donc avoir plus d'une place possible. Le lecteur qui désire trouver une illustration d'une structure particulière devra consulter le glossaire, car, si cette structure est figurée, les numéros des planches correspondantes y seront donnés sous la rubrique appropriée. Toutes les structures ne sont pas représentées, mais toutes les plus courantes le sont, ainsi que la plupart, mais non la totalité, des formes rares.

Zu den Tafeln

Die Tafeln sind im Sinne der in der Einleitung beschriebenen und zusammengefaßten Klassifikation angeordnet worden. Da jedoch einzelne Tafeln verschiedene Strukturen oder Strukturmerkmale zeigen, können sie an verschiedenen Stellen zu finden sein. Der Leser, der die Abbildung einer bestimmten Struktur sucht, sollte im Verzeichnis der Fachausdrücke nachschlagen. Er wird dort zu dem gesuchten Stichwort die Nummern der zugehörigen Tafeln finden. Nicht alle Strukturen sind abgebildet, wohl aber die häufigsten, und von den selteneren Strukturen werden die meisten gezeigt.

Láminas

Las láminas están ordenadas de acuerdo a la clasificación esbozada y sintetizada en la Introducción. Sin embargo alguna lámina en particular puede indicar diversas características y por ello su ubicación aparecerá alternada con otras. El lector que desee localizar una lámina de una estructura en particular, debe consultar el Glosario donde, si ella está ilustrada, hallará a continuación del término mismo, el número de la lámina requerida. No todas la estructuras se encuentran ilustradas, pero sí las más comunes; del mismo modo pueden consultarse la mayoría de las formas raras.

PLATE 1A

Pleistocene glacial varves. Silt is light colored, clay is dark colored. Silt layers grade abruptly into overlying clay. Note lateral continuity, cyclic character, and subequal thickness of beds. South Hadley Falls, Hampshire County, Massachusetts, U.S.A. (Photograph by R. F. FLINT).

Glaziale Warven (Pleistozän). Siltlagen hell, Tonlagen dunkel. Abrupter Übergang von Silt zu Ton. Beachte horizontale Beständigkeit, zyklische Ausbildung, aber ungleiche Mächtigkeit der einzelnen Lagen.

Varves glaciaires d'âge Pleistocène. L'aleurite apparait en clair, l'argile en foncé. Les couches d'aleurite passent brusquement aux argiles sus-jacentes. Noter la continuité latérale, le caractère cyclique et l'épaisseur presque uniforme des couches.

Varves glaciarios formados por capas limosas (claras) que pasan bruscamente a capas arcillosas (oscuras). Nótese la continuidad lateral, el carácter cíclico y el espesor casi uniforme de las capas. Pleistoceno.

PLATE 1B

Graded bedding in Archean graywacke. Note sharp basal contact of sandstone (light) grading upward to argillite (dark). Note continuity and uniform thickness of beds; note unequal thickness of same. Near Ruby Island, East Bay, Minnitaki Lake, District of Kenora, Ontario, Canada.

Gradierte Schichtung in archäischer Grauwacke. Beachte scharf abgesetzte Basis des hellen Sandsteins und nach oben hin Übergang in dunklen Tonstein. Beachte kontinuierlich gleiche Mächtigkeit einzelner verschieden mächtiger Schichten.

Couches granoclassées dans une grauwacke d'âge Archéen. Noter le contact basal très marqué du grès (en clair) qui passe vers le haut à une argilite (en foncé). Noter aussi la continuité et l'épaisseur uniforme de chaque couche et la variation d'épaisseur des différentes couches.

Estratificación gradada en grauvacas arcaicas. Nótese la base bien definida de cada ciclo (niveles claros) y la gradación hacia arriba a argillitas (niveles oscuros); además resalta la continuidad lateral y uniformidad del espesor de los estratos, con excepción de los superiores.

PLATE 2

Uniform, even bedding in halite (white) with anhydritic clay material (black). "Liniensalz" of
Zechstein 3 (Permian). Beds average approximately 8 cm thick. Ronnenberg Potash Mine, near
Hannover, Germany (Photograph by RICHTER-BERNBURG).

Gleichmäßige, ebene Schichtung in Steinsalz (hell), wechsellagernd mit anhydritischem Ton
(schwarz). Lagen durchschnittlich ungefähr 8 cm mächtig. Liniensalz, Zechstein 3 (Perm).

Stratification uniforme et régulière d'une halite (blanche) avec une argile à anhydrite (noire)
«Liniensalz» du Zechstein 3 (Permien). Epaisseur moyenne des couches, environ 8 cm.

Estratificación uniforme y regular en halita (clara), con material arcilloso anhidrítico (oscuro)
intercalado. Espesor aproximado de las capas 8 cm. "Liniensalz" de Zechstein 3, Pérmico.

PLATE 3

Calcareous Paleocene flysch formed by thin, alternating beds of marl and siliceous limestone. The hard siliceous beds resist weathering but the marly mudstones are easily eroded. Note extreme persistence of single beds and absence of lenticular bedding or scour-and-fill structures. Extreme east end of beach of San Telmo, near Zumaja, Vascongadas Province, Spain (DE LLARENA 1954, pl. 2, Fig. 1).

Kalkiger Flysch (Paleozän). Wechsellagerung dünner Mergel- und kieseliger Kalkstein-Schichten. Während die verkieselten Lagen der Verwitterung trotzen, wittern die mergeligen Tonsteine leicht aus. Beachte außerordentliche Beständigkeit einzelner Lagen und Abwesenheit von linsenförmiger Schichtung und Erosionsrinnen.

Flysch calcaire d'âge Paléocène, formé de minces couches alternantes de marne et de calcaire siliceux. Les couches siliceuses dures sont résistantes mais les argiles marneuses sont facilement érodées. Noter la continuité de chaque couche et l'absence de stratification lenticulaire et de structures de creusement et remplissage.

Flysh formado por estratos finos y alternados de margas y calizas silíceas. Los últimos resisten la meteorización, mientras que los estratos margosos son fácilmente erosionados. Nótese la persistencia en la continuidad de cada estrato, la falta de estratificación lenticular y de estructuras de relleno de cauce. Paleoceno.

PLATE 4

Uniform and regular alternation of limestone (light) and shale (dark). Note unusual continuity of beds. Height of cliff is approximately 25 meters. S zone (Carboniferous). Harness Slade, Castlemarine, Pembrokeshire, England (Geological Survey of Great Britain).

Gleichmäßige, regelmäßige Wechsellagerung von Kalkstein (hell) und Schieferton (dunkel). Beachte ungewöhnliche Beständigkeit der Schichten. Höhe der Klippe ungefähr 25 m. S-Zone (Karbon).

Alternance uniforme et régulière de calcaire (clair) et d'argile schisteuse (foncée). Noter la continuité remarquable des couches. La falaise a environ 25 m de haut. Zone S (Carbonifère).

Alternancia uniforme y regular de calizas (claras) y lutitas (oscuras). Nótese la persistencia poco común en la continuidad de los estratos. Altura aproximada del afloramiento, 25 metros. Zona S, Carbonífero.

PLATE 5 A

Rhythmic bedding in a flysch sequence, Los Molles Formation (Tertiary). Note continuity and uniform thickness of sandy (light) layers. Near Guaquen, Aconcagua Province, Chile. (Photograph by G. CECIONI.)

Rhythmische Lagerung in Flysch-Schichten; Los Molles Formation (Tertiär). Beachte konstante Mächtigkeit und Beständigkeit der hellen sandigen Lagen.

Stratification répétée dans une série de flysch, formation de Los Molles (Tertiaire). Remarquer a continuité et l'épaisseur uniforme des couches sableuses (claires).

Estratificación rítmica en flysch. Nótese la continuidad y espesor uniforme de los estratos areniscosos (claros). Formación Los Molles, Terciario.

PLATE 5 B

Thin and even-bedded siltstone and shale of Tradewater Formation (Pennsylvanian) above Bell Coal at the old Bell Mines. 3750 F.S.L. 5200 F.E.L. M-18, Crittenden County, Kentucky, U.S.A.

Dünner, gleichmäßig geschichteter Siltstein und Schieferton der Tradewater Formation (Pennsylvanian).

«Siltstone» et argile schisteuse en lits minces et réguliers de la formation de Tradewater (Pennsylvanien).

Estratificación fina y regular en limolitas y lutitas. Formación Tradewater, Pennsylvaniano.

PLATE 6A

Overturned graded siltstone (light) and slate (dark). Top to right. Note prominent slaty cleavage which is refracted upon passing through beds of varying competency. St. Francis Group (Ordovician), Moe River, Compton County, Quebec, Canada (Geological Survey of Canada).

Gradierter Siltstein (hell) und Tonschiefer (dunkel); überkippt. Schichtoberfläche zeigt nach rechts. Beachte prominente Schiefrigkeit, die in Lagen unterschiedlicher Kompetenz gebrochen wird. St. Francis Group (Ordovizium).

«Siltstone» granoclassée et renversée (claire) et ardoise (foncée). Haut à droite de la photo. Noter le clivage très marqué de l'ardoise, qui se réfracte plus ou moins suivant la rigidité des couches traversées. Ordovicien.

Estratos de limolitas gradadas (claras) y pizarras (oscuras), volcados; techo del conjunto hacia la derecha. Nótese el conspicuo clivaje secundario, refractado a través de estratos de distinta competencia. Grupo St. Francis, Ordovícico.

PLATE 6B

Uniform, rhythmically bedded sandstone of the Tyee Formation (Tertiary). Graded bedding is indicated by sharp lower surfaces of the more resistant sandstone beds and their indefinite upper surfaces, best seen in lower half of outcrop. Near the top of the exposure, the siltstones in the upper parts of the rhythmites have been eroded by the currents that transported the detritus of the overlying bed. Along Siletz River, five miles north of Siletz, Lincoln County, Oregon,U.S.A. (SNAVELY and WAGNER, 1963, Fig. 7).

Gleichförmiger, rhythmisch geschichteter Sandstein. Tyee Formation (Tertiär). Gradierte Schichtung wird an der scharfen unteren Begrenzung der Sandlagen und ihrem allmählichen Übergang nach oben deutlich, gut sichtbar in der unteren Aufschlußhälfte. Im oberen Teil des Aufschlusses sind die Siltlagen der rhythmischen Schichten von der Strömung ausgewaschen, die den Detritus der darüberliegenden Schicht transportierte.

Grès rhythmique à litage uniforme de la formation de Tyee (Tertiaire). Le granoclassement est indiqué par le fait que les niveaux inférieurs des couches de grès plus résistantes sont plus nets, et leur niveaux supérieurs moins bien définis; ceci se voit le mieux dans la moitié inférieure de la coupe. Vers le haut de la coupe, les «siltstones» des parties supérieures des séquences manquent, ayant été emportées par les courants qui transportaient les matériaux de la couche sus-jacente en formation.

Estratificación rítmica y uniforme en areniscas gradadas. La estratificación gradada queda indicada por las superficies inferiores bien definidas de los estratos areniscosos más resistentes; las superiores están mal definidas. Estas características tienen mejor desarrollo en la mitad inferior del afloramiento. Cerca del techo, las limolitas de los niveles superiores de las ritmitas han sido erodadas por las corrientes que han transportado los detritos depositados como capas superiores. Formación Tyee, Terciario.

PLATE 7

Rhythmically interbedded sand- or siltstone and shale (Devonian). Note uniformity and continuity of bedding. Thick sandstone bed shows ball-and-pillow development. West shore of Susquehanna River, opposite Sunbury, Northumberland County, Pennsylvania, U.S.A.

Rhythmisch wechsellagernde Sand- oder Siltsteine und Schieferton (Devon). Beachte Gleichförmigkeit und Kontinuität der Schichtung. Die gebankten Sandsteine zeigen Ballenstruktur.

Grès ou «siltstone», et argile schisteuse à stratification en alternance répétée (Dévonien). Noter l'uniformité et la continuité de la stratification. La couche épaisse de grès montre un début de structure en boules et coussins.

Interestratificación rítmica de areniscas o limolitas y lutitas; nótese la uniformidad y continuidad de la estratificación. El estrato areniscoso grueso presenta desarrollo de estructura almohadillada. Devónico.

PLATE 8A (left)

Regular, uniform bedding in "Linien-Anhydrite" of Zechstein 2 (Permian). Germany (Photograph by G. RICHTER-BERNBURG).

Regelmäßige, gleichförmige Schichtung im Linienanhydrit des Zechsteins 2 (Perm).

Stratification régulière et uniforme dans la «Linien-Anhydrite» du Zechstein 2 (Permien).

Estratificación regular y uniforme. "Linien-Anhydrite", Zechstein 2, Pérmico,.

PLATE 8B (right)

Regularly laminated, cyclical bedding in Pleistocene silt and clay. Paired summer silt (light) and homogeneous winter clay (dark) constitute a varve. At Twin Falls Dam, Teefy Township, Cochrane District, Ontario, Canada (Geological Survey of Canada).

Gleichmäßig feinschichtige, zyklische Schichtung in Silt und Ton (Pleistozän). Eine einzelne Warve besteht aus einer hellen Sommerlage (Silt) und einer dunklen homogenen Winterlage (Ton).

Stratification régulièrement feuilletée et à alternance répétée dans des limons et argiles pleistocènes. Un limon d'été (clair) associé en paire à une argile homogène d'hiver (foncée) forment une varve.

Laminación cíclica y regular. Cada ciclo, formado por limo (claro) depósitado durante el verano y arcilla (oscura) en el invierno, constituye un varve. Pleistoceno.

PLATE 9A

Bedding contrasts in basal Carboniferous limestone. Even, uniform, thin-bedded sandy limestone overlying ill-sorted, poorly-stratified, coarse conglomerate. Bosworlos, Newfoundland, Canada (Photograph by D. M. BAIRD).

Kontrastreiche Schichtung in basalem Kalkstein (Karbon). Im Hangenden ebene, gleichförmige, dünne Schichtung in sandigem Kalkstein. Schlecht sortiertes und kaum geschichtetes, grobes Konglomerat im Liegenden.

Contrastes dans la stratification d'un calcaire carbonifère inférieur. Lits réguliers, uniformes et minces de calcaire sableux surmontant un conglomérat grossier mal trié et à stratification obscure.

Contraste en la estratificación; caliza arenosa con estratificación fina, uniforme y regular, sufrayacente a un conglomerado grueso pobremente seleccionado y estratificado. Carbonífero pasal.

PLATE 9B

Cyclic sedimentation in red clastics of Catskill Formation (Devonian). Each cycle begins abruptly with coarser, light-colored sandstone and grades upward into dark red siltstone and shale. East side of Susquehanna River at Peters Mountain, Dauphin County, Pennsylvania, U.S.A.

Repetitionsschichtung in roten, klastischen Gesteinen der Catskill Formation (Devon). Jeder Zyklus beginnt unvermittelt mit grobem, hellem Sandstein und geht nach oben in dunkelroten Siltstein und Schieferton über.

Sédimentation rhythmique dans des sédiments clastiques rouges de la formation de Catskill (Dévonien). Chaque cycle débute brusquement par un grès plus grossier et plus clair, et passe vers le haut à des «siltstones» et argiles schisteuses d'un rouge plus foncé.

Sedimentación cíclica en sedimentos clásticos rojos. Cada ciclo comienza bruscamente con arenisca clara y pasa en forma gradual hacia arriba, a limolita y lutita rojas oscuras. Formación Catskill, Devónico.

PLATE 10A

Even bedding in St. Peter Sandstone (Ordovician). SW 1/4 SW 1/4 SW 1/4 sec. 20, T. 7 N., R. 2 W. Calhoun County, Illinois, U.S.A. (Illinois Geological Survey).

Ebene Schichtung im St. Peter-Sandstein (Ordovizium).

Stratification régulière dans le grès de St. Peter (Ordovicien).

Estratificación regular en la arenisca St. Peter (Ordovícico).

PLATE 10B

Thick "massive" sandstone with some horizontal bedding in lower part. Height of exposure is 12 meters. Lance Formation (Cretaceous). SE 1/4 NW 1/4 sec. 15, T. 21 N., R. 101 W., Sweetwater County, Wyoming, U.S.A. (Jersey Production Research Company).

Massiger, grobgebankter Sandstein der Lance Formation (Kreide). Parallelschichtung im unteren Teil. Aufschlußhöhe 12 m.

Grès épais, massif, à stratification plus nette dans la partie inférieure. Hauteur de l'affleurement, 12 m. Formation de Lance (Crétacé).

Arenisca "masiva" de considerable espesor, con estratificación más visible en los niveles inferiores; altura del afloramiento 12 metros. Formación Lance, Cretácico.

PLATE 11 A

Regular interbedded fine-grained sandstone and shale. Note "wavy" rippled surfaces of more resistant sandstone beds just above hammer handle. Gardeau Formation (Devonian). Along New York State Route 63, approximately 1 mile north of Dansville, Sparta Township, Livingston County, New York, U.S.A.

Regelmäßig wechsellagernde Feinsand- und Schiefertonlagen. Beachte die „gewellten", gerippelten Oberflächen der widerstandsfähigeren Sandsteinlagen oberhalb des Hammerstiels. Gardeau Formation (Devon).

Couches régulièrement intercalées de grès fin et d'argile schisteuse. Noter les surfaces onduleuses des couches de grès plus résistantes juste au-dessus du manche du marteau. Formation de Gardeau (Dévonien).

Areniscas finas y lutitas regularmente interestratificadas. Nótese la superficie ondulante de las capas areniscosas más resistentes, justamente arriba del mango del martillo. Formación Gardeau, Devónico.

PLATE 11 B

Whin, wavy bedding, of ripple mark origin in sandstone. Aux Vases Sandstone (Mississippian) SW 1/4 SW 1/4 NE 1/4, sec. 36, T. 5 S., R. 9 W., Randolph County, Illinois, U.S.A. (Illinois Geological Survey).

Dünne, gewellte, aus Rippelmarken bestehende Schichtung im Aux Vases-Sandstein (Mississippian).

Grès à stratification fine, onduleuse, provenant de rides de plage. Grès de Aux Vases (Mississipien).

Estratificación ondulada, fina, de origen ondulítico. Arenisca Aux Vases, Mississipiano.

PLATE 12

Interbedded limestone (light) and shale (dark). Limestone beds show some lensing but have good continuity. Base of Lower Lias (Jurassic). Lyme Regis, Dorset, England (Geological Survey of Great Britain).

Kalkstein (hell) wechsellagernd mit Schieferton (dunkel). Die Kalksteinlagen keilen gelegentlich aus, sind aber im ganzen ziemlich beständig. Basis des unteren Lias (Jura).

Calcaire (clair) et argile schisteuse (foncée) intercalés. Les lits calcaires tendent vers un aspect lenticulaire mais la continuité latérale reste bonne. Base du Lias inférieur (Jurassique).

Calizas (claras) y lutitas (oscuras) interestratificadas; las primeras muestran cierta tendencia a la lenticulación pero presentan buena continuidad. Base del Lías Inferior.

PLATE 13 A

Interbedded calcilutite and black calcareous shale. Calcilutite layers (white) show pinch and swell (sedimentary boudinage) and short fractures (tectonic). Martinsburg Shale (Ordovician). On U.S. Highway 122 near Shoemakersville, Berks County, Pennsylvania, U.S.A.

Kalkpelit (hell) wechsellagernd mit kalkigem Schieferton (dunkel). Beachte An- und Abschwellen der Kalkpelitlagen (Sediment-Boudinage). Fiederspalten sind tektonischen Ursprungs. Martinsburg-Schiefer (Ordovizium).

Calcilutite (calcaire pélitique) et schiste calcaire noir en lits intercalés. Les couches de calcilutite (blanches) sont boudinées et montrent de courtes fractures d'origine tectonique. Formation de Martinsburg (Ordovicien).

Calcilutitas (claras) y lutitas calcáreas interestratificadas. Las primeras presentan estructura moniliforme y pequeñas fracturas tectónicas. Lutita Martinsburg, Ordovícico..

PLATE 13 B

Irregularly interbedded limestone (A) and dolomite (B). Note bifurcation of dolomitic layers and isolation of pod-like bodies of calcilutite. Conococheague Limestone (Cambrian). About 2 miles northeast of Clear Spring, Washington County, Maryland, U.S.A.

Unregelmäßige Wechsellagerung von Kalkstein (A) und Dolomit (B). Beachte sich gabelnde Dolomitlagen, die isolierte Kalkpelit-Linsen umgeben. Conococheague-Kalkstein (Kambrium).

Couches irrégulièrement intercalées de calcaire (A) et de dolomie (B). Noter la bifurcation des couches dolomitiques qui isolent de petites lentilles de calcilutite. Calcaire de Conococheague (Ordovicien).

Calizas (A) y dolomías (B) irregularmente interestratificadas. Nótese la bifurcación de la dolomía y los cuerpos aislados lentiformes calcipelíticos. Caliza Conococheague, Cámbrico.

PLATE 14

Mottled bedding; dolomite (dark) and limestone (light). Origin of pattern unknown; perhaps secondary. Stonehenge Limestone Member, Beekmantown Group (Ordovician). Along Western Maryland Railroad 0.5 mile east of Charlton, Washington County, Maryland, U.S.A.

Fleckige Schichtung aus dunklem Dolomit und hellem Kalkstein. Ursprung des fleckigen Musters unbekannt; vielleicht sekundär. Stonehenge-Kalkstein Member, Beekmantown Group (Ordovizium).

Dolomie (foncée) et calcaire (clair) à stratification mouchetée d'origine inconnue, peut-être secondaire. Terme calcaire de la formation de Stonehenge, groupe de Beekmantown (Ordovicien).

Estratificación moteada de origen desconocido, probablemente secundario; dolomía (oscura) y caliza (clara). Miembro Stonehenge, Grupo Beekmantown, Ordovícico.

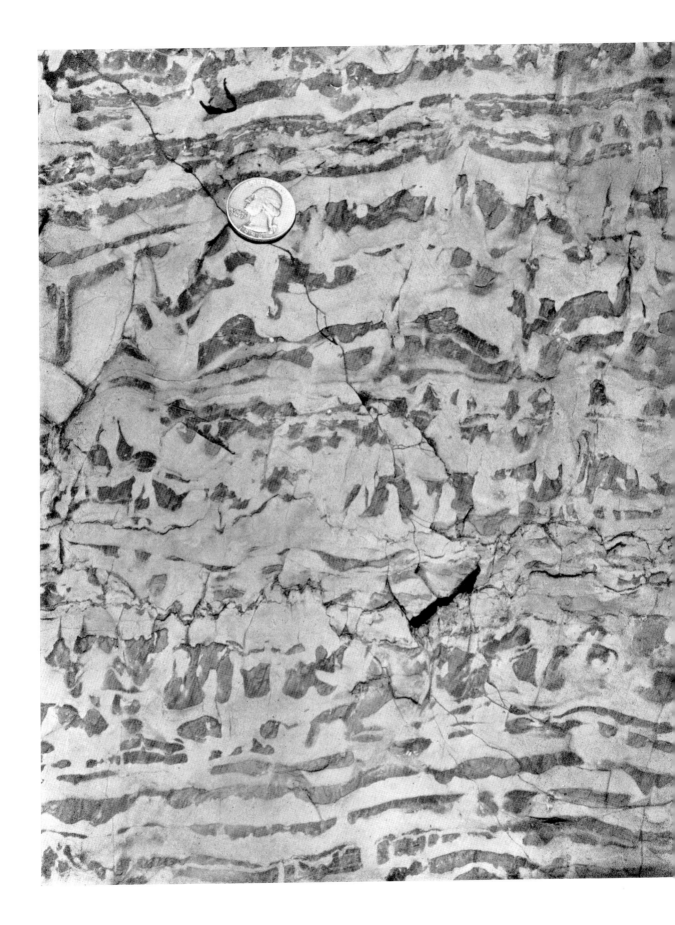

PLATE 15 A

Complex bedding in limestone conglomerate (A) and pseudoconglomerate (B) of boudinage (?) origin. Beekmantown Group (Ordovician). Along tracks of Western Maryland Railroad, about three-quarters mile east of Charlton, Washington County, Maryland, U.S.A.

Verwickelte Schichtung in Kalkstein-Konglomerat (A) und Pseudokonglomerat (B). Letzteres vielleicht Sediment-Boudinage (?). Beekmantown Group (Ordovizium).

Alternance complexe d'un conglomérat calcaire (A) et d'un pseudoconglomérat (B) probablement dû au boudinage. Groupe de Beekmantown (Ordovicien).

Estratificación compleja en conglomerado calizo (A) y pseudo-conglomerado (B) de origen tensional (?). Grupo Beekmantown, Ordovícico.

PLATE 15 B

Nodular structure in limestone (Cretaceous). Exposed in road cut along U.S. Highways 67 and 377, between Blanket and Brownwood, Parker County, Texas, U.S.A.

Knollenstruktur in Kalkstein (Kreide).

Calcaire noduleux (Crétacé).

Estructura nodular en caliza; Cretácico.

PLATE 16A   (top)

Incipient nodular bedding in dolomite with irregular shale parting. Twice natural scale. Joliet Limestone (Silurian). Lincoln quarry, Joliet, Will County, Illinois, U.S.A. (Illinois Geological Survey).

Beginnende knollige Schichtung in Dolomit mit unregelmäßigen Schieferton-Lagen. Zweifach natürliche Größe. Joliet-Kalkstein (Silur).

Début de texture en nodules dans une dolomie avec schistes éffilochés. Grossi deux fois. Calcaire de Joliet (Silurien).

Dolomía con estratificación nodular incipiente y lutita con laminación muy fina e irregular; ×2. Caliza Joliet, Silúrico.

PLATE 16B (left)

Small carbonate nodules in calcareous underclay (Pennsylvanian). Note lack of bedding. Illinois, U.S.A. (NORMAN, 1959, pl. 2).

Kleine Karbonatknollen in kalkigem Basal-Ton (Pennsylvanian). Beachte Abwesenheit von Schichtung.

Petits nodules calcaires dans une «underclay» calcaire (Pennsylvanien). Noter l'absence de litage.

Nódulos calcáreos pequeños en sotoarcilla calcárea. Nótese la falta de estratificación. Pennsylvaniano.

PLATE 16C (right)

Nodular bedding. Irregular carbonate nodules in calcareous shale. Natural size. Bankston Limestone (Pennsylvanian). Illinois, U.S.A. (Illinois Geological Survey).

Knollige Schichtung. Unregelmäßige Karbonatknollen in kalkhaltigem Schieferton. Natürliche Größe. Bankston-Kalkstein (Pennsylvanian).

Texture noduleuse. Nodules calcaires irréguliers dans un schiste calcaire. Grandeur nature. Calcaire de Bankston (Pennsylvanien).

Estratificación nodular; nódulos calcáreos irregulares en lutita calcárea. Tamaño natural. Caliza Bankston, Pennsylvaniano.

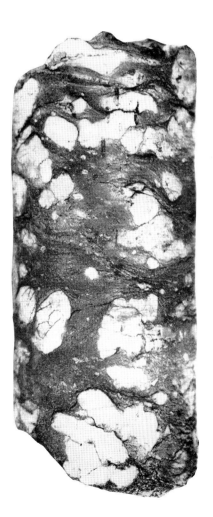

PLATE 17A (left)

Flaser bedding. Light colored cross-bedded lenses of siltstone interbedded with shale. Natural size (Pennsylvanian). Reprinted from "Sedimentary Rocks", 2nd edition, 1957, by F. J. PETTI-JOHN by permission of HARPER and Row. Copyright 1957, by HARPER and BROTHERS.

Flaserschichtung. Wechsellagerung von hellen, schräggeschichteten Siltstein-Linsen und Schieferton. Natürliche Größe (Pennsylvanian).

«Flaser bedding» (structure lenticulaire). Lentilles claires de «siltstone» en stratification croisée alternant avec une argile schisteuse. Grandeur nature (Pennsylvanien).

Estratificación flaser. Lentes (claros) planoconvexos de limolita con laminación entrecruzada, intercalados en lutita. Tamaño natural. Pennsylvaniano.

PLATE 17B (middle)

Flaser bedding. Lenses and pods of cross-bedded, light colored siltstone interbedded with shale. Natural size. Carbondale Formation (Pennsylvanian). Gallatin County, Illinois, U.S.A. (Illinois Geological Survey).

Flaserschichtung. Helle, schräggeschichtete Siltstein-Linsen und -Schmitzen in Schieferton. Natürliche Größe. Carbondale Formation (Pennsylvanian).

«Flaser bedding». Lentilles, arrondies ou allongées, entre-croisées, de «siltstone» claire, alternant avec une argile schisteuse. Grandeur nature. Formation de Carbondale (Pennsylvanien).

Estratificación flaser. Lentes (claros) de limolita con laminación entrecruzada, intercalados en lutita. Tamaño natural. Formación Carbondale, Pennsylvaniano.

PLATE 17C (right)

Lenses, pods, and stringers of siltstone interbedded with shale. Note siltstone filled burrow. Natural size. Tar Springs Sandstone (Mississippian). J. H. FORESTER, No. 1 fee, sec. 5, T. 6 S., R. 1 W., Perry County, Illinois, U.S.A. (Illinois Geological Survey).

Linsen, Schmitzen und Bänder aus Siltstein wechsellagernd mit Schieferton. Beachte mit Silt gefüllten Bau. Natürliche Größe. Tar Springs-Sandstein (Mississippian).

Lentilles arrondies ou allongées, et filets de «siltstone», alternant avec une argile schisteuse. Noter le tube rempli de «siltstone». Grandeur nature. Grès de Tar Springs (Mississipien).

Lentes y guías de limolita, intercaladas en lutita. Estructura tubular con relleno limolítico. Tamaño natural. Arenisca Tar Spring, Mississippiano.

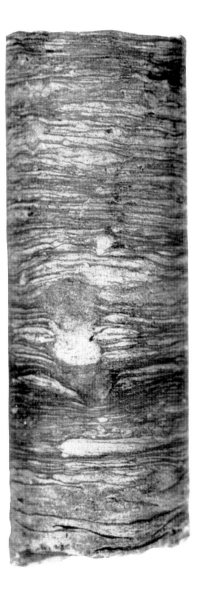

PLATE 18A (left)

Regularly interlaminated siltstone and shale. Natural size. Tar Springs Sandstone (Mississippian) J. H. FORESTER, No. 1 fee, sec. 5, T. 6 S., R. 1 W., Perry County, Illinois, U.S.A. (Illinois Geological Survey).

Regelmäßig wechsellagernde, feinschichtige Siltsteine und Schiefertone. Natürliche Größe. Tar Springs-Sandstein (Mississippian).

«Siltstone» et schiste en couches minces régulièrement alternées. Grandeur nature. Grès de Tar Springs (Mississipien).

Interlaminación regular de limolita y lutita; tamaño natural. Arenisca Tar Spring, Mississippiano.

PLATE 18B (middle)

Fine sandstone and siltstone interbedded with irregular, inclined, shale partings (Pennsylvanian). Natural size. Illinois, U.S.A. (Illinois Geological Survey).

Feinsand- und Siltstein mit unregelmäßigen, geneigten Schieferton-Zwischenlagen (Pennsylvanian). Natürliche Größe.

Grès fin et «siltstone» en stratification alternante avec des délits schisteux très fins, irréguliers et inclinés (Pennsylvanien). Grandeur nature.

Arenisca fina y limolita interestratificadas, con láminas muy finas lutíticas, irregulares e inclinadas. Tamaño natural. Pennsylvaniano.

PLATE 18C (right)

Pods and lenses of light-colored siltstone interbedded with shale. Natural size. Tar Springs Sandstone (Mississippian). J. H. FORESTER, No. 1 fee, sec. 5, T. 6 S., R. 1 W., Perry County, Illinois, U.S.A. (Illinois Geological Survey).

Schmitzen und Linsen aus Siltstein (hell) in Schieferton. Natürliche Größe. Tar Springs-Sandstein (Mississippian).

Lentilles allongées ou arrondies de «siltstone» claire alternant avec une argile schisteuse. Grandeur nature. Grès de Tar Springs (Mississipien).

Lentes limolíticos (claros) y lutita interestratificados. Tamaño natural. Arenisca Tar Spring, Mississippiano.

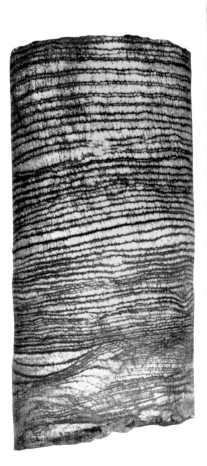

PLATE 19A

Erosional channel in sandstone. Note massive bed above channel. Tar Springs Sandstone (Mississippian). NE 1/4 NE 1/4 NW 1/4 sec. 18, T. 15 S., R. 7 E., Pope County, Illinois, U.S.A. (Illinois Geological Survey).

Erosionsrinne im Sandstein des Tar Springs-Sandsteins (Mississippian). Beachte massige Bankung im Hangenden der Rinne.

Chenal d'érosion dans un grès. Noter la couche massive qui surmonte le chenal. Grès de Tar Springs (Mississipien).

Cauce erosional en arenisca. Nótese el estrato masivo sobre el cauce. Arenisca Tar Spring Mississippiano.

PLATE 19B

Channel in uniformly bedded sandstones and siltstones of Tradewater Formation (Pennsylvanian). 5,200 F.E.L., 10,800 F.N.L., H-30, Drakesboro Quadrangle, Muhlenberg County, Kentucky, U.S.A. (Photograph by J. A. SIMON).

Erosionsrinne in gleichmäßig geschichteten Sand- und Siltsteinen der Tradewater Formation (Pennsylvanian).

Chenal dans des grès et «siltstones» à stratification uniforme. Formation de Tradewater (Pennsylvanien).

Cauce en areniscas y limolitas con estratificación uniforme. Formación Tradewater, Pennsylvaniano.

PLATE 20A

Scour channel, filled with nonlaminated calcarenite, truncating underlying laminated calcilutite. Cynthiania Group (Ordovician). 2400 F.W.L., 1500 F.S.L. CC-65, Pendelton County, Kentucky, U.S.A.

Erosionsrinne in feingeschichtetem Kalkpelit, gefüllt mit massigem Kalksandstein. Cynthiania Group (Ordovizium).

Rigole affouillée puis comblée par une calcarénite non stratifiée, qui recoupe la calcilutite litée sous-jacente. Groupe de Cynthiana (Ordovicien).

Cauce rellenado por calcarenita truncando a una calcilutita laminada. Grupo Cynthiana, Ordovícico.

PLATE 20B

Thin and evenly laminated siltstone (A), small erosional channel (B), and ripple cross-lamination (C). Sandstone above Indiana Coal V in the Petersburgh Formation (Pennsylvanian). NE 1/4 NE 1/4 sec. 17, T. 2 S., R. 7 W., Pike County, Indiana, U.S.A.

Eben- und feingeschichteter Siltstein (A), kleine Erosionsrinne (B) und Rippelschichtung (C). Sandstein im Hangenden von Indiana Kohleflöz V. Petersburgh Formation (Pennsylvanian).

«Siltstone» (A) finement et régulièrement litée, petite rigole d'érosion (B) et stratification croisée de rides (C). Grès qui surmonte le Indiana Coal V, dans la formation de Petersburgh (Pennsylvanien).

Limolita con laminación fina y uniforme (A), pequeño cauce de erosión (B) y migro-óndulas entrecruzadas (C). Arenisca sobre el manto "Indiana Coal V", Formación Petersburgh, Pennsylvaniano.

PLATE 21 A

Edge of washout of Indiana Coal III (Pennsylvanian). Coal bed (A) and underclay (B) replaced by sandstone (C) with unconformable base. Some rafted stringers of coal (D) at base of sandstone. Center NE 1/4 NE 1/4 sec. 13, T. 2 N., R. 6 W., Daviess County, Indiana, U.S.A.

Rand eines Priels Indiana Coal III (Pennsylvanian). Sandstein (C) mit diskordanter Lagerung an die Stelle von Kohleflöz (A) und Basiston (B) getreten. Beachte losgerissene Kohlefetzen an der Sohle des Sandsteins (D).

Rive d'un chenal affouillé dans Indiana Coal III (Pennsylvanien). La couche de charbon (A) et l'argile de mur (B) sont remplacées par un grès (C) discordant. Filets de charbon déplacés (D) à la base du grès.

Borde de cauce. Manto de carbón (A), sotoarcilla (B), arenisca en discordancia (C) y algunas guías de carbón (D) en su bas. "Indiana Coal III", Pennsylvaniano.

PLATE 21 B

Discontinuous layers of coal interbedded with sandstone. Note lack of vertical ordering and lateral continuity. Compare with plate 4. High wall of abandoned strip mine near washout of Indiana Coal V. Petersburgh Formation (Pennsylvanian) SE 1/4 NE 1/4 sec. 15, T. 2 N., R. 7 W., Pike County, Indiana, U.S.A.

Unterbrochene Kohle- und Sandsteinlagen. Beachte Fehlen vertikaler Gliederung und seitlicher Beständigkeit. Vergleiche mit Tafel 4. Senkrechte Wand im alten Tagebau, Nähe Priel (Indiana Coal V). Petersburgh Formation (Pennsylvanian).

Couches de charbon discontinues alternant avec du grès. Noter l'arrangement vertical désordonné et l'absence de continuité latérale. Contraster avec la planche 4. Paroi d'une mine à ciel ouvert abandonnée, près d'un chenal érodé, Indiana Coal V de la formation de Petersburgh (Pennsylvanien).

Mantos discontínuos de carbón intercalados en arenisca. Nótese la falta de ordenamiento vertical y de continuidad lateral. Compárese con lámina 4. "Indiana Coal V" de la Formación Petersburgh, Pennsylvaniano.

PLATE 22

Erosional contact in sand of McNairy Member of the Gulfian Series (Cretaceous). Note irregular base and clay pebbes in overlying sand (weathered dark) and cross-bedding in underlying sand light). Bruceton quadrangle (14,900 F.S.L., 16,800 F.W.L.), Benton County, Tennessee, U.S.A.

Erosionskontakt im Sandstein des McNairy Member (Gulfian Series, Kreide). Beachte umregelmäßige Basis und Tongerölle in darüberliegendem Sand (dunkel verwittert) und Schrägschichtung im liegenden Sandstein (hell).

Contact dû à l'érosion dans le sable du terme de McNairy de la série de Gulfian (Crétacé). Noter la base irrégulière, et les cailloux d'argile, dans le sable supérieur (devenu foncé par altération), et la stratification croisée du sable inférieur (clair).

Contacto erosional en arenas. Nótese la base irregular y la grava de arcilla del depósito (oscuro) suprayacente a una arena (clara) con estratificación entrecruzada. Miembro McNairy, Serie Gulf, Cretácico.

PLATE 23

Massive appearing sandstone and radiographs showing actual bedding. Note uniform bedding (A and D), cross-bedding (B and C) and disturbed bedding (E). (HAMBLIN, 1962, Fig. 1).

Massig erscheinender Sandstein und Röntgenaufnahmen desselben mit sichtbarer Schichtung. Beachte gleichförmige Schichtung (A und D), Schrägschichtung (B und C) und gestörte Schichtung (E).

Grès d'apparence massive, dont les radiographies révélent la stratification. Noter les exemples de stratification uniforme (A et D), croisée (B et C) et dérangée (E).

Radiografía de una arenisca con apariencia masiva, revelando la verdadera estratificación. Nótese la estratificación uniforme (A y D), entrecruzada (B y C) y disturbada (E).

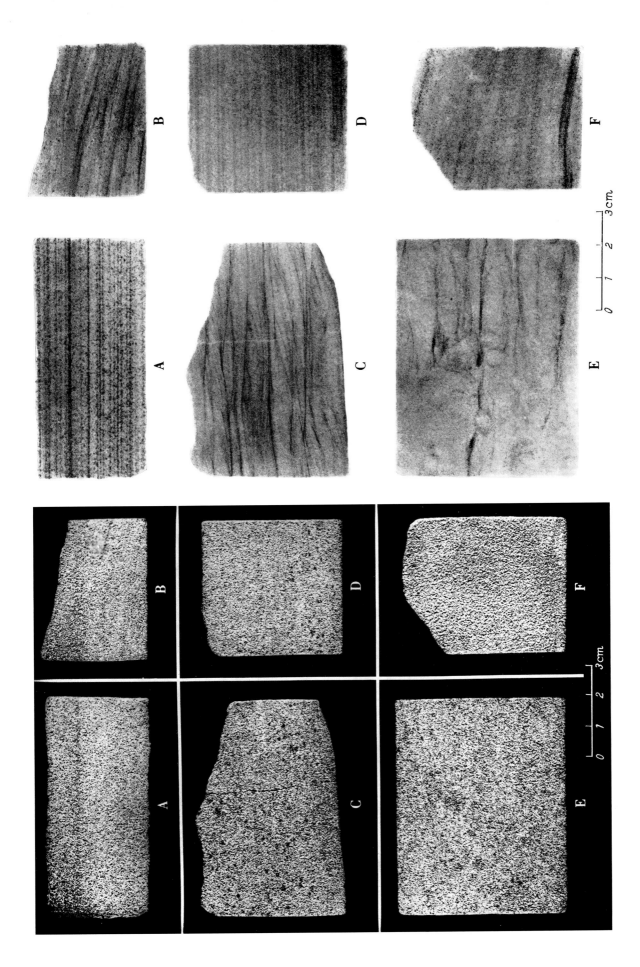

PLATE 24 A

Beds of variable thickness in dolomite. Thick massive regular bed overlying thinner beds (Silurian). At Sunset Point, approximately 7 miles southeast of Galena, Jo Daviess County, Illinois, U.S.A. (Illinois Geological Survey).

Dolomitschichten verschiedener Mächtigkeit. Massig gebankte Schichten im Hangenden, dünne Lagen im Liegenden (Silur).

Couches d'épaisseur variable dans une dolomie. Couche épaisse, massive et régulière surmontant des lits plus minces (Silurien).

Dolomía en estratos de espesores variables. Un estrato masivo de espesor considerable yace sobre estratos delgados. Silúrico.

PLATE 24 B

Massive mudstone. Note absence of fissility and general absence of bedding. Bloomsburg Member of Wills Creek Formation (Silurian). Along tracks of Western Maryland Railroad, Round Top, Washington County, Maryland, U.S.A.

Massiger Pelit. Beachte Fehlen von Teilbarkeit und Schichtung. Bloomsburg Member der Wills Creek Formation (Silur).

Mudstone massive. Noter l'absence de fissilité et l'absence générale de stratification. Terme de Bloomsburg de la formation de Wills Creek (Silurien).

Limolita masiva; nótese la falta de fisilidad y de estratificación, característica poco común en este tipo de sedimentita. Miembro Bloomsburg, Formación Wills Creek, Silúrico.

PLATE 25

Thick, uniform massive bedding in Moehave Formation (Triassic). NW 1/4 SE 1/4 sec. 15, T. 41 S., R. 10 W., Washington County, Utah, U.S.A. (Photograph by W. K. HAMBLIN).

Mächtige, gleichförmig massige Schichtung. Moehave Formation (Trias).

Couches massives, épaisses et uniformes dans la formation de Moehave (Trias).

Estratos masivos de considerable espesor y uniformes. Formación Moehave, Triásico.

PLATE 26

Tillite of Gowganda Formation (Cobalt Series, Huronian). Essentially a bedding plane surface showing weak alignment of elongate rock fragments. Rock surface has been striated or scratched by Pleistocene glaciers. Highway 17, about 4 miles south of Latchford, District of Timiskaming. Ontario, Canada.

Tillit der Gowganda Formation (Cobalt Series, Huron). Blick auf eine nahezu horizontale Schichtfläche. Beachte schwache Einregelung der Gesteinsfragmente. Die Gesteinsoberfläche zeigt Schrammen und Striemen von pleistozänen Gletschern.

Tillite de la formation de Gowganda (Série Cobalt, Huronien). La surface d'affleurement est dans le plan de stratification, presque horizontale, et montre l'alignement peu prononcé des fragments rocheux allongés. La surface de la roche a été striée ou égratignée par les glaciers pleistocènes.

Tilita de la Formación Gowganda. La superficie expuesta es paralela al plano de depositación y aproximadamente horizontal. El ejemplar contiene fragmentos elongados con cierta orientación, carece de estratificación y presenta raspaduras o estrias originadas por atrición de un glaciar del Pleistoceno. Serie Cobalt, Huroniano.

PLATE 27

Sand waves of two sizes in tidal creek. Sand waves formed by ebb current, from upper right to lower left. Bay of Arcachon, near Le Teich, Gironde, France (Photograph by L. M. J. U. VAN STRAATEN).

Sandwellen zweier verschiedener Größen in einem Priel. Sandwellen wurden von Ebbströmung geformt, die von rechts oben nach links unten floß.

Vagues de sable de deux tailles différentes dans un étier. Les vagues de sable ont été formées par le courant de jusant, allant d'en haut à droite à bas à en gauche.

Ondas de arena de dos longitudes formadas por corrientes de bajamar. Corriente desde el ángulo superior derecho al inferior izquierdo. Actual.

PLATE 28

Sand waves in modern carbonate sands. Cat Cay sand belt, approximately one mile wide, northwest corner of Great Bahama Bank, Bahama Islands (Shell Development Company).

Sandwellen in rezenten Karbonatsanden.

Vagues de sable dans des sables calcaires actuels.

Ondas de arena calcárea; Actual.

PLATE 29A

Air photograph of sand waves on island in Mississippi River. Wave crests are 20 to 60 feet (7 to 20 m) apart. Current from left to right, Waterproof, Louisiana, U.S.A. (U.S. Army Engineers, Vicksburg District)

Luftaufnahme von Sandwellen auf einer Insel im Mississippi. Die Wellenkämme sind von 7 bis 20 m voneinander entfernt. Strömungsrichtung von links nach rechts.

Ile dans le cours du Mississippi montrant des vagues de sable. La distance entre les crêtes des vagues est de 7 à 20 m. Courant de gauche à droite.

Ondas de arena en una isla del Río Mississippi (vista aérea). La distancia entre crestas oscila entre 7 y 20 m. Corriente de izquierda a derecha. Actual.

PLATE 29B

Cross-bedding in point bar of Vermilion River. Current from left to right. SW 1/4 SW 1/4 sec. 30, T. 18 N., R. 9 W., Vermillion County, Indiana, U.S.A.

Schrägschichtung in Sandbank am Gleithang des Vermilion-Flusses. Strömungsrichtung von links nach rechts.

Stratification entrecroisée dans un bourrelet arqué de la rivière Vermilion. Sens du courant de gauche à droite.

Estratificación entrecruzada en un espolón aluvial del Río Vermilion; corriente de izquierda a derecha.

PLATE 30A

Cross-bedding in limestone of Miami Oolite (Pleistocene). Current from left to right. Cocoanut Grove, Dade County, Florida, U.S.A. (Shell Development Company).

Schrägschichtung im Miami-Oolit (pleistozäner Kalkstein). Strömungsrichtung von links nach rechts.

Calcaire à stratification entrecroisée dans l'oolithe de Miami (Pleistocène). Sens du courant de gauche à droite.

Estratificación entrecruzada en caliza oolítica; corriente de izquierda a derecha. Pleistoceno.

PLATE 30B

Cross-bedding in sandstone of Caseyville Formation (Pennsylvanian). Current from right to left. Hammer head rests on contact of cross-bedded layer with true bedding. NE 1/4 NE1/4 NW 1/4 sec. 21, T. 7 S., T. 5 W., Randolph County, Illinois, U.S.A. (Illinois Geological Survey).

Schrägschichtung im Sandstein der Caseyville Formation (Pennsylvanian). Strömungsrichtung von rechts nach links. Hammer an der Grenzfläche zwischen schräg- und normalgeschichtetem Sandstein.

Grès à stratification oblique de la formation de Caseyville (Pennsylvanien). Sens du courant de droite à gauche. La tête du marteau marque le contact d'une couche oblique avec une couche à stratification vraie.

Estratificación entrecruzada en areniscas. La masa del martillo está apoyada en el contacto entre las capas frontales y el plano inferior de la unidad sedimentaria que las contiene. Corriente de derecha a izquierda. Formación Caseyville, Pennsylvaniano.

PLATE 31

Well developed tabular cross-bedding. Note uniformity of inclination. Rough Rock Sandstone (Upper Carboniferous) at Thievely Scout, near Burnley, Lancashire, England (SHACKLETON, 1962, pl. 8).

Gut entwickelte, tafelige Schrägschichtung. Beachte gleichförmiges Einfallen. Rough Rock Sandstein (Oberkarbon).

Stratification oblique tabulaire bien marquée. Noter l'inclinaison uniforme. Grès de Rough Rock (Carbonifère supérieur).

Estratificación entrecruzada tabular bien desarrollada. Nótese la uniformidad de la inclinación. Arenisca Rough Rock, Carbonífero Superior.

PLATE 32A

Cross-bedding of variable direction in modern sand dunes. Great Sand Dunes National Monument, Colorado, U.S.A. (Photograph by G. P. MERK).

Schrägschichtung wechselnder Einfallsrichtung in rezenten Sanddünen.

Stratification entrecroisée de direction variable dans des dunes de sable actuelles.

Estratificación entrecruzada con inclinación en direcciones variables; arena de duna. Actual.

PLATE 32B

Cross-bedding in Purslane Sandstone, Pocono Group (Mississippian). Note tangential foresets. Current from right to left. Approximately 6 miles west of Hancock on U.S. Highway 40 at Sideling Hill, Washington County, Maryland, U.S.A.

Schrägschichtung im Purslane-Sandstein, Pocono Group (Mississippian). Beachte tangentiale Leeblätter. Strömungsrichtung von rechts nach links.

Stratification entrecroisée dans le grès de Purslane, groupe de Pocono (Mississipien). Noter les couches frontales qui deviennent tangentielles. Sens du courant de droite à gauche.

Estratificación entrecruzada en arenisca. Nótese las capas frontales dispuestas tangencialmente a la base de cada unidad. Corriente de derecha a izquierda. Arenisca Purslane. Grupo Pocono, Mississippiano.

PLATE 33 A

Cross-section of large cross-bedded trough. Current obliquely into hillside. Navajo Sandstone (Jurassic). SE 1/4 SE 1/4 SE 1/4 sec. 32, T. 40 S., R. 17 W., Washington County, Utah. (Photograph by W. K. HAMBLIN).

Profil einer größeren, schräggeschichteten Mulde. Strömungsrichtung schräg in Aufschlußwand hinein. Navajo-Sandstein (Jura).

Coupe d'une grande auge à stratification entrecroisée. Le courant se dirigeait obliquement vers l'intérieur de la colline. Grès Navajo (Jurassique).

Corte transversal de un cauce amplio con estratificación entrecruzada; corriente oblícua al faldeo. Arenisca Navajo, Jurásico.

PLATE 33 B

Cross-bedding in Pleasantview Sandstone (Pennsylvanian). NE 1/4 NE 1/4 SW 1/4 sec. 31, T. 2 N., R. 1 E., Schuyler County, Illinois, U.S.A. (Illinois Geological Survey).

Schrägschichtung im Pleasantview-Sandstein (Pennsylvanian).

Stratification entrecroisée dans le grès de Pleasantview (Pennsylvanien).

Estratificación entrecruzada. Arenisca Pleasantview, Pennsylvaniano.

PLATE 34A

Thick tabular cross-bedding. Current from right to left. Sandstone of Caseyville Formation (Pennsylvanian). NE 1/4 NW 1/4 NW 1/4 sec. 2, T. 11 S., R. 7 E., Hardin County, Illinois, U.S.A. (Illinois Geological Survey).

Mächtige, tafelige Schrägschichtung. Strömungsrichtung von rechts nach links. Sandstein der Caseyville Formation (Pennsylvanian).

Stratification oblique tabulaire. Le courant allait de droite à gauche. Grès de la formation de Caseyville (Pennsylvanien).

Estratificación entrecruzada tabular de considerable espesor; corriente de derecha a izquierda. Formación Caseyville, Pennsylvaniano.

PLATE 34B

Top view of foresets of large cross-bedded layer overlain by thin-bedded, horizontal sandstone. Current from lower right to upper left. Degonia Sandstone (Mississippian). NE 1/4 NE 1/4 NE 1/4 sec. 23, T. 11 S., R. 1 W., Union County, Illinois, U.S.A. (Illinois Geological Survey).

Aufsicht auf Leeblätter einer mächtigen, schräggeschichteten Lage mit dünnschichtigem, horizontalem Sandstein im Hangenden. Strömungsrichtung von rechts unten nach links oben. Degonia-Sandstein (Mississippian).

Couches frontales, vues de dessus, d'un grand banc oblique surmonté par un grès en lits minces, horizontaux. Le courant allait d'en bas à droite vers la gauche en haut. Grès de Degonia (Mississipien).

Vista en planta de una amplia unidad sedimentaria formada por capas frontales, infrayacente a una arenisca finamente estratificada. Corriente desde el ángulo inferior derecho al superior izquierdo. Arenisca Degonica, Mississippiano.

PLATE 35 A

Cross-bedding in calcarenite. Ste. Genevieve Limestone (Mississippian). W. K. HAMBLIN stands on top of one unit; another lower cross-bedded unit shown in foreground. NW 1/4 NE 1/4 NE 1/4 sec. 34, T. 2 S., R. 10 W., Monroe County, Illinois, U.S.A. (Illinois Geological Survey).

Schrägschichtung in Kalksandstein. Ste. Genevieve Kalkstein (Mississippian). W. K. HAMBLIN steht auf der einen Einheit; im Vordergrund darunter eine andere schräggeschichtete Einheit.

Stratification oblique dans le calcaire calcarénitique de Ste. Genevieve (Mississipien). W. K. HAMBLIN se tient au sommet d'une des assises.

Estratificación entrecruzada en calcarenita. W. K. HAMBLIN se encuentra de pie en el techo de una unidad suprayacente a la que se observa en primer plano. Caliza Ste. Genevieve, Mississippiano.

PLATE 35 B

Cross-bedding in Loyalhanna Limestone (Mississippian). Note overlying thick "massive" limestone bed with uniform, even lamination. Conemaugh Gorge, near Johnstown, Cambria County, Pennsylvania, U.S.A. (Photograph by L. D. MECKEL jr.).

Schrägschichtung im Loyalhanna-Kalkstein (Mississippian). Beachte mächtigen, massigen Kalkstein mit ebener, einheitlicher Feinschichtung im Hangenden.

Stratification entrecroisée dans le calcaire de Loyalhanna (Mississipien). Noter le banc de calcaire épais, massif, à litage uniforme et régulier, qui se trouve au sommet.

Estratificación entrecruzada en caliza, infrayacente a un grueso estrato (caliza) "masivo" con laminación uniforme y regular. Caliza Loyalhanna, Mississippiano.

PLATE 36

Cross-bedding in the Navajo Sandstone (Jurassic). SW 1/4 SW 1/4 SW 1/4 sec. 20, T. 41 S., R. 9 W., Zion National Park, Kane County, Utah, U.S.A. (Photograph by W. K. HAMBLIN).

Schrägschichtung im Navajo-Sandstein (Jura).

Stratification entrecroisée dans le grès de Navajo (Jurassique).

Estratificación entrecruzada. Arenisca Navajo, Jurásico.

PLATE 37

Cross-bedding in calcareous sandstone ot Loyalhanna Limestone (Mississippian). Current from right to left. Abandoned quarry, north of U.S. Highway 30, at crest of Laurel Hill, Westmoreland County, Pennsylvania, U.S.A. (Photograph by R. W. Adams).

Schrägschichtung im Loyalhanna-Kalksandstein (Mississippien). Strömungsrichtung von rechts nach links.

Stratification entrecroisée dans le grès calcaire des calcaires de Loyalhanna (Mississipien). Le courant allait de droite à gauche.

Estratificación entrecruzada en arenisca calcárea. Corriente de derecha a izquierda. Caliza Loyalhanna, Mississippiano.

PLATE 38A

Micro-cross-lamination of Schilfsandstein (Triassic). Current from lower right to upper left. Burrer quarry at Maulbronn, Württemberg, Germany. (Photograph by PAUL WURSTER.)

Kleindimensionale Schrägschichtung in triassischem Schilfsandstein. Strömungsrichtung von rechts unten nach links oben.

Microstratification croisée de Schilfsandstein (Trias). Le plan de la photo est le plan de stratification. Le courant allait de la droite en bas a la gauche en haut.

Laminación micro-entrecruzada; corriente desde el ángulo inferior derecho al superior izquierdo. Schilfsandstein, Triásico.

PLATE 38B

Siltstone showing oversteepened and deformed current ripple cross-lamination. Current from right to left. Note laminations overturned in down-current direction. Martinsburg Shale (Ordovician). Near Grantville, Dauphin County, Pennsylvania, U.S.A. (Photograph by E. F. McBRIDE).

Siltstein mit übersteilter und verformter Strömungsrippel-Schrägschichtung. Strömungsrichtung von rechts nach links. Beachte die nach stromabwärts überkippten Feinschichten. Martinsburg-Schiefer (Ordovizium).

«Siltstone» à fin litage entrecroisé dû à la migration de rides de courant. Noter les feuillets redressés et déformés. Le courant allait de droite à gauche. Les lits sont déversés dans le sens du courant. Argile schisteuse de Martinsburg (Ordovicien).

Limolita con estructura entrecruzada originada por migración de óndulas, cuyas láminas presentan marcada pendiente y deformación. Nótese las láminas volcadas en sentido de la corriente (de derecha a izquierda). Limolita de la Formación Martinsburg, Ordovícico.

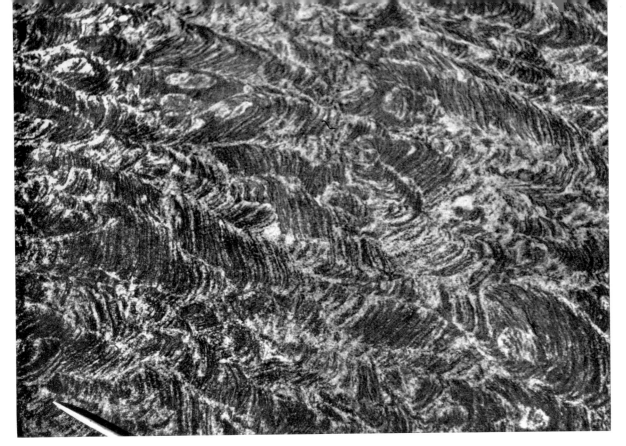

PLATE 39

Ripple cross-lamination in river alluvium. Ripple laminae rhythmically superimposed with development of pseudobedding in upper half of picture, 123,5 mile canyon, Colorado River, Arizona, U.S.A. (McKEE, 1938, Fig. 4-D).

Rippel-Schrägschichtung im Alluvium des Colorados. Sich rhythmisch überlagernde Rippelschichten mit Pseudoschichtung in der oberen Bildhälfte.

Feuilletage entrecroisé de rides dans les alluvions de la rivière du Colorado. Les feuillets, dûs à la migration des rides, sont superposés en alternance répétée, avec formation de pseudo-stratification dans la moitié supérieure de la photo.

Migro-óndulas entrecruzadas en aluvio del Río Colorado. Ondulas laminadas superpuestas con desarrollo de pseudo-estratificación en la mitad superior de la lámina. Actual.

PLATE 40A

Inclined bedding in outwash sands and gravels of Lemont drift (Pleistocene). Ice front to right, NE 1/4 NW 1/4 sec. 24; T. 37 N., R. 12 E., Cook County, Illinois, U.S.A. (Illinois Geological Survey).

Geneigte Schichtung in Auswaschsanden und Geröllen der Lemont Drift (Pleistozän). Eisfront lag rechts.

Stratification inclinée dans les sables et graviers fluvio-glaciaires des dépôts de Lemont (Pleisto-cène). Front du glacier vers la droite.

Estratificación inclinada en detrito fluvio-glaciario; frente del glaciar a la derecha. Arenas y gravas del drift Lemont, Pleistoceno.

PLATE 40B

Inclined bedding with long uniform foresets in Lafayette gravel (Pliocene). Current from right to left. On Kentucky State Highway 121, 0.5 miles southwest of New Concord, Calloway County, Kentucky, U.S.A.

Geneigte Schichtung mit langen, gleichmäßigen Leeblättern im pliozänen Lafayette-Schotter. Strömungsrichtung von rechts nach links.

Stratification inclinée avec couches frontales uniformes sur une grande largeur dans les graviers de Lafayette (Pliocène). Courant allant de droite à gauche.

Estratificación inclinada formada por capas frontales largas y uniformes. Corriente de derecha a izquierda. Grava Lafayette, Plioceno.

PLATE 41 A

Dipping flank beds of Thornton reef (Silurian). Illinois Tollway, near quarry of the Materials Service Corporation at Thornton, Cook County, Illinois, U.S.A. (Illinois Geological Survey).

Geneigte Flankenschichten am Thornton-Riff (Silur).

Strates inclinées sur le flanc du récif de Thornton (Silurien).

Estratificación inclinada en el flanco de un arrecife. Arrecife Thornton, Silúrico.

PLATE 41 B

Inclined reef-flank bedding in Edwards Limestone (Cretaceous). Inclined beds dip to the left away from biohermal reef at right. Compare with plate 41 A. Middle Bosque River, three miles southeast of Crawford, McLennan County, Texas, U.S.A.

Geneigte Riff-Flankenschichtung im Edwards-Kalkstein (Kreide). Die geneigten Schichten fallen vom Bioherm-Riff rechts nach links hin ein. Vergleiche mit Tafel 41 A.

Stratification inclinée sur le flanc d'un récif dans les calcaires de Edwards (Crétacé). Les couches inclinées plongent vers la gauche, en s'éloignant du bioherme que se trouverait à droite. Comparer avec 41 A.

Estratificación inclinada con pendiente hacia la izquierda, en el flanco de un arrecife. El bioherma se encuentra a la derecha, fuera del marco fotográfico. Compárese con lámina 41 A. Caliza Edwards, Cretácico.

PLATE 42A

"Potomac Marble" or limestone conglomerate. Note bedding, parallel to length of specimen, fragment size grading coarse to fine upwards, and weak orientation of elongate fragments parallel to bedding. Taneytown facies, Newark Series (Triassic), Maryland, U.S.A.

„Potomac Marble" (Kalkstein-Konglomerat). Beachte Schichtung parallel zur Länge des Hand-stücks, Größe der Fragmente, die von grob (unten) nach fein (oben) hin übergehen, und die schwache Ausrichtung länglicher Fragmente parallel zur Schichtung.

«Potomac Marble», ou conglomérat calcaire. La stratification est parallèle au grand côté de l'échantillon, les éléments allant de grossiers à fins vers le haut; noter aussi l'orientation des fragments allongés qui tend à être parallèle à la stratification.

Conglomerado calizo ("Potomac Marble"). Nótese: la estratificación paralela a la longitud de la muestra; el tamaño de los fragmentos gradando hacia arriba, de grueso a fino; y la débil orienta-ción de los fragmentos elongados paralelos a la estratificación. Facies Taneytown, Serie Newark, Triásico.

PLATE 42B

Graded bedding in Martinsburg Shale (Ordovician). Three graded units, each show weak lamina-tions at top. Note load casts and weak current ripple-bedding in uppermost bed. Near Middle-town, Orange County, New York, U.S.A. (McBRIDE, 1962, Fig. 7).

Gradierte Schichtung im Martinsburg-Schiefer (Ordovizium). Drei gradierte Einheiten, je im Oberteil schwach feinschichtig; beachte Belastungsmarken und schwache Strömungs-Rippel-schichtung in der obersten Lage.

Couches granoclassées dans l'argile schisteuse de Martinsburg (Ordovicien). Trois unités grano-classées dont chacune montre une amorce de litage au sommet; noter les empreintes de charge (poches) et la stratification de rides de courant peu accentuée dans la couche supérieure.

Estratificación gradada. Tres unidades con gradación a pelitas débilmente laminadas. El estrato superior presenta calcos de carga y estratificación de óndulas ácueas de escaso desarrollo. Lutita Martinsburg, Ordovícico.

PLATE 43 A

Imbrication of limestone slabs in creek. Current from right to left. Junction of State Highway 125 and Whiteoak Creek, Lewis Township, Brown County, Ohio, U.S.A.

Dachziegel-Lagerung aus Kalksteinplatten. Strömungsrichtung von rechts nach links.

Plaques de calcaire imbriquées dans un cours d'eau. Courant de droite à gauche.

Lajas de caliza imbricadas. Corriente de derecha a izquierda. Actual.

PLATE 43 B

Detailed view of chert gravel shown in plate 40 B. Note general alignment of long dimensions of pebbles parallel to pocket knife (parallel to bedding) and weak up-current imbrication. Current flowed from right to left.

Nah-Ansicht der Hornstein-Schotter aus Tafel 40 B. Beachte allgemeine Ausrichtung der Gerölle mit ihrer Längsachse parallel zum Taschenmesser (parallel zur Schichtung). Schwache Dachziegel-Lagerung stromaufwärts. Strömungsrichtung von rechts nach links.

Détail du gravier de chailles de la planche 40 B. Noter l'arrangement généralement parallèle à la stratification (indiquée par le couteau), des plus grandes dimensions des cailloux. Légère imbrication vers l'amont. Le courant allait de droite à gauche.

Vista en detalle de la lámina 40 B, de una grava compuesta por ftanita. Nótese la orientación general paralela al cortaplumas, en la elongación de los rodados (paralela a la estratificación) y la débil imbricación en sentido opuesto a la corriente (de derecha a izquierda).

PLATE 44 A

Flat, soft, mud pebbles on tidal flat. Weak imbrication by current from right to left. Jade Busen, Lower Saxony, Germany (Senckenberg am Meer, L 8445).

Abgeflachte, weiche Schlickgerölle auf Wattoberfläche. Schwach ausgebildete Dachziegel-Lagerung, bedingt durch Strömung von rechts nach links.

Galets mous aplatis sur une slikke. Légère imbrication par le courant allant de droite à gauche.

Zona intercotidal con cantos rodados chatos, plásticos y débilmente imbricados. Corriente de derecha a izquierda. Actual.

PLATE 44 B

Casts from a shale-pebble conglomerate in Hardinsburg sandstone (Mississippian). Note weak alignment of pebble casts. Specimen approximately 1 meter long. SW 1/4 NE 1/4 NE 1/4 sec. 32, T. 13 S., R. 4 E., Johnson County, Illinois, U.S.A. (Illinois Geological Survey).

Abdrücke eines Konglomerates von Tonschiefer-Geröllen im Hardinsburg-Sandstein (Mississippian). Beachte schwache Ausrichtung der Geröllabdrücke. Handstück ungefähr 1 m lang.

Moulages de cailloux schisteux dans un conglomérat dans les grès de Hardinsburg (Mississipien). Noter la légère tendance à l'alignement des moulages. Longueur de l'échantillon: environ 1 mètre.

Calco en arenisca de un conglomerado constituído por grava de lutita débilmente orientada. Longitud aproximada de la muestra, 1 m. Arenisca Hardinsburg, Mississippiano.

PLATE 45 A

Quartz pebble conglomerate in sandstone of Caseyville Formation (Pennsylvanian). Quartz pebbles lie in plane of bedding. Long axes of pebbles show weak orientation. SE 1/4 SW 1/4 NW 1/4 sec. 22, T. 7 S., R. 5 W., Randolph County, Illinois, U.S.A. (Illinois Geological Survey).

Konglomerat aus Quarzgeröllen im Sandstein der Caseyville Formation (Pennsylvanian). Die Quarzgerölle liegen in der Ebene der Schichtung. Ihre Längsachsen zeigen schwache Einregelung.

Conglomérat de cailloux quartzeux dans les grès de la formation de Caseyville (Pennsylvanien). Les cailloux de quartz sont dans le plan de stratification et leurs grands axes sont plus ou moins orientés.

Intercalación conglomerádica cuarzosa en arenisca. Nótese la débil orientación en la elongación de los clastos, que reposan en un plano de estratificación. Formación Caseyville, Pennsylvaniano.

PLATE 45 B

Oriented *Donax* shells and sand lineation. Ebb current from top to bottom. Beach at "Little Shell" on Mustang Island, San Patricio County, Texas, U.S.A. (Photograph by E. F. McBride).

Orientierte *Donax*gehäuse und Sandriefung. Ebbstrom von oben nach unten.

Coquilles de *Donax* orientées, et linéation du sable. Le courant de jusant va du haut en bas de la photo.

Conchillas (*Donax*) orientadas y marcas de lineación en arena. Reflujo desde el borde superior de la lámina al inferior. Actual.

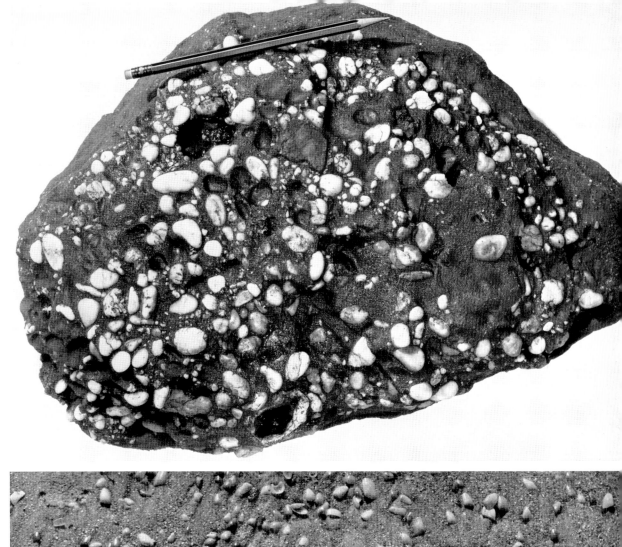

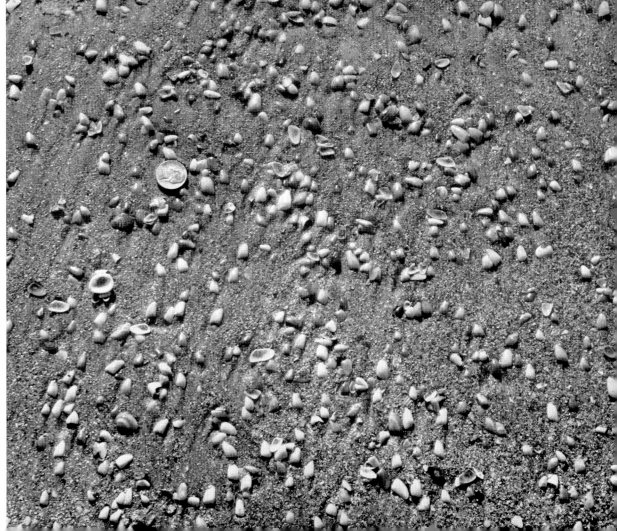

PLATE 46A

Well-oriented parafusulinids in limestone of Antler Peak Limestone (Permian). Average length of parafusulinids is approximately 15 mm. Northeast corner of T. 42 N., R. 32 E., Humbolt County, Nevada, U.S.A. (U.S. Geological Survey).

Gut eingeregelte Parafusuliniden im Kalkstein der Antler Peak-Schichten (Perm). Durchschnittslänge der Parafusuliniden ungefähr 15 mm.

Parafusulines bien orientées dans un calcaire de la formation d'Antler Peak (Permien). Longueur moyenne des parafusulines, environ 15 mm.

Caliza con parafusulínidos marcadamente orientados. Longitud promedio de los organismos 15 mm. Caliza Antler Peak, Pérmico.

PLATE 46B

Oolites (×9), slightly etched, in Ste. Genevieve Limestone (Mississippian). Note preferred orientation of elongate oolites. Anna Quarries, Inc., near Anna, Union County, Illinois, U.S.A. (Illinois Geological Survey).

Leicht angeätzte Ooide (×9) im Ste. Genevieve-Kalkstein (Mississippian). Beachte bevorzugte Ausrichtung der länglichen Ooide.

Oolithes (×9), légèrement burinées, dans le calcaire de Ste. Genevieve (Mississipien). Noter l'orientation des oolithes allongées.

Oolites (×9) levemente corroídas, Nótese la orientación de las formas elongadas. Caliza Ste. Genevieve, Mississippiano.

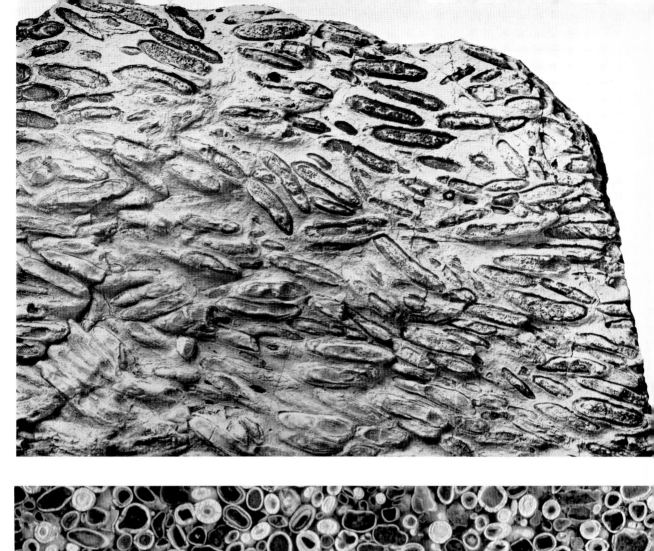

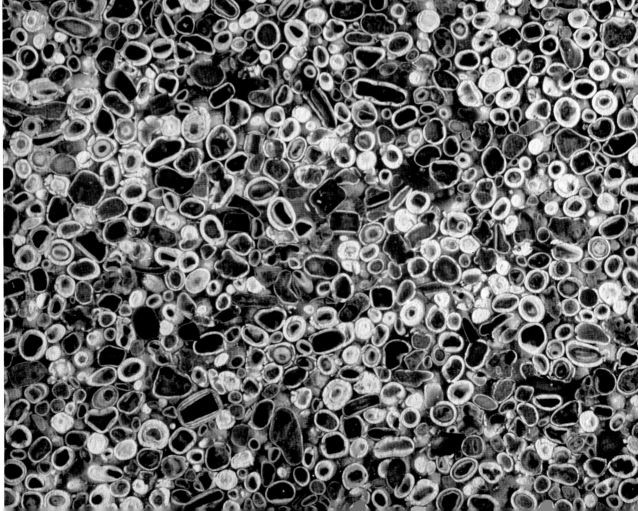

PLATE 47A

Vertical section of Devonian limestone showing outlines of brachiopod shells, crinoid stems and other fossil debris interbedded in calcilutite (natural size). Weak currents failed to orient strongly bivalve brachiopod shells. Compare with plate 47B. Near Milan, Rock Island County, Illinois, U.S.A. (Illinois Geological Survey).

Vertikalschnitt durch einen devonischen Kalkstein; Umrisse von Brachiopodenschalen, Crinoidenstämmen und anderen in Kalkpelit eingelagerten Fossilfragmenten (natürliche Größe). Die Strömung war zu schwach, um die durchweg doppelschaligen Brachiopoden einzuregeln. Vergleiche mit Tafel 47B.

Coupe verticale dans un calcaire dévonien, montrant les contours de coquilles de brachiopodes, de tiges de crinoides et d'autres débris fossiles, interstratifiés dans une calcilutite (grandeur nature). Les courants faibles n'ont pas pu orienter nettement les coquilles bivalves des brachiopodes. Comparer avec 47B.

Conchillas de braquiópodos, tallos de crinoídeos y otros restor fósiles, en caliza y calcipelita interestratificadas. Una débil corriente ha sido suficiente para orientar netamente las conchillas de braquiópodos. Tamaño natural. Compárese con lámina 47B. Devónico.

PLATE 47B

Top of limestone bed showing the stable position of current-oriented brachiopod shells to be convex up. Cincinnatian Series (Ordovician). Near Seaman, Adams County, Ohio, U.S.A.

Oberseite einer Kalksteinbank. Die von der Strömung eingeregelten Brachiopoden weisen in stabiler Lage mit ihrer konvexen Seite nach oben. Cinncinnatian Series (Ordovizium).

Plan de stratification d'un calcaire, vu de dessus, montrant qu'en position stable les coquilles de brachiopodes orientées par le courant ont leur partie convexe vers le haut. Série de Cincinnati (Ordovicien).

Techo de una caliza con conchillas de braquiópodos en posición estable, orientadas por corrientes con sus caras convexas hacia arriba. Serie Cincinnatiana, Ordovícico.

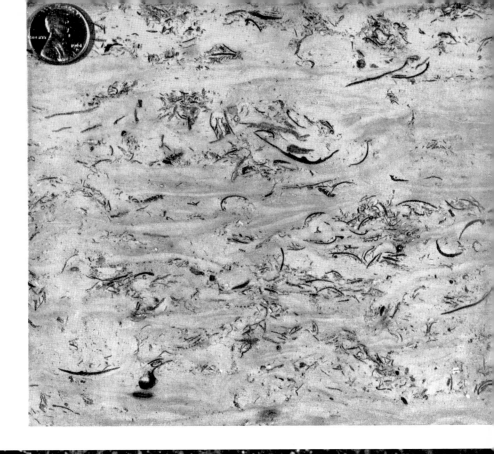

PLATE 48

Laminated limestone. Laminations probably result from an algal growth structure ("*weedia*"-type stromatolite). Note irregularities such as bifurcation, local disruption and arching laminations. Conococheague Limestone (Cambrian). Approximately 2 miles northeast of Clear Spring, Washington County, Maryland, U.S.A.

Feingeschichteter Kalkstein, in dem die Feinschichten wahrscheinlich durch Algenwachstum bedingt sind (Stromatolithen vom „*Weedia*"-Typ). Beachte Unregelmäßigkeiten wie Gabelung, örtliche Unterbrechungen und bogige Feinschichtung. Conococheague-Kalkstein (Ordovizium).

Calcaire feuilleté. Les feuillets résultent probablement de la croissance d'une algue (stromatolithe du type «*Weedia*»). Noter les irrégularités telles que bifurcation, disruption locale et feuillets bombés. Calcaire de Conococheague (Ordovicien).

Caliza laminada, probablemente originada por actividad biológica de algas; estructura estromatolítica, tipo *weedia*. Nótese algunas láminas bifurcadas, arqueadas y discontínuas. Caliza Conococheague, Ordovícico.

PLATE 49A

Stromatolitic dolomite showing several stromatolite columns. Denault Formation (Precambrian),
Marion Lake, Labrador, Canada (Photograph by W. HILLER).

Stromatolithischer Dolomit mit mehreren pfeilerartigen Stromatolithen. Denault Formation
(Präkambrium).

Dolomie stromatolithique avec plusieurs colonnes de stromatolithes. Formation de Denault
Précambrien).

Dolomía estromatolítica con algunas columnas estromatolíticas. Formación Denault, Precámbrico.

PLATE 49B

Stromatolite: close-up of upper bedding plane surface of massive biohermal bed of plate 50B.
Small stromatolitic heads (*proliferum* type) have weathered into relief. Conococheague Lime-
stone (Cambrian). Along Western Maryland Railroad, about one half mile east of Big Spring
Station, Washington County, Maryland, U.S.A.

Stromatolithen in der Schichtfläche einer massigen Biohermbank. Nahaufnahme von Tafel 50B.
Reliefartige Herauswitterung von kleinen Stromatolithenköpfen („*proliferum*"-Typ). Conococheague-Kalkstein (Kambrium).

Stromatolithe: Détail du plan de stratification, vu de dessus, du bioherme massif de la planche
50B. Les petits stromatolithes (du type *proliferum*) ont été laissé en relief par l'érosion. Calcaires
de Conococheague (Cambrien).

Estromatolita. Vista en planta y en detalle del techo de un estrato masivo biohermal (lám. 50B),
con pequeñas "cabezas" estromatolíticas (tipo *proliferum*) realzadas por meteorización. Caliza
Conococheague, Cámbrico.

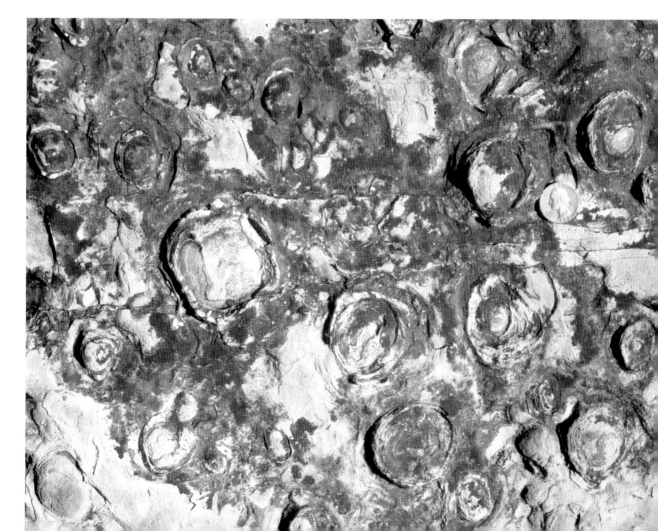

PLATE 50A

Casts and molds of rudistid pelecypods in massive, poorly bedded limestone. Edwards Limestone (Cretaceous). Along Middle Bosque River Valley, three miles southwest of Crawford, McLennan County, Texas, U.S.A.

Ausgüsse und Abdrücke von Lamellibranchiaten (Rudisten) in massigem, schlecht geschichtetem Kalkstein. Edwards-Kalkstein (Kreide).

Moules et moulages de rudistes dans un calcaire massif, à stratification peu marquée. Calcaire d'Edwards (Crétacé).

Calcos y moldes de pelecípodos (rudistas) en caliza masiva levemente estratificada, Caliza Edwards, Cretácico.

PLATE 50B

Massive stromatolitic bioherm. The lower beds are dolomitic; the bioherm is mainly structureless limestone. At the very top of the bed are closely packed stromatolites of *proliferum* type (see pl. 49B). Conococheague Limestone (Cambrian). Along Western Maryland Railway approximately 0.5 mile east of Big Spring Station, Washington County, Maryland, U.S.A.

Massiges stromatolithisches Bioherm. Die unteren Schichten sind dolomitisch, das Bioherm im großen und ganzen strukturloser Kalkstein. Den obersten Teil der Bank bilden gedrängt wachsende Stromatolithen vom *Proliferum*-Typ (s. Tafel 49B). Conococheague-Kalkstein (Kambrium).

Bioherme stromatolithique massif. Les couches inférieures sont dolomitiques; le bioherme consiste surtout en calcaire sans structure définie. A son sommet s'entassent des stromatolithes du type *proliferum* (cf. pl. 49B). Calcaire de Conococheague (Cambrien).

Bioherma estromatolítico en parte masivo, compuesto principalmente por caliza de base dolomítica. En el techo, la estructura algácea (tipo *proliferum*) se encuentra estrechamente agrupada (véase lám. 49B). Caliza Conococheague, Cámbrico.

PLATE 51

Organic framework of coralites with finer grained calcareous matrix that contains some skeletal debris. Note stylolitic contact between some coralites (Devonian). Nat. Assoc. Petrol., No. 1, Marion County Coal Co., Sec. 32, T. 2 N., R. 1 E., Marion County, Illinois, U.S.A. (Illinois Geological Survey).

Organisches Gerüst aus Koralliten, mit feinkörniger Kalkmatrix, die Skeletfragmente enthält. Beachte stylolithischen Kontakt zwischen einigen Koralliten (Devon).

Charpente de squelettes coralliens, avec pâte calcaire à grain plus fin contenant quelques débris de squelettes. Noter le contact à stylolithes entre certains coraux (Dévonien).

Armazón orgánico compuesto por coralitos ligados por una mátriz calcárea de grano fino que engloba algunos restos orgánicos. Nótese el contacto estilolítico entre algunos coralitos. Devónico.

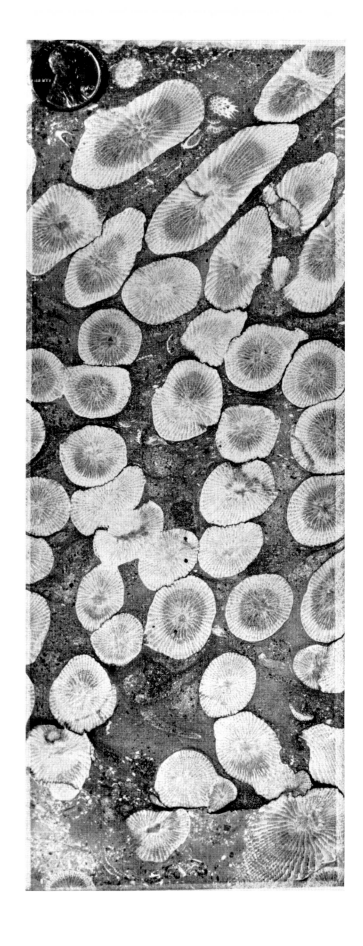

PLATE 52A

Load casts on underside of thin-bedded sandstone. Tar Springs Sandstone (Mississippian). SW 1/4 SW 1/4 NW 1/4 sec. 30, T. 7 S., R. 6 W., Randolph County, Illinois, U.S.A. (Illinois Geological Survey).

Belastungsmarken an der Unterseite dünngeschichteten Sandsteins. Tar Springs-Sandstein (Mississippian).

Empreintes de charge à la face inférieure d'une couche de grès mince. Grès de Tar Springs (Mississipien).

Calcos de carga en la superficie inferior de una arenisca finamente estratificada. Arenisca Tar Spring, Mississippiano.

PLATE 52B

Large crenulated load casts on underside of sandstone. Aux Vases Sandstone (Mississippian). SW 1/4 NW 1/4 NW 1/4 sec. 3, T. 5 S., R. 9 W., Monroe County, Illinois, U.S.A. (Illinois Geological Survey).

Schichtunterseite mit großen, gerunzelten Belastungsmarken. Aux Vases-Sandstein (Mississippian).

Grandes empreintes de charge plissotées à la face inférieure d'un grès. Grès de Aux Vases (Mississipien).

Calcos de carga amplios y corrugados, en la superficie inferior de una arenisca. Arenisca Aux Vases, Mississippiano.

PLATE 53 A

Bottom irregularities, ascribed by ARAI (1957, pl. 1, Fig. 1) to load casting, at base graded conglomeratic bed. Akahira Formation (Tertiary). Ushikubi Pass, Ogano-machi, Chichibu-gum, Saitama Prefecture, Japan.

Unregelmäßige Strukturen an der Sohlfläche einer gradierten konglomeratischen Schicht, von ARAI (1957, Tafel 1, Abb. 1) als Belastungsmarken gedeutet. Akahira Formation (Tertiär).

Irrégularités de fond, attribuées par ARAI à des figures de charge, à la base d'un conglomérat granoclassé. Formation de Akahira (Tertiaire).

Base irregular de un estrato conglomerádico gradado, atribuída por ARAI (1957, pl. 1, Fig. 1) a deformación por carga. Formación Akahira, Terciario.

PLATE 53 B

Flame structure at base of graywacke bed. Mud plumes overturned downcurrent. Current from right to left. Martinsburg Shale (Ordovician), Hamburg, Berks County, Pennsylviana, U.S.A.

Flammenstruktur an der Unterseite einer Grauwackenbank. „Flammenspitzen" nach stromabwärts überkippt. Strömungsrichtung von rechts nach links. Martinsburg-Schiefer (Ordovizium).

«Flame structure» à la base d'une grauwacke. Panaches de boue renversés dans le sens du courant qui allait de droite à gauche. Argile schisteuse de Martinsburg (Ordovicien).

Base de una grauvaca con estructura flamiforme. Las crestas han sido volcadas en sentido de la corriente (de derecha a izquierda). Lutita Martinsburg, Ordovícico.

PLATE 54A

Load casts with low relief and random pattern. Cypress Sandstone (Mississippian). NW 1/4 SE 1/4 sec. 23, T. 14 S., R. 4 E., Massac County, Illinois, U.S.A. (Illinois Geological Survey).

Belastungsmarken mit geringem Relief und wahlloser Anordnung. Cypress-Sandstein (Mississippian).

Empreintes de charge désordonnées et à relief peu accentué. Grès de Cypress (Mississipien).

Calcos de carga no orientados y de escaso relieve. Arenisca Cypress, Mississippiano.

PLATE 54B

Torose load casts (?) (CROWELL, 1955, p. 1360; pl. 2, Fig. 4). Lusk Shale (Pennsylvanian). NW 1/4 SE 1/4 NE 1/4 sec. 28, T. 10 S., R. 2 W., Union County, Illinois, U.S.A. (Illinois Geological Survey).

Wulstige Belastungsmarken (?). Lusk-Schiefer (Pennsylvanian).

Empreintes de charge (?) cordées et digitées. Argile schisteuse de Lusk (Pennsylvanien).

Calcos de carga (?) polilobulados. Lutita Lusk, Pennsylvaniano.

PLATE 55 A

Small flute casts. Current parallel to pencil from left to right. McAras Brook Formation (Mississippian?). Near McAras Brook, Antigonish County, Nova Scotia, Canada (Photograph by W. HILLER).

Kleine Strömungswülste. Strömungsrichtung von links nach rechts, parallel zum Bleistift. McAras Brook-Formation (Mississippian).

Petits «flute casts». Le crayon indique le sens du courant (de gauche à droite). Formation de McAras Brook (Mississipien).

Turboglifos pequeños. Corriente de derecha a izquierda, paralela al lápiz. Formación McAras Brook, Mississippiano?

PLATE 55 B

Small, closely spaced and overlapping elongate flute casts on underside of sandstone of Caseyville Formation (Pennsylvanian). Current from top to bottom. NE 1/4 NW 1/4 sec. 31, T. 11 S., R. 2 E., Johnson County, Illinois, U.S.A. (POTTER and GLASS, 1958, pl. 4 B).

Kleine, dicht gescharte und sich überlappende, längliche Strömungswülste an der Sohlfläche einer Sandsteinbank der Caseyville Formation (Pennsylvanian). Strömungsrichtung von oben nach unten.

Face inférieure d'un grès montrant de petits «flute casts» rapprochés et allongés et qui se recouvrent. Courant de haut en bas. Formation de Caseyville (Pennsylvanien).

Base de una arenisca con turboglifos pequeños estrechamente agrupados e imbricados en dirección de sus elongaciones. Corriente desde el borde superior de la lámina al inferior. Formación Caseyville, Pennsylvaniano.

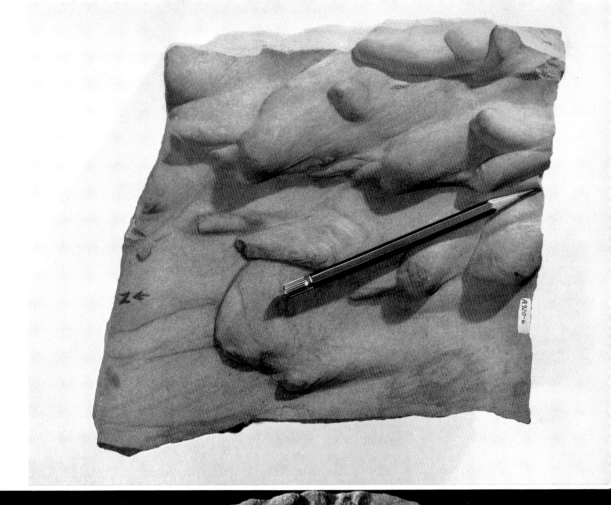

PLATE 56A

Flute casts of several sizes. Current from lower right to upper left. Atoka Formation (Pennsylvanian). On U.S. Highway 59, 1.2 miles south of Hodgen, Le Flore County, Oklahoma, U.S.A. (Jersey Production Research Company).

Strömungswülste verschiedener Größe. Strömungsrichtung von rechts unten nach links oben. Atoka Formation (Pennsylvanian).

«Flute casts» de tailles diverses. Le courant allait d'en bas à droite à en haut à gauche. Formation d'Atoka (Pennsylvanien).

Turboglifos de diversos tamaños. Dirección de la corriente desde el ángulo inferior derecho al superior izquierdo. Formación Atoka, Pennsylvaniano.

PLATE 56B

Small flute casts with development of relatively blunt up-current end. Current from lower right to upper left. Atoka Formation (Pennsylvanian). Approximately 1.2 miles south of Hodgen, Le Flore County, Oklahoma, U.S.A. (Jersey Production Research Company).

Kleine Strömungswülste; das stumpfe obere Ende zeigt stromaufwärts. Strömungsrichtung von rechts unten nach links oben. Atoka Formation (Pennsylvanian).

Petits «flute casts» qui se terminent brusquement vers l'amont du courant. Formation d'Atoka (Pennsylvanien). Le courant allait d'en bas à droite à en haut à gauche.

Turboglifos pequeños con extremos relativamente romos en sentido opuesto a la corriente. Sentido de la corriente desde el ángulo inferior derecho al superior izquierdo. Formación Atoka, Mississippiano.

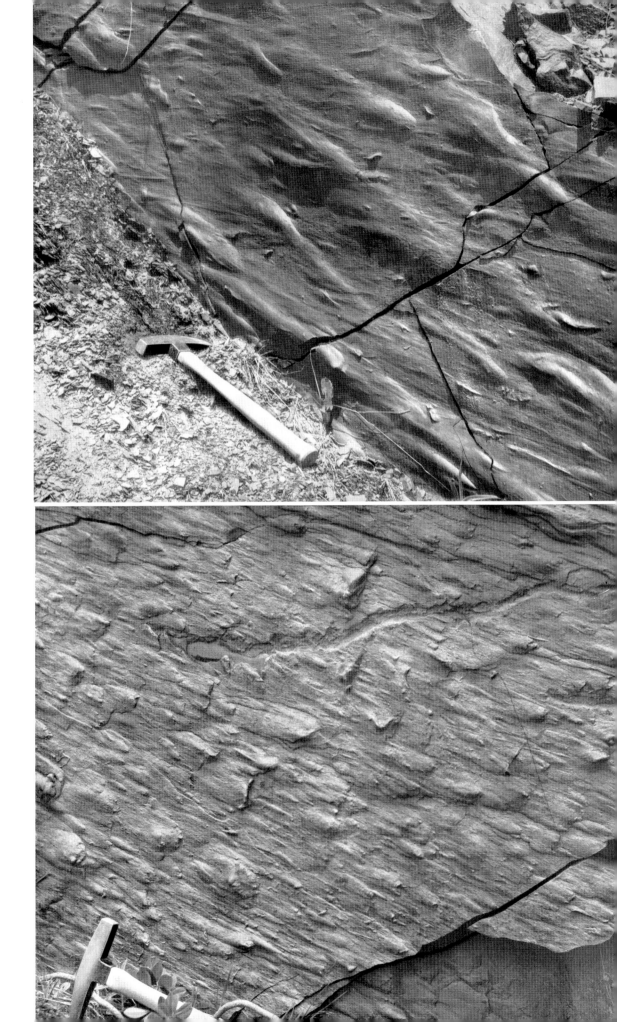

PLATE 57A

Simple flute casts and large groove cast (A) at lower edge of specimen. Current from right to left. Gardeau Formation (Devonian). Intersection of State Highway 235 and Coffee Hill Road, West Sparta Township, Livingston County, New York, U.S.A. (Photograph by W. HILLER).

Einfache Strömungswülste und große Rillenmarke (A) an der unteren Handstückkante. Strömungsrichtung von rechts nach links. Gardeau Formation (Devon).

«Flute casts» simples avec rainure assez importante (A) à la partie inférieure de l'échantillon. Courant de droite à gauche. Formation de Gardeau (Dévonien).

Turboglifos simples y amplio calco de surco (A) en el borde inferior de la muestra. Corriente de derecha a izquierda. Formación Gardeau, Devónico.

PLATE 57B

Flute casts on base of graded bed in the Tyee Formation (Tertiary). Note arrangement of flutes in rows and overlapping character. Current moved from bottom to top of picture. Green Mountain, 5 miles south of Valsetz, Polk County, Oregon, U.S.A. (U.S. Geological Survey).

Strömungswülste an der Sohle einer gradierten Schicht. Beachte Aufreihung und Überlappung der Wülste. Strömungsrichtung von unten nach oben. Tyee Formation (Tertiär).

«Flute casts» à la base d'une couche granoclassée dans la formation de Tyee (Tertiaire). Noter que les moulages sont en rangées et se recouvrent.

Turboglifos en la base de un estrato gradado, dispuestos en hileras e imbricados Corriente desde el borde inferior de la lámina al superior. Formación Tyee, Terciario.

PLATE 58

Large closely-spaced, and in part overlapping, fan-shaped to linguiform flute casts on base of thick sandstone bed. Current from left to right. Smithwick Formation (Pennsylvanian). Three miles east of Marble Falls, Burnett County, Texas, U.S.A. (Photograph by E. F. McBride).

Große, dicht gescharte und zum Teil sich überlappende, fächer- bis zungenförmige Strömungswülste an der Sohle einer mächtigen Sandsteinbank. Strömungsrichtung von links nach rechts. Smithwick Formation (Pennsylvanian).

Grands «flute casts», rapprochés et qui se recouvrent, à la face inférieure d'une épaisse assise de grès, allant de formes en éventail à un type linguiforme. Courant de gauche à droite. Formation de Smithwick (Pennsylvanien).

Turboglifos grandes con aspecto de abanicos aluviales o linguiformes, estrechamente agrupados y en parte imbricados, en la base de una arenisca de considerable espesor. Corriente de izquierda a derecha. Formación Smithwick, Pennsylvaniano.

PLATE 59A

Three large drag grooves with flute marks and squamiform load casts. Note rhythmic development of flute marks and squamiform casts perpendicular to grooves. Rhythmic pattern of
flutes and loads called "dinosaur leather" (CHADWICK, 1948, p. 1315). Current from lower left
to upper right. Normanskill Shale (Ordovician). On U.S. Highway 9 W just north of overpass
crossing Interstate 87, Greene County, New York, U.S.A. (Photograph by E. CHOWN).

Drei große Schleifrillen mit Strömungswülsten und schuppenförmigen Belastungsmarken.
Beachte rhythmische Abfolge von Strömungswülsten und schuppenförmigen Belastungsmarken
in rechtem Winkel zu den Rillen. Dieses rhythmische Muster wurde von CHADWICK (1948,
S. 1315) als „Dinosaurier Leder" bezeichnet. Strömungsrichtung von links unten nach rechts
oben. Normanskill-Schiefer (Ordovizium).

Trois grandes cannelures de traînage avec «flute marks» et empreintes de charge en écailles.
Noter l'alternance de «flute marks» et d'empreintes en écaille arrangées perpendiculairement aux
cannelures. Cette alternance constitue la «peau de dinosaure» de CHADWICK. Le courant allait
d'en bas à gauche à en haut à droite. Schistes de Normanskill (Ordovicien).

Tres calcos de arrastre grandes, turboglifos y calcos de carga escamiformes. Nótese el desarrollo
rítmico de los dos últimos ("cuero de dinosaurio"), perpendiculares a los calcos de arrastre.
Corriente desde el ángulo inferior izquierdo al superior derecho. Lutita Normanskill, Ordovícico.

PLATE 59B

Close up of flute marks and squamiform load casts shown in plate 59A. Current from left to
upper right. Normanskill Shale (Ordovician). On U.S. Highway 9 W just north of overpass
above Interstate 87, Greene County, New York, U.S.A. (Photograph by E. CHOWN).

Nahaufnahme von Strömungswülsten und schuppenförmigen Belastungsmarken aus Abbildung 59A. Strömungsrichtung von links unten nach rechts oben. Normanskill-Schiefer
(Ordovizium).

Détail des «flute marks» et empreintes de charge en écailles de la planche 59A. Le courant
allait d'en bas à gauche à en haut à droite.

Vista en detalle de turboglifos y calcos de carga escamiformes, correspondiente a la lámina 59A.
Corriente desde el ángulo inferior izquierdo al superior derecho. Lutita Normanskill, Ordovícico.

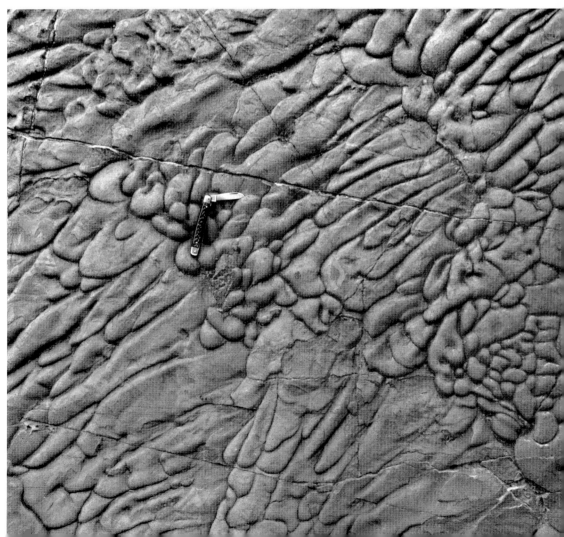

PLATE 60A

Groove casts (drag marks). Note superimposed striations on several of the groove casts and discordant direction of brush cast at A. Tar Springs Sandstone (Mississippian). 17-G-25, near Crofton, Christian County, Kentucky, U.S.A. (Photograph by DALE FARRIS).

Rillenmarken (Schleifmarken). Etliche Strömungswülste von Riefen überprägt. Beachte diskordante Richtung der Quastenmarke bei A. Tar Springs-Sandstein (Mississippian).

Cannelures (empreintes de traînage). Remarquer les stries superposées sur les cannelures, parallèlement à leur direction, et l'orientation discordante de l'éraflure en A. Grès de Tar Springs (Mississipien).

Calcos de surcos (calcos de marcas de arrastre). Nótese las estriaciones superpuestas en algunos de los calcos y un calco de empuje con distinta orientación en A. Arenisca Tar Spring, Mississippiano.

PLATE 60B

Large twisted flute (?) cast. Atoka Formation (Pennsylvanian). On U.S. Highway 59, 1.2 miles south of Hodgen, Le Flore County, Oklahoma, U.S.A. (Jersey Production Research Company).

Großer wirbeliger Strömungswulst (?). Atoka Formation (Pennsylvanien).

«Flute cast» de grande taille, tordu. Formation d'Atoka (Pennsylvanien).

Turboglifo (?) amplio de aspecto retorcido. Formación Atoka, Pennsylvaniano.

PLATE 61

Flute casts (A) groove casts (B), and prod casts (C). Note two sets of groove casts, one parallel to flutes and one at an acute angle. Note multiple striations of the diagonal groove marks, the deflection of one, and the obliteration of this earlier set by flutes. Current from right to left. Hatch Formation (Devonian), Conesus Lake, Livingston County, New York (Photograph by W. Hiller).

Strömungswülste (A), Rillenmarken (B) und Stoßmarken (C). Rillenmarken treten in zwei Sätzen auf, von denen der eine parallel, der andere in spitzem Winkel zu den Strömungswülsten verläuft. Beachte vielfache Riefung der diagonalen Rillenmarken, die Ablenkung einer einzelnen und die Tilgung des ersten Satzes durch Strömungswülste. Strömungsrichtung von rechts nach links. Hatch Formation (Devon).

«Flute casts» (A), cannelures (B) et «prod casts» (C). Il y a deux groupes de cannelures, l'un parallèle aux «flute casts», l'autre à angle aigu. Noter les nombreuses stries le long des cannelures en diagonale, dont l'une est déviée, et le fait que ce premier groupe de cannelures est oblitéré par les «flute casts». Le courant allait de droite à gauche. Formation de Hatch (Dévonien).

Turboglifos (A), calcos de surcos (B) y calcos de punzamiento (C). Nótese dos juegos de calcos de surcos, uno paralelo a los turboglifos y el otro en ángulo agudo; calcos de surcos estriados diagonales; la deflexión y la obliteración por turboglifos. Corriente de derecha a izquierda. Formación Hatch, Devónico.

PLATE 62

Groove (A) and chevron cast (B) on the under surface of sandstone layer. Sherburne Formation (Devonian). Taughannock Falls, north of Ithaca, Tompkins County, New York, U.S.A. (Reprinted with permission from DUNBAR and RODGERS, Principles of Stratigraphy, copyright 1957 by John Wiley and Sons.)

Rille (A) und Fiedermarke (B) an der Sohlfläche einer Sandsteinschicht. Sherburne Formation (Devon).

Moulages de cannelures (A) et empreintes en chevron (B) à la face inférieure d'un grès. Formation de Sherburne (Dévonien).

Calco de surco (A) y calco espigado (B) en la superficie inferior de una arenisca. Formación Sherburne. Devónico.

PLATE 63 A

Load-casted groove casts (A) and vibration marks (B). Atoka Formation (Pennsylvanian). On U.S. Highway 59, 2 miles south of Hodgen, Le Flore County, Oklahoma, U.S.A. (Jersey Production Research Company).

Durch Belastung überformte Rillenmarken (A) und Schwingungsmarken (B). Atoka Formation (Pennsylvanian).

Moulage de cannelures déformées par surcharge (A) et empreintes de vibration (B). Formation d'Atoka (Pennsylvanien).

Calco de surcos ahondados por efecto de carga (A) y marcas de vibración (B). Formación Atoka, Pennsylvaniano.

PLATE 63 B

Load-casted groove casts. Line of movement parallel to groove, Atoka Formation (Pennsylvanian). Along U.S. Highway 59, 2 miles south of Hodgen, Le Flore County, Oklahoma, U.S.A. Jersey Production Research Company).

Durch Belastung überformte Rillenmarken. Bewegungssinn parallel zur Rille. Atoka Formation Pennsylvanian).

Moulages de cannelures déformées par surcharge. Direction du courant parallèle aux cannelures. Formation d'Atoka (Pennsylvanien).

Calco de surcos ahondados por efecto de carga, paralelos a la dirección de la corriente. Formación Atoka, Pennsylvaniano.

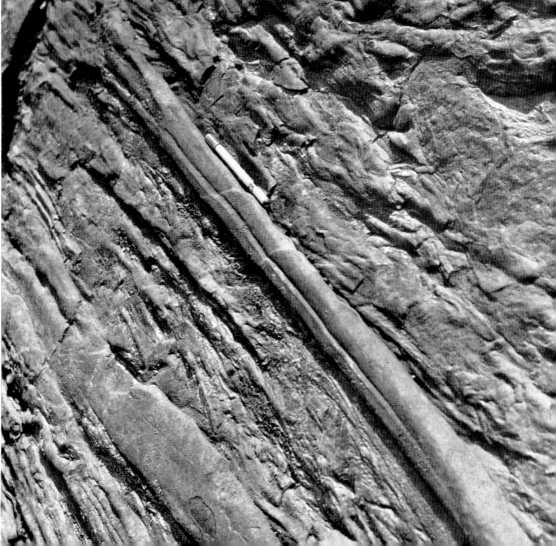

PLATE 64A

Groove casts (A) and other sole marks of undetermined origin, on limestone of Cincinnatian Series (Ordovician). Current direction parallel to pencil. Five Mile Creek, Anderson Township, Hamilton County, Ohio, U.S.A. (Photograph by DALE FARRIS).

Rillenmarken (A) und andere unbestimmte Schichtflächen-Marken auf einer Kalksteinbank der Cincinnatian Series (Ordovizium). Strömungsrichtung parallel zum Bleistift.

Cannelures (A) et autres empreintes d'origine inconnue, sur la face inférieure d'un calcaire du Cincinnatien (Ordovicien). Direction du courant parallèle au crayon.

Calcos de surcos (A) y otras marcas de base de origen desconocido, en caliza. Corriente paralela al lápiz. Serie Cincinnatiana, Ordovícico.

PLATE 64B

Sole showing a few poorly developed groove casts and many miscellaneous smaller markings. Atoka Formation (Pennsylvanian). Approximately 1.2 miles south of Hodgen, Le Flore County, Oklahoma, U.S.A. (Jersey Production Research Company).

Schichtunterseite mit wenigen, schlecht entwickelten Rillenmarken und allerlei anderen kleineren Marken. Atoka Formation (Pennsylvanian).

Cannelures peu marquées et autres empreintes diverses, plus petites, à la face inférieure d'une couche. Formation d'Atoka (Pennsylvanien).

Calcos de surcos débilmente desarrollados y calcos de distintas marcas más pequeñas, en la superficie inferior de un estrato. Formación Atoka. Pennsylvaniano.

PLATE 65 A

Groove casts (slide or drag marks?). Martinsburg Shale (Ordovician). Three-quarters of a mile north of Lenhartsville, Berks County, Pennsylvania, U.S.A. (McBride, 1962, Fig. 13).

Rillenmarken (Gleit- oder Schleifmarken?). Martinsburg-Schiefer (Ordovizium).

Cannelures (empreintes de glissement, ou de traînage?) Argile schisteuse de Martinsburg (Ordovicien).

Calcos de surcos (marcas de deslizamiento o de arrastre?). Lutita Martinsburg, Ordovícico.

PLATE 65 B

Cast of slide mark in sole of turbidite. Note parallelism of markings. Movement parallel to markings. Kulm graywacke (Lower Carboniferous). Southwest of Braunshausen, Sauerland, Germany (Photograph by Ph. H. Kuenen).

Ausguß einer Gleitmarke auf der Unterseite eines Turbidits. Streng parallele Ausrichtung der Marken. Bewegungssinn parallel zu den Marken. Kulm-Grauwacke (Unterkarbon).

Moulage d'une empreinte de glissement à la face inférieure d'une turbidite (grauwacke). Noter le parallélisme des empreintes. Direction du mouvement parallèle aux empreintes. Grauwacke de Kulm (Carbonifère inférieur).

Calco de deslizamiento en la superficie inferior de una turbidita. Nótese el paralelismo de la estructura, también paralela al movimiento. Grauvaca Kulm, Carbonífero Inferior.

PLATE 66

Groove casts (A), prod casts (B) and brush marks (C). Current from bottom to top. Note multiple secondary striations on larger groove casts and abrupt termination of largest groove. Gardeau Formation (Devonian). Patterson Gulley, northwest of Dansville, Sparta Township, Livingston County, New York, U.S.A. (Photograph by W. HILLER).

Rillenmarken (A), Stoßmarken (B) und Quastenmarken (C). Strömungsrichtung von unten nach oben. Beachte vielfache sekundäre Riefung der größeren Rillenmarken und das unvermittelte Absetzen der größten Rillenmarke. Gardeau Formation (Devon).

Cannelures (A), prod casts (B) et éraflures (C). Le courant allait de bas en haut. Noter les nombreuses stries secondaires sur les cannelures les plus importantes, et l'extrémité abrupte de la plus grande de toutes. Formation de Gardeau (Dévonien).

Calcos de surcos (A), de punzamiento (B) y de roce (C). Nótese las múltiples estriaciones secundarias sobre los calcos de surcos más grandes y la erminación abrupta del mayor. Corriente desde el borde inferior de la lámina al superior. Formación Gardeau, Devónico.

PLATE 67

Large asymmetrical groove cutting symmetrical ripples. Groove approximately at right angles to ripple crests. Fish River Series, Nama System (Precambrian). South-West Africa (Photograph by I. W. HÄLBICH).

Große unsymmetrische Rille, symmetrische Rippeln ungefähr rechtwinkelig durchsetzend. Fish River Series, Nama System (Präkambrium).

Grande cannelure asymétrique traversant des rides symétriques. La cannelure est presque perpendiculaire aux crêtes des rides. Groupe de Fish River, Nama (Précambrien).

Calco de surco grande y asimétrico, en ángulo aproximadamente recto a las crestas simétricas de una ondulita. Serie Fish R ver, Sistema Nama.

PLATE 68

Brush marks (A), prod casts (B), "cabbage leaf" or frondescent cast (C), imprints of fish vertebrae (skip casts) (D), drag marks (E). Current flowed from left to right. Krosno beds (Oligocene), Wetlina, Poland (Photograph from DZULYNSKI and SLACZKA, 1958, pl. 30).

Quastenmarken (A), Stoßmarke (B), Fächermarke (C), Eindrücke von Fischwirbeln (Hüpfmarken) (D), Schleifmarken (E). Strömungsrichtung von links nach rechts. Krosno-Schichten (Oligozän).

Eraflures (A), prod casts (B), moulage en feuille de chou ou en frondes, empreintes laissées par des vertèbres de poisson roulées par le courant (D) (skip casts), et empreintes de traînage (E). Le courant allait de gauche à droite. Couches de Krosno (Oligocène).

Calcos de empuje (A), calcos de punzamiento (B), calco caulifoliado (C), calcos de marcas producidas por vértebras de peces (D) y calcos de marcas de arrastre (E). Corriente de izquierda a derecha. Capas Krosno, Oligoceno.

PLATE 69A

Multiple brush mark. Current from left to right. Menilite beds, Upper Eocene, Rudawka Rymanowska, central Carpathians, Poland (Photograph from DZULYNSKI and SANDERS, 1962, pl. 15 B).

Vielfache Quastenmarke. Strömungsrichtung von links nach rechts. Meniliten-Schichten, Obereozän.

«Brush-mark» multiple. Courant de gauche à droite. Couches de Menilite (Eocène supérieur).

Calcos de empuje múltiples. Corriente de izquierda a derecha. Capas Menilite, Eoceno Superior.

PLATE 69B

"Rill-casts". Current from right to left. Krosno beds (Oligocene). Rudawka Rymanowska, central Carpathian Mountains, Poland (Photograph from DZULYNSKI and SANDERS, 1962, pl. 2A).

„Gefurchte Strömungswülste". Strömungsrichtung von rechts nach links. Krosno-Schichten (Oligozän).

«Rigoles». Courant de droite à gauche. Couches de Krosno (Oligocène).

"Rill-casts". Corriente de derecha a izquierda. Capas Krosno, Oligoceno.

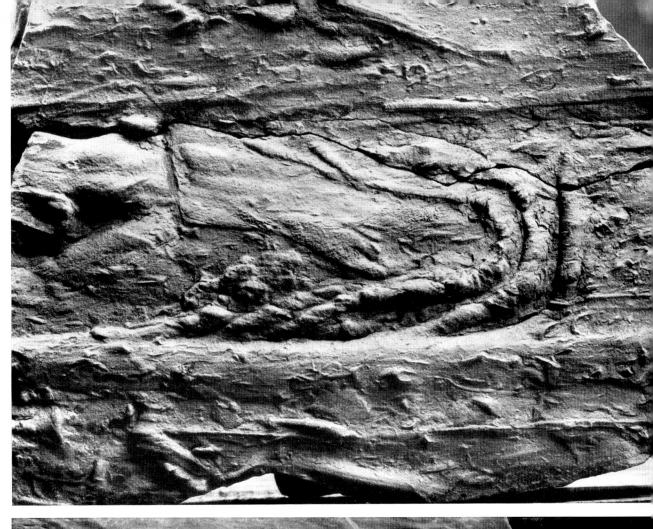
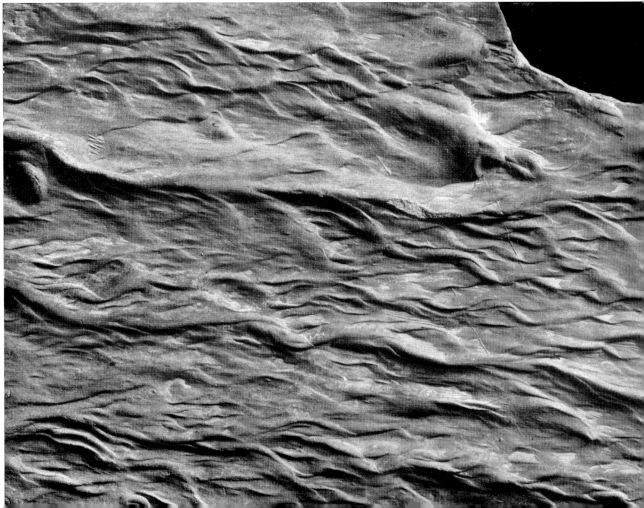

PLATE 70A

Lebensspuren on underside of sandstone. Haymond Formation (Pennsylvanian). Approximately 12 miles northeast of Marathon, Brewster County, Texas, U.S.A. (Photograph by E. F. McBride).

Lebensspuren auf der Unterseite eines Sandsteins. Haymond Formation (Pennsylvanian).

Pistes à la face inférieure d'un grès. Formation de Haymond (Pennsylvanien).

Icnofósiles en la superficie inferior de una arenisca. Formación Haymond, Pennsylvaniano.

PLATE 70B

Ichnofossils, probably burrows ("Arthrophycus"). Tuscarora Quartzite (Silurian). Blue Mountain Gap, U.S. Highway 11-15, Perry County, Pennsylvania, U.S.A.

Ichnofossilien, vermutlich Bauten (,,Arthrophycus"). Tuscarora-Quarzit (Silur).

Pistes fossiles, probablement moulages de galeries («Arthrophycus»). Quartzite de Tuscarora (Silurien).

Icnofósiles, probablemente "Arthrophycus". Cuarcita Tuscarora, Silúrico.

PLATE 71

Casts of burrows (biohieroglyphs) on underside of siltstone. Waverly Group (Mississippian). Five miles north of Alma, Ross County, Ohio, U.S.A. (Photograph by DALE FARRIS).

Abdrücke von Bauten (Bio-Hieroglyphen) auf Schichtunterseite eines Siltsteins. Waverly Group (Mississippian).

Moulages de tubes (biohiéroglyphes) à la face inférieure d'une «siltstone» (microgrès). Couches de Waverly (Mississipien).

Estructura tubular (biohieroglifo) en la base de una limonita. Grupo Waverly, Mississippiano.

PLATE 72A

Weakly-developed, shallow "scour casts" and diverse burrow casts (biohieroglyphs). Atoka Formation (Pennsylvanian). On U.S. Highway 59, 1.2 miles south of Hodgen, Le Flore County, Oklahoma, U.S.A. (Jersey Production Research Company).

Schwach ausgebildete flache Kolkmarken und verschiedene Bio-Hieroglyphen (Ausgüsse von Bauten). Atoka Formation (Pennsylvanian).

Empreintes d'affouillement peu marquées, et divers moulages de tubes (biohiéroglyphes). Formation d'Atoka (Pennsylvanien).

Turboglifos escasamente desarrollados y estructura tubular (biohieroglifos). Formación Atoka, Pennsylvaniano.

PLATE 72B

Sole markings of undetermined nature on underside of Berea Sandstone (Mississippian). Upper Twin Creek, Niles Township, Scioto County, Ohio, U.S.A.

Schichtflächen-Marken unbestimmten Ursprungs auf Schichtunterseite. Berea-Sandstein (Mississippian).

Empreintes d'origine inconnue à la face inférieure du grès de Berea (Mississipien).

Marcas de base de origen desconocido. Arenisca Berea, Mississippiano.

PLATE 73 A

Multiple glacial striations of Pleistocene ice in Hayes River Group (Precambrian). Principal direction of ice movement approximately parallel to hammer handle. East end of Hyers Island, Oxford Lake, Manitoba, Canada (Geological Survey of Canada).

Vielfache glaziale Kritzen (Pleistozän) auf Schichten der Hayes River Group (Präkambrium). Haupt-Bewegungsrichtung des Eises ungefähr parallel zum Hammerstiel.

Stries glaciaires multiples formées par un glacier pleistocène sur les couches de Hayes River (Précambrien). La direction principale de mouvement du glacier était à peu près parallèle au manche du marteau.

Abundantes estriaciones glaciarias producidas durante el Pleistoceno sobre rocas del Grupo Hayes River, Precámbrico. Desplazamiento principal del hielo, aproximadamente paralelo al mango del martillo.

PLATE 73 B

Fine-textured, uniformly-oriented glacial striations on Livingston Limestone Member of Bond Formation (Pennsylvanian). NW 1/4 sec. 21, T. 18 N., R. 13 W., Vermilion County, Illinois, U.S.A. (Illinois Geological Survey).

Fein strukturierte, einheitlich orientierte glaziale Schrammen auf Livingston-Kalkstein (Bond Formation, Pennsylvanian).

Stries glaciaires de structure fine et d'orientation uniforme, sur le calcaire de Livingston de la formation de Bond (Pennsylvanien).

Estriaciones glaciarias finas de orientación uniforme, en caliza. Miembro Livingston, Formación Bond, Pennsylvaniano.

PLATE 74

Boulder pavement with embedded striated and facetted till stones. Note tendency of long axes and striations to be parallel to one another. Dwyka Conglomerate (Permian). Near Gibeau, South-West Africa (Photograph by J. W. Hälbich).

Blockschicht mit eingelagerten, geschrammten und facettierten Geschiebe-Blöcken. Beachte parallele Ausrichtung der Längsachsen und Schrammen. Dwyka-Konglomerat (Perm).

Pavage de blocs avec cailloux morainiques à facettes). et ries. Noter que les grands axes des cailloux, et les stries, tendent à être parallèles les uns aux autres. Conglomérat de Dwyka (Permien).

Pavimento formado por fragmentos tilíticos estriados y facetados. Nótese la tendencia al paralelismo entre los ejes mayores de los clastos y las estriaciones. Conglomerado Dwyka, Pérmico.

PLATE 75

Roches moutonnées on dolomite of Epworth Formation (Precambrian). Note abrupt lee-side termination. Ice moved from left to right. Fifty-five miles south of Coronation Gulf, lat. 66° 56′, long. 114° 03′, District of MacKenzie, Northwest Territories, Canada (Geological Survey of Canada).

Rundhöcker aus Dolomit. Epworth Formation (Präkambrium). Beachte unvermittelten Abbruch an der Leeseite. Eis bewegte sich von links nach rechts.

Roches moutonnées sur une dolomie de la formation d'Epworth (Précambrien). Noter leur terminaison avale abrupte. Le glacier allait de gauche à droite.

Dolomía aborregada con terminaciones abruptas a sotavento del desplazamiento de la masa de hielo. Formación Epworth, Precámbrico.

PLATE 76A

Parting lineation in sandstone. Current parallel to hammer. Haymond Formation (Pennsylvanian). Pecos County, Texas, U.S.A. (Photograph by E. F. McBride).

Strömungs-Streifung in Sandstein. Strömungsrichtung parallel zum Hammer. Haymond Formation (Pennsylvanian).

Linéation de délit dans un grès. Courant parallèle au marteau. Formation de Haymond (Pennsylvanien).

Lineación por corriente en arenisca; corriente paralela al martillo. Formación Haymond, Pennsylvaniano.

PLATE 76B

Current parting on two bedding plane surfaces of Salt Wash Sandstone Member of Morrison Formation (Jurassic). Line of movement parallel to hammer handle. Near Big Hole Mine, Blanding District, San Juan County, Utah, U.S.A. (STOKES, 1953, Fig. 2A).

Strömungs-Streifung auf zwei Schichtflächen im Salt Wash Sandstein. Morrison Formation (Jura). Strömungsrichtung parallel zum Hammerstiel.

Délit par le courant le long de deux plans de stratification du grès de Salt Wash, formation de Morrison (Jurassique). Direction du courant parallèle au manche du marteau.

Lineación por corriente en dos planos de estratificación. Miembro Salt Wash (areniscas), Formación Morrison, Jurásico.

PLATE 77

Asymmetrical, transverse ripple marks with relatively continuous crests on tidal flat. Hyogo-ku, Kobe City, Kinki District, Japan (Fuji Film Company).

Unsymmetrische Transversalrippeln mit relativ kontinuierlichen Kämmen, im Watt.

Rides asymétriques, transversales, à crêtes assez continues, sur l'estran.

Ondula transversa con crestas asimétricas y relativamente continuas, en una zona intercotidal. Actual.

PLATE 78A

Asymmetrical current ripples on point bar of Vermilion River. Current from upper left to lower right. NE 1/4 SW 1/4 SW 1/4 sec. 30, T. 18 N., R. 9. W., Vermillion County, Indiana, U.S.A.

Unsymmetrische Strömungsrippeln in Sandbank des Vermilion-Flusses. Strömungsrichtung von oben links nach unten rechts.

Rides de courant asymétriques, sur un bourrelet arqué de la Vermilion River. Courant allant d'en haut à gauche à en bas à droite.

Ondula asimétrica en un espolón aluvial del Río Vermilion. Corriente desde el ángulo superior izquierdo al inferior derecho. Actual.

PLATE 78B

Asymmetrical ripple marks on modern marine beach. Wave length approximately 20 cm. Great Western Beach, Scarborough, Cumberland County, Maine, U.S.A. (Photograph by ROBERT DOW).

Unsymmetrische Rippelmarken an rezentem Meeresstrand. Wellenlänge ungefähr 20 cm.

Rides de plage asymétriques sur une plage marine actuelle. Longueur d'onde des rides, 20 cm environ.

Ondula asimétrica en una playa marina. Longitud de onda 20 cm aproximadamente. Actual.

PLATE 79

Three sand waves on which are superimposed sand waves of smaller magnitude (ripple marks) on point bar of Vermilion River. Current from right to left parallel to shovel and lineation. NE 1/4 SW 1/4 SW 1/4 sec. 30, T. 18 N., R. 9 W., Vermillion County, Indiana, U.S.A.

Drei große Sandwellen; von Sandwellen geringerer Größe (Rippelmarken) überlagert. Sandbank im Vermilion-Fluß. Strömungsrichtung von rechts nach links, parallel zu Spaten und Riefung.

Trois vagues de sable géantes, avec des rides superposées d'amplitude moindre, sur un bourrelet arqué de la Vermilion River. Courant de droite à gauche, parallèle à la pelle.

Tres amplias ondas de arena con superposición de otras de menor magnitud, en un espolón aluvial del Río Vermilion. Corriente de derecha a izquierda, paralela a la pala y a la lineación. Actual.

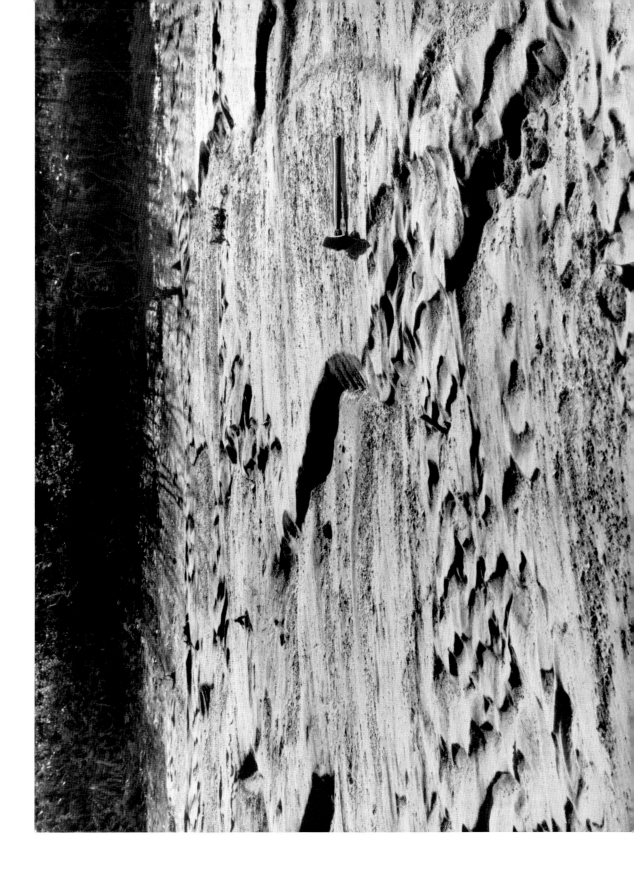

PLATE 80

Uniform transverse ripple marks. Kanto District, Japan (Fuji Film Company).

Gleichförmige Transversalrippeln.

Rides de plages transversales uniformes.

Ondula transversa y uniforme. Actual.

PLATE 81 A

Transverse ripple marks with development of secondary ripple marks on crests. Mud covered sand of tidal flat. Note worm excrement. Lower Saxony, Germany (Senckenberg am Meer, L 8329).

Transversalrippeln mit sekundären Rippeln auf den Kämmen. Mit Schlamm bedeckte Sandfläche im Watt. Beachte Wurmexkremente.

Rides de plage transversales, avec rides secondaires sur les crêtes. Estran dont la vase est recouverte de sable. Remarquer les tortillons d'excréments de vers arénicoles.

Ondula transversa con desarrollo de otra secundaria sobre sus crestas. Arena cubierta por fango en una zona intercotida. Nótese los excremento de vermes. Actual.

PLATE 81 B

Uniform, transverse ripple marks. Note trail. Note also larger granules in ripple troughs. Tidal flat, Mellum Island, Lower Saxony, Germany (Senckenberg am Meer, L 2919).

Gleichförmige Transversalrippeln. Beachte zunehmende Korngröße in Rippeltälern und Spur.

Rides transversales uniformes. Noter la présence d'une piste d'origine animale, et de grains plus gros dans les creux.

Ondula transversa uniforme. Nótese la concentración de granos mayores en las depresiones y la presencia de un rastro. Zona intercotidal. Actual.

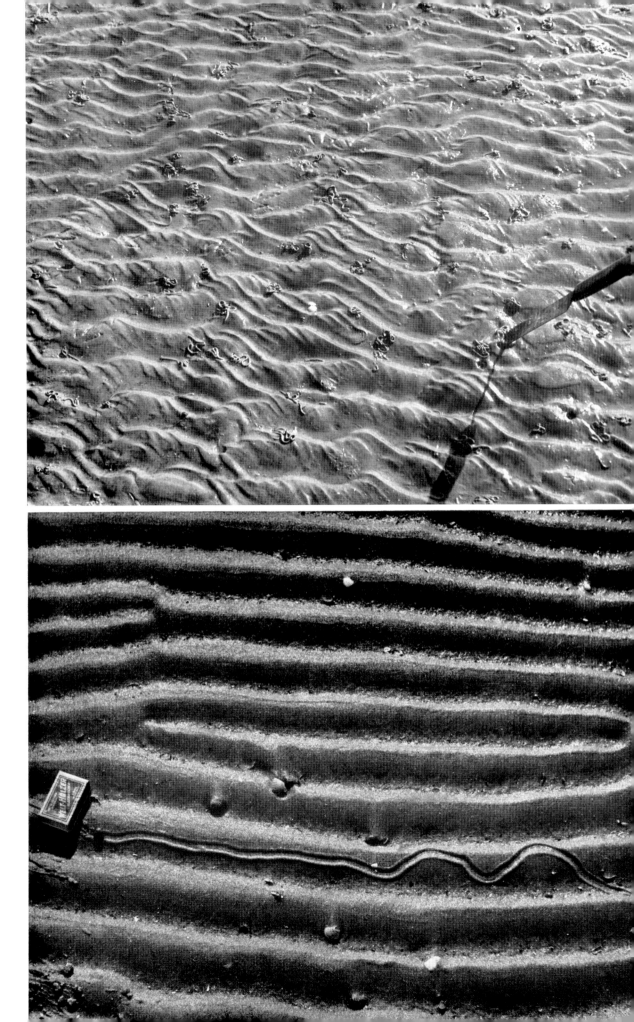

PLATE 82A

Ripple-marked sand of tidal flat. Note mud-filled ripple troughs which in places form connected and isolated crescents. Jade Busen, Lower Saxony, Germany (Senckenberg am Meer, L 8340).

Gerippelte Sandlagen im Watt. Beachte Schlickfüllung in Rippeltälern, die stellenweise zusammenhängende und isolierte Bögen bildet.

Coupe dans le sable marqué de rides d'un estran actuel. Noter les creux des rides, remplis de vase, qui peuvent former des croissants, soit réunis les uns aux autres, soit isolés.

Ondula en arena de una zona intercotidal. Nótese el relleno de fango que en ciertas partes tiene formas de medias lunas aisladas o conectadas. Actual.

PLATE 82B

Transverse ripple marks (A), linguoid ripple marks (B), and small current crescents (C) on tidal flat. Current from upper left to lower right. Mellum Island, Lower Saxony, Germany (Senckenberg am Meer, L 8285).

Transversalrippeln (A), zungenförmige Rippeln (B) und kleine Strömungsbögen (C) im Watt. Strömungsrichtung von links oben nach rechts unten.

Ride transversale (A), ride linguiforme (B), et petits croissants dûs au courant (C), sur l'estran. Courant allant d'en haut à gauche à en bas à droite.

Ondula transversa (A), óndula linguoide (B) y pequeñas marcas en herradura (C). Corriente desde el ángulo superior izquierdo al inferior derecho. Zona intercotidal, Actual.

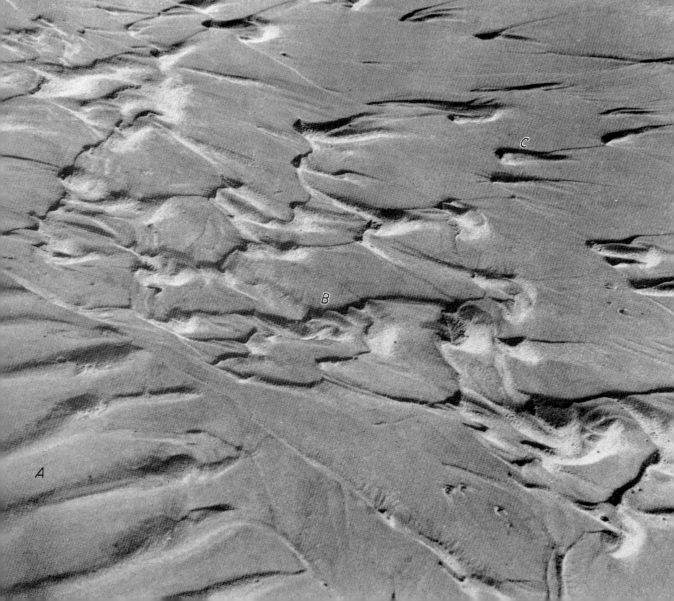

PLATE 83 A

Aeolian dune and ripple marks. Ripples formed by wind from right to left. Monahans State Park, Ward County, Texas, U.S.A. (Photograph by E. F. McBride).

Winddüne mit Rippelmarken. Windrichtung, für die Rippeln verantwortlich, von rechts nach links.

Dune et rides éoliennes. Les rides ont été formées par un vent soufflant de droite à gauche.

Formación de óndula en una duna. Sentido del viento de derecha a izquierda. Actual.

PLATE 83 B

Ripple marks with branching pattern, on modern marine beach. Great Western Beach, Scarborough, Cumberland County, Maine, U.S.A. (Photograph by Robert Dow).

Sich verzweigende Rippelmarken an rezentem Meeresstrand.

Rides ramifiées sur une plage marine actuelle.

Ondula ramificada en una playa marina. Actual.

PLATE 84A

Linguoid or cuspate ripple marks on thin-bedded sandstone of Caseyville Formation (Pennsylvanian). Current from left to right. Johnson County, Illinois, U.S.A. (Illinois Geological Survey).

Zungen- oder sichelförmige Rippelmarken auf dünnschichtigem Sandstein. Caseyville Formation (Pennsylvanian). Strömungsrichtung von links nach rechts.

Rides arquées ou en feston, sur grès finement lité de la formation de Caseyville (Pennsylvanien).

Ondula linguoide en arenisca finamente estratificada. Corriente de izquierda a derecha. Formación Caseyville, Pennsylvaniano.

PLATE 84B

Linguoid ripple marks in sand of tidal flat. Current from left to right. Mellum Island, Lower Saxony, Germany (Senckenberg am Meer, L 7985).

Zungenförmige Rippelmarken in Wattensand. Strömungsrichtung von links nach rechts.

Rides de plage arquées en croissant dans le sable de l'estran. Courant de gauche à droite.

Ondula linguoide. Corriente de izquierda a derecha. Zona intercotidal, Actual.

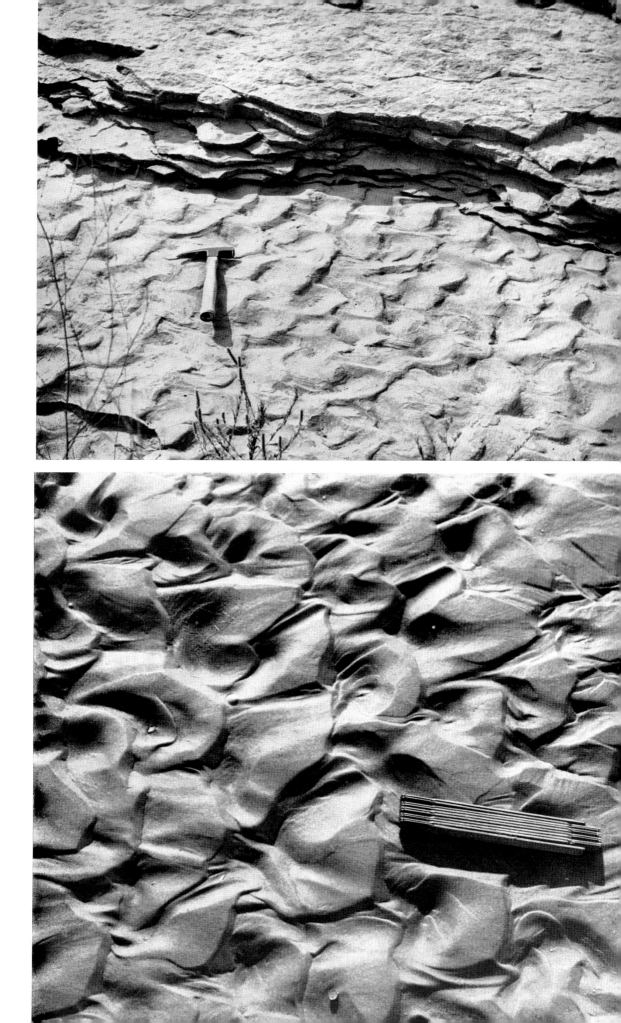

PLATE 85 A

Asymmetrical ripple marks in sandstone. Current from upper right to lower left. Tar Springs Sandstone (Mississippian). SW 1/4 SW 1/4 NE 1/4 sec. 30, T. 7 S., R. 6 W., Randolph County, Illinois, U.S.A. (Illinois Geological Survey).

Unsymmetrische Rippelmarken im Sandstein des Tar Springs-Sandstein (Mississippian). Strömungsrichtung von rechts oben nach links unten.

Rides asymétriques sur un grès. Le courant allait d'en haut à droite à en bas à gauche. Tar Springs (Mississipien).

Ondula asimétrica. Corriente desde el ángulo superior derecho al inferior izquierdo. Arenisca Tar Spring, Mississippiano.

PLATE 85 B

Regular, nearly symmetrical ripple marks. Karoo Series (Permian). South Africa (Photograph by W. HILLER).

Regelmäßige, nahezu symmetrische Rippelmarken. Karru Series (Perm).

Rides régulières, presque parfaitement symétriques. Couches de Karoo (Permien).

Ondulita regular aproximadamente simétrica. Serie Karoo, Pérmico.

PLATE 86A

Interference ripple marks. Hampshire Formation (Devonian). Near junction of State Highway 45 and U.S. Highway 50, east of Romney, Hampshire County, West Virginia, U.S.A. (Photograph by N. L. McIver).

Interferenzrippeln. Hampshire Formation (Devon).

Rides complexes à interférence. Formation de Hampshire (Dévonien).

Ondulitas de interferencia. Formación Hampshire. Devónico.

PLATE 86B

Symmetrical, branching ripple marks in argillaceous siltstone, cut by minute calcite veinlets of later tectonic origin (Carboniferous). Conche, Newfoundland, Canada (Photograph by D. M. Baird).

Symmetrische, sich verzweigende Rippeln in tonigem Siltstein (Karbon). Winzige, mit Kalkspat gefüllte Klüfte sind tektonischen Ursprungs.

Rides symétriques, divisées, dans un grès fin argileux, recoupées par de fines veines de calcite d'origine tectonique postérieure (Carbonifère).

Ondulita simétrica ramificada y venillas calcíticas de origen tectónico, en limolita arcillosa. Carbonífero.

PLATE 87A

Interference ripple marks on sandstone (Pennsylvanian). Cape Breton Island, Nova Scotia, Canada (KINDLE, 1917, pl. 30a).

Interferenzrippeln in Sandstein (Pennsylvanian).

Rides d'interférence sur un grès (Pennsylvanien).

Ondulitas de interferencia en arenisca. Pennsylvaniano.

PLATE 87B

Small, symmetrical, transverse ripple marks. Note smaller medial ripples between normal crests. Sandstone of Keweenawan Series (Precambrian). SE 1/4 NE 1/4 SE 1/4 sec. 10, T. 48 N., R. 1 E., Gogebic County, Michigan, U.S.A. (Photograph by W. H. HAMBLIN).

Kleine symmetrische Transversalrippeln in Sandstein. Beachte kleinere Kämme zwischen den Normalrippeln. Keweenawan Series (Präkambrium).

Petites rides symétriques, transversales. Noter les petites rides intermédiaires entre les crêtes de hauteur normale. Grès de Keweenawan (Précambrien).

Ondulitas pequeñas, simétricas y transversas. Nótese la ondulita más pequeña entre las crestas de la mayor.

PLATE 88

Ripple marks and animal trails on underside of thin-bedded sandstone of Clore Limestone (Mississippian). NW 1/4 SE 1/4 SE 1/4 sec. 2, T. 8 S., R. 6 W., Randolph County, Illinois, U.S.A. (Illinois Geological Survey).

Rippelmarken und Lebensspuren. Unterseite eines dünnschichtigen Sandsteins. Clore-Kalkstein (Mississippian).

Rides de plage et pistes d'origine animale à la face inférieure d'un grès finement lité. Calcaires de Clore (Mississipien).

Ondulita y rastros fósiles en la base de una arenisca finamente estratificada. Caliza Clore, Mississippiano.

PLATE 89A

Transverse ripple marks in calcarenite of Cincinnatian Series (Ordovician). Ohio Coordinate System, South Zone, X: 1,614,700 Y: 315,512, Pleasant Township, Brown County, Ohio, U.S.A.

Transversalrippeln in Kalksandstein. Cincinnatian Series (Ordovizium).

Rides transversales dans une calcarénite du Cincinnatien (Ordovicien).

Ondulita transversa en calcarenita. Serie Cincinnatian, Ordovícico.

PLATE 89B

Large ripple marks in the Mississagi Quartzite (Huronian, Precambrian) of the Blind River area. Wave length is approximately 75 cm and amplitude is 15 cm. Striker Township, Ontario, Canada (McDowell, 1957, p. 11).

Große Rippelmarken im Mississagi-Quarzit (Huronian, Präkambrium). Wellenlänge ungefähr 75 cm, Amplitude 15 cm.

Rides de grande taille, dans la quartzite de Mississagi (Huronien, Précambrien), de la région de Blind River. Longueur d'onde environ 75 cm, amplitude environ 15 cm.

Ondulita amplia (longitud de onda 75 cm y amplitud 15 cm, aproximadamente). Cuarcita Mississagi, Huroniano, Precámbrico.

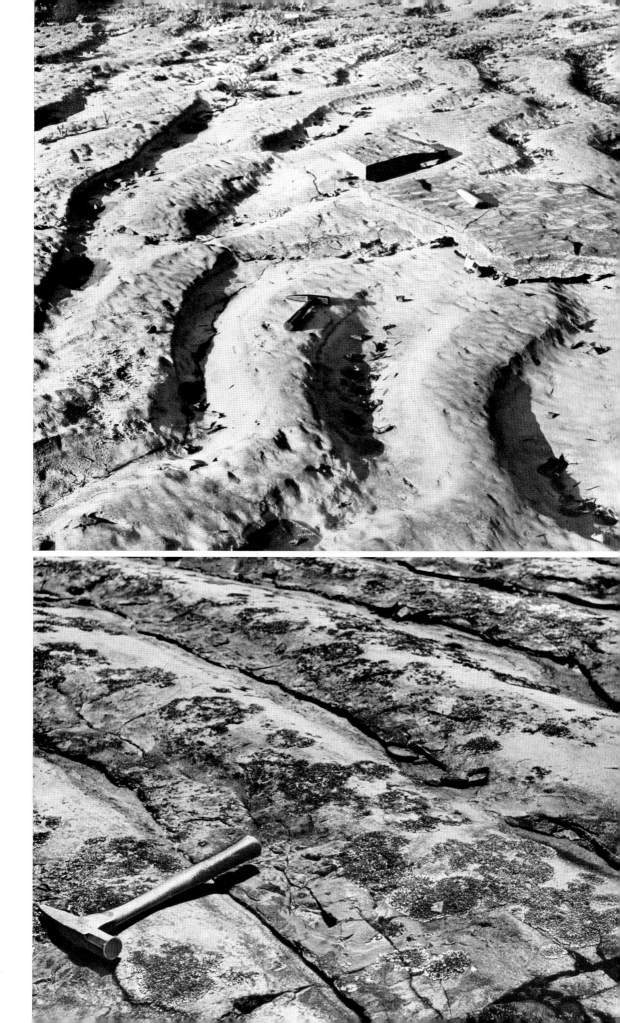

PLATE 90A

Large, separated ripples of bioclastic limestone in shale, Cincinnatian Series (Ordovician). Ohio Coordinate System, South Zone, X: 1,593,200; Y: 294,900, Pleasant Township, Brown County, Ohio, U.S.A.

Große, voneinander getrennte Rippeln aus bioklastischem Kalkstein in Schieferton; Cincinnatian Series (Ordovizium).

Grandes rides espacées, de calcaire bioclastique dans une argile schisteuse. Couches cincinnatiennes (Ordovicien).

Ondulita con crestas amplias y separadas, en una caliza bioclástica intercalada en sedimentos lutíticos. Serie Cincinnatiana, Ordovícico.

PLATE 90B

Large, separated ripples of bioclastic limestone in shale, Cincinnatian Series (Ordovician). Compare with plate 90A. Ohio Coordinate System, South Zone, X: 1,593,200; Y: 294,900, Pleasant Township, Brown County, Ohio, U.S.A.

Nah-Ansicht von Tafel 90A.

Détail de 90A.

Ondulita con crestas amplias y separadas, en una caliza bioclástica intercalada en sedimentos lutíticos. Compárese con lámina 90A. Serie Cincinnatiana, Ordovícico.

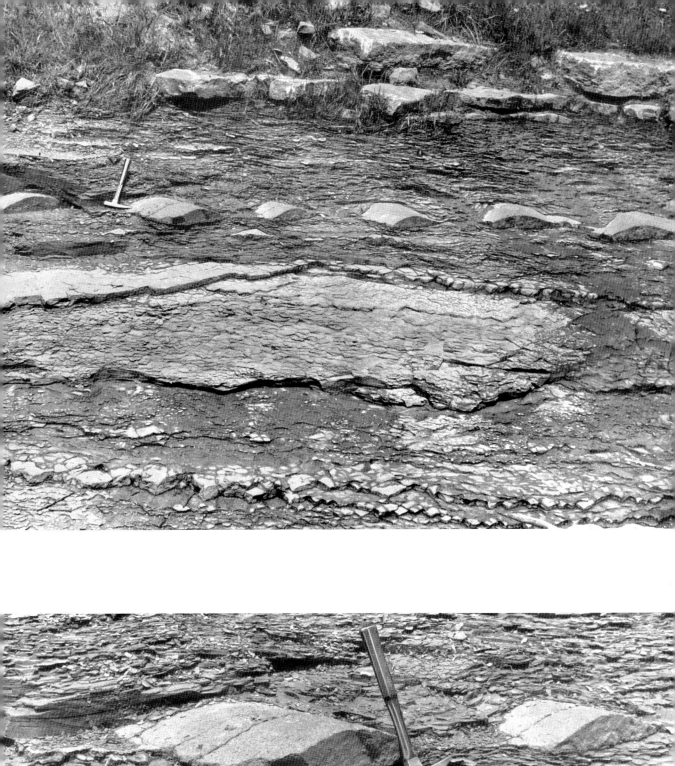

PLATE 91 A

Ripple marks in very shallow, elongate, subparallel troughs. Ripples generally strike transverse to trough axis. Current from lower right to upper left parallel to hammer and trough axiis Degonia Sandstone (Mississippian). SE 1/4 SW 1/4 SE 1/4 sec. 2, T. 11 S., R. 2 W., Unon. County, Illinois, U.S.A.

Rippeln in sehr flachen, langgestreckten, subparallelen Trögen. Streichen der Rippeln im allgemeinen transversal zu den Trogachsen. Strömungsrichtung von rechts unten nach links oben, parallel zu Hammer und Trogachsen. Degonia-Sandstein (Mississippian).

Rides dans des dépressions très peu marquées, allongées, plus ou moins parallèles. La plupart des rides sont perpendiculaires à l'axe de la dépression. Le courant allait d'en bas à droite à en haut à gauche, parallèlement au marteau et à l'axe de la dépression. Grès de Degonia (Mississipien).

Ondulitas transversas en cauces superficiales, elongados y subparalelos. Corriente desde el ángulo inferior derecho al superior izquierdo, paralela al martillo y al eje longitudinal de los cauces. Arenisca Degonia, Mississippiano.

PLATE 91 B

Current crescents, formed around pebbles, and rill marks on sandy beach. Ebb current from right to left. Along Chesapeake and Delaware Canal, two miles west of St. Georges, Newcastle County, Delaware, U.S.A.

Um Gerölle herum geformte Luvgräben und Rieselmarken auf Sandstrand. Ebbstrom-Richtung von rechts nach links.

Cupules en croissant autour de cailloux, et rigoles sur une plage de sable. Courant de jusant de droite à gauche.

Marcas en herradura formadas alrededor de rodados y marcas de escurrimiento, en arena de playa. Corriente de reflujo de derecha a izquierda. Actual.

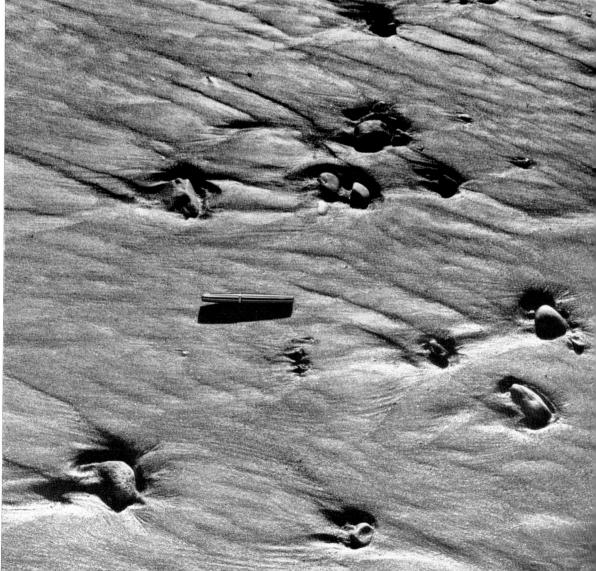

PLATE 92A

Large, distributary rill pattern on marine beach sand. Current from left to right (BAIRD, 1962, Fig. 9).

Breites, sich verzweigendes Rieselmuster auf marinem Sandstrand. Strömungsrichtung von links nach rechts.

Grand réseau de rigoles à bras divergents sur une plage de sable marin.

Marcas de escurrimiento grandes, de diseño distributivo; corriente de izquierda a derecha. Actual.

PLATE 92B

Current crescent casts, around shale pebbles, and primary current lineation on bedding plane of sandstone. Current from upper right to lower left parallel to pencil. Flagstone, McMaster University, Hamilton, Ontario, Canada.

Um Schiefertongerölle herum geformte Luvgräben und primäre Strömungsriefung. Schichtfläche eines plattigen Sandsteins. Strömungsrichtung von rechts oben nach links unten, parallel zum Bleistift.

Cupules en croissant (moulages) autour de moulages de cailloux schisteux, et arrangement linéaire, contemporain du dépôt, dans le plan de stratification d'un grès (dallage). Courant allant d'en haut à droite à en bas à gauche, parallèlement au crayon.

Calcos en herradura formados alrededor de rodados de lutita y lineación por corriente, en un plano de estratificación de una arenisca lajosa. Corriente desde el ángulo superior derecho al inferior izquierdo, paralela al lápiz.

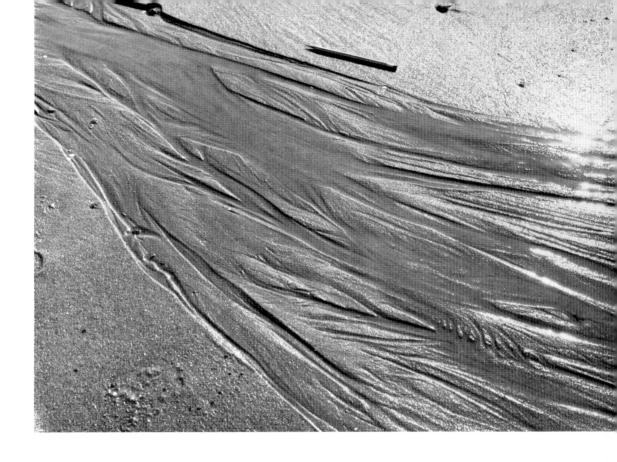

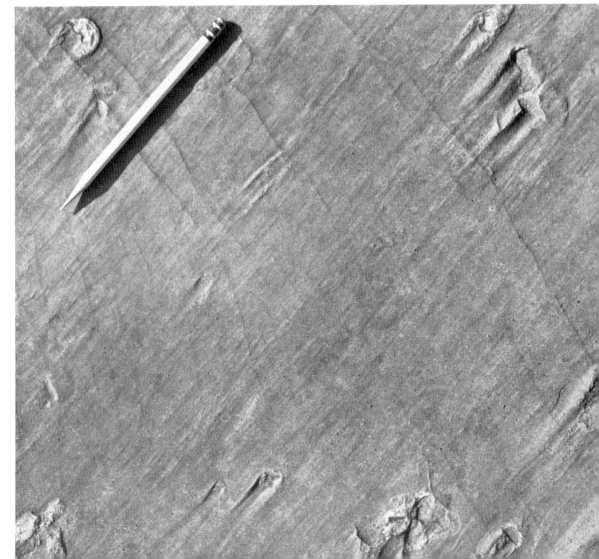

PLATE 93 A

Current crescents produced by deflection of current around shale fragments. Current from right to left. Compare with plate 91 B. Athabaska Formation (Precambrian). East of Turnor Point, Lake Athabaska, Saskatchewan, Canada (FAHRIG, 1961, pl. 1, p. 26).

Strömungskämme, durch Ausspülungen um Schieferbrocken herum geformt. Strömungsrichtung von rechts nach links. Athabaska Formation (Präkambrium). Vergleiche mit Tafel 91 B.

Cupules en croissant produites quand le courant est dévié autour de débris de schiste. Courant de droite à gauche. Cf. planche 91 B. Formation d'Athabaska (Précambrien).

Calcos en herradura formados alrededor de fragmentos lutíticos; las marcas han sido originadas por deflexión de la corriente. Sentido de la corriente de derecha a izquierda. Compárese con ámina 91 B. Formación Athabaska, Precámbrico.

PLATE 93 B

Current crescent casts on underside of sandstone of Juniata Formation (Ordovician). Current from top to bottom. On State Highway 74 at Waggoners Gap, Cumberland County, Pennsylvania, U.S.A. (Photograph by N. L. McIVER).

Strömungskämme auf der Unterseite einer Sandsteinschicht. Juniata Formation (Ordovizium). Strömungsrichtung von oben nach unten.

Cupules en croissant à la face inférieure d'un grès de la formation de Juniata (Ordovicien). Courant de bas en haut.

Calcos en herradura en la base de una arenisca. Corriente desde el borde superior de la lámina al inferior. Formación Juniata, Ordovícico.

PLATE 94

Modern desiccation mark with rain imprints. Note progressive drying and greater density of desiccation cracks in drier (light) area. Fort Dodge, Webster County, Iowa, U.S.A. (Photograph by D. M. BAIRD).

Rezente Trockenrisse mit Regentropfen-Eindrücken. Beachte fortgeschrittene Austrocknung und dichtere Scharung der Trockenrisse in der hellen Bildseite.

Empreintes de dessiccation, avec empreintes de gouttes de pluie. Noter le dessèchement progressif, avec densité plus grande de fentes de retrait dans la région plus sèche (plus claire).

Grietas de desecación y marcas de lluvia. Nótese la desecación progresiva y la mayor densidad de las grietas en el área (clara) más seca. Actual.

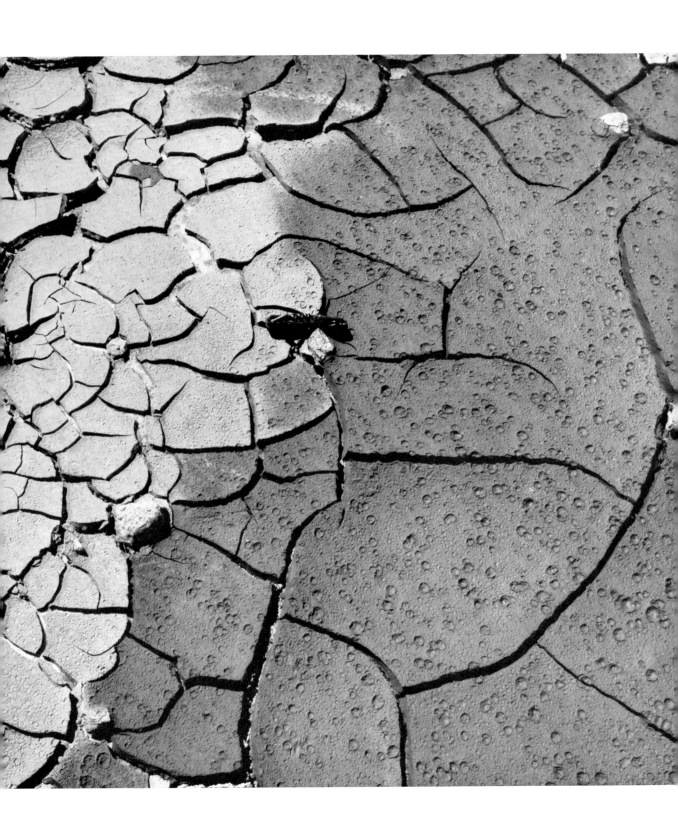

PLATE 95

Modern mud cracks showing formation of mud curls or flakes. West Bay, near Eypemouth, Bridport, Dorset, England (Geological Survey of Great Britain).

Rezente Trockenrisse mit eingerollten Tonhäuten.

Fentes de dessiccation actuelles, montrant la formation de pellicules enroulées et d'écailles.

Grietas de desecación y formación de rizos o láminas de fango. Actual.

PLATE 96A

Irregular modern molds of ice crystals. Note variable size and tendency of molds to be straight and intersect. Some irregular desiccation cracks are also present in upper left and lower right corners. Clay pit approximated 2.5 miles southwest of Newcastle, New Castle County, Delaware, U.S.A.

Unregelmäßige Abdrücke von Eiskristallen, rezent. Beachte unterschiedliche Größe der im allgemeinen geraden und sich überschneidenden Abdrücke. Unregelmäßige Trockenrisse oben links und unten rechts.

Moules actuels irréguliers de cristaux de glace. Noter la taille variable des moules qui tendent à être rectilignes et à se croiser. Quelques fentes irrégulières de dessiccation se voient aussi dans les coins supérieur gauche et inférieur droit.

Moldes irregulares de cristales de hielo, de variadas dimensiones, rectos e intersectados. En el ángulo superior izquierdo e inferior derecho se observan algunas grietas de desecación irregulares. Actual.

PLATE 96B

Infilled desiccation cracks in mudstone of Thurso Sandstone. Old Red Sandstone (Devonian). Clairdon shore, east of Thurso, Caithness, Scotland (Geological Survey of Great Britain).

Ausgefüllte Trockenrisse (Netzleisten) im Pelit des Thurso-Sandsteins. Old Red-Sandstein (Devon).

Fentes de dessiccation colmatées dans une argile schisteuse des grès de Thurso. Vieux-grès-rouges (Dévonien).

Grietas de desecación rellenadas, en limolita. Arenisca Thurso, Old Red-Sandstone, Devónico.

PLATE 97

Mud cracks in argillaceous limestone of Wills Creek Formation (Silurian). Along tracks of Western Maryland Railroad, Round Top, Washington County, Maryland, U.S.A.

Trockenrisse in tonigem Kalkstein der Wills Creek Formation (Silur).

Fentes de dessiccation dans un calcaire argileux de la formation de Wills Creek (Silurien).

Grietas de desecación en caliza arcillosa. Formación Wills Creek, Silúrico.

PLATE 98 A

Casts of salt crystals in shaley siltstone. Wupatki Member, Moenkopi Formation (Triassic). New Cameron, Coconino County, Arizona, U.S.A. (McKee, 1954, pl. 10C).

Salzkristall-Pseudomorphosen in tonigem Siltstein. Wupatki Member der Moenkopi Formation (Trias).

Moulages de cristaux de sel dans une argile schisteuse. Terme de Wupatki, formation de Moenkopi (Trias).

Calcos de cristales de sal, en lutita. Miembro Wupatki, Formación Moenkopi, Triásico.

PLATE 98 B

Rhythmic pattern of beach cusps in medium-sized gravel. Alum Bay, Isle of Wight, Great Britain (Kuenen, 1948, pl. 1-B).

Rhythmisches Muster von Strandhörnern in Schotter mittlerer Korngröße.

Arrangement rythmique de croissants de plage dans un gravier de taille moyenne.

Cuspilitos formados por grava mediana. Actual.

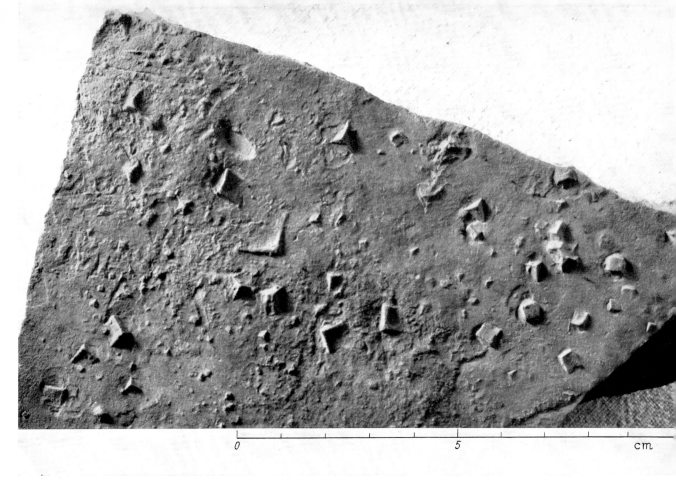

PLATE 99

Dinosaur tracks in dolomite of Glen Rose Limestone (Cretaceous). Davenport Ranch, West Verde Creek about 9 miles southwest of Bandera, Bandera County, Texas, U.S.A. (Reprinted from *Science*, 1961, by permission; Shell Oil Company photograph).

Dinosaurier-Fährten im Dolomit des Glen Rose-Kalksteins (Kreide).

Empreintes de pas de dinosaures dans une dolomie; calcaires de Glen Rose (Crétacé).

Huellas de dinosaurio en dolomía. Caliza Glen Rose, Cretácico.

PLATE 100A

Deformation structure (ball-and-pillow?) in glacial sand and gravel (Pleistocene). Ontario, Canada (Photograph by D. M. BAIRD).

Verformungsstruktur (Ballenstruktur?) in glazialen Sanden und Schottern (Pleistozän).

Figures de déformation (coussins?) dans des sables et graviers glaciaires (Pleistocène).

Estructura de deformación (estructura almohadillada?) en arenas y gravas glaciarias. Pleistoceno.

PLATE 100B

Cross-section of large pillow structure in Thorold Sandstone (Silurian). New Jolly road cut, Hamilton, Ontario, Canada.

Schnitt durch eine große Kissenstruktur im Thorold-Sandstein (Silur).

Coupe d'un grand coussin dans le grès de Thorold (Silurien).

Estructura almohadillada. Arenisca Thorold, Silúrico.

PLATE 101 A

Large sandstone pillow in basal part of ball-and-pillow zone (Devonian). In small quarry, on U.S. Highway 11-15, near Port Trevorton, Snyder County, Pennsylvania, U.S.A.

Großer Sandsteinballen im Unterteil einer Ballenstruktur-Zone (Devon).

Coussin de grès de grande taille, dans la partie basale d'une zone en coussins (Dévonien).

Estructura almohadillada en la base de una arenisca deformada por carga. Devónico.

PLATE 101 B

Ball-and-pillow structure in sandstone of Chemung Formation (Devonian). Note that structure is confined to a specific bed, larger more prominant pillows and better-defined pillows at base, gradual passage upward into normal sandstone, and predominance of down-facing convexities. Along Juniata River about 3 miles north of Amity Hall, Perry County, Pennsylvania, U.S.A.

Ballenstruktur im Sandstein der Chemung Formation (Devon). Die Struktur beschränkt sich auf eine einzelne Schicht. Größere, besser ausgebildete Ballen im Unterteil; allmählicher Übergang nach oben in ungestörten Sandstein. Konvexflächen überwiegend nach unten gerichtet.

Structure en boules-et-coussins («pseudo-nodules») dans les grès de Chemung (Dévonien). Noter que cette structure est confinée à une couche donnée, avec des coussins plus grands et plus marqués à la base, et passage graduel vers le haut à un grès normal; les surfaces convexes sont généralement tournées vers le bas.

Extructura almohadillada en arenisca. Nótese la estructura limitada a un estrato; su mejor desarrollo y definición en la base; el pasaje gradual hacia arriba a una arenisca normal; y la convexidad inferior de las formas, más marcada. Formación Chemung. Devónico.

PLATE 102A

Ball-and-pillow structure in limestone (calcisiltite). Note large size of pillows at base becoming smaller upward and passing into undisturbed beds. Internal laminations conform to margins of pillows. Cincinnatian Series (Ordovician). Ohio Coordinate System, South Zone, X: 1,665,498, Y: 243,120, Spring Township, Adams County, Ohio, U.S.A.

Ballenstruktur in Kalkstein (Kalksiltstein). Beachte enorme Größe der basalen Ballen, die nach oben hin in kleinere Ballen und schließlich in ungestörtes Gestein übergehen. Innere Fein-schichtung zeichnet die Ballenumrisse ab. Cincinnatian Series (Ordovizium).

Structure en boules-et-coussins dans un calcaire (calcisiltite). Noter que les coussins sont de grande taille à la base, deviennent progressivement plus petits vers le haut et passent à des couches indemnes. La lamination interne est conforme aux contours des coussins. Série de Cincinnati (Ordovicien).

Estructura almohadillada en caliza (calcipelita). El tamaño de estas formas en la base, disminuye hacia arriba hasta estratos no deformados; la laminación y superficie de estas masas son con-cordantes. Serie Cincinnatiana, Ordovícico.

PLATE 102B

Ball-and-pillow structure in limestone (calcisiltite) of Cincinnatian Series (Ordovician). Note large size of pillows at base becoming smaller upward and passing into undisturbed beds. Ohio Coordinate System, South Zone, X: 1,665,500, Y: 243,100, Spring Township, Adams County, Ohio, U.S.A.

Ballenstruktur in Kalkstein (Kalksiltstein) der Cincinnatian Series (Ordovizium). Beachte enorme Größe der Ballen, die nach oben hin allmählich in kleinere Ballen und schließlich in ungestörtes Gestein übergehen.

Structure en boules-et-coussins dans un calcaire (calcisiltite) de la série de Cincinnati (Ordo-vicien). Noter que les coussins, de grande taille à la base, deviennent progressivement plus petits vers le haut avant de passer à des couches indemnes.

Estructura almohadillada en caliza (calcipelita). El tamaño de estas formas en la base, disminuye hacia arriba hasta estratos no deformados. Serie Cincinnatiana, Ordovícico.

PLATE 103 A

Ball-and-pillow structure in sandstone. Note restriction to well-defined bed overlain and underlain by undisturbed strata. Thorold Formation (Silurian). New Jolly road cut, Hamilton, Ontario, Canada.

Ballenstruktur in Sandstein. Struktur ist auf eine einzige Schicht beschränkt. Im Hangenden und Liegenden ungestörtes Gestein. Thorold Formation (Silur).

Structure en boules-et-coussins dans un grès. Noter que les pseudo-nodules sont confinés à un niveau bien défini, intercalé dans des couches indemnes. Formation de Thorold (Silurien).

Estructura almohadillada en arenisca, limitada a un estrato entre otros dos normales (no deformados). Formación Thorold, Silúrico.

PLATE 103 B

Ball-and-pillow structure in sandstone of Waverly Group (Mississippian). Deformed layer approximately 1.3 meters thick. Note lateral persistence of deformed layer. On U.S. Highway 23, approximately 0.5 miles north of Waverly, Ross County, Ohio, U.S.A.

Ballenstruktur im Sandstein der Waverly Group (Mississippian). Verformte Bank ungefähr 1,3 m mächtig. Beachte seitliche Beständigkeit der verformten Schicht.

Structure en boules-et-coussins dans un grès du groupe de Waverly (Mississipien). La couche déformée a une épaisseur d'environ 1,30 m. Noter son extension latérale.

Estructura almohadillada en arenisca. Nótese la continuidad lateral del estrato deformado (espesor aproximado 1.3 m). Grupo Wavely, Mississippiano.

PLATE 104A

Large sandstone ball in interbedded sandstones and shales of Lusk Shale (Pennsylvanian). Note differential compaction over ball. Along Interstate Highway 57, NE 1/4 NE 1/4 NE 1/4 sec. 27, T. 11 S., R. 1 E., Union County, Illinois, U.S.A. (Illinois Geological Survey).

Großer Sandsteinballen in Sandstein, wechsellagernd mit Schieferton. Lusk-Schiefer (Pennsylvanian). Beachte unterschiedliche Verdichtung oberhalb des Ballens.

Boule de grès de grande taille, dans des couches alternantes de grès et d'argile schisteuse des schistes de Lusk (Pennsylvanien). Noter l'effet de compaction différentielle autour et au-dessus de la boule.

Estructura almohadillada en areniscas y lutitas interestratificadas. Nótese los efectos de la compactación diferencial sobre la estructura. Lutita Lusk, Pennsylvaniano.

PLATE 104B

Broken sandstone ball in horizontal lying sediments. Note differential compaction over ball. Lusk Shale (Pennsylvanian). Along Interstate Highway 57, NE 1/4 NE 1/4 NE 1/4 sec. 27, T. 11 S., R. 1 E., Union County, Illinois, U.S.A. (Illinois Geological Survey).

Zerbrochener Sandsteinballen in Horizontalschichten. Beachte unterschiedliche Verdichtung oberhalb des Ballens. Lusk-Schiefer (Pennsylvanian).

Boule de grès brisée, dans des couches horizontales. Noter l'effet de compaction différentielle au-dessus de la boule. Schistes de Lusk (Pennsylvanien).

Estructura almohadillada quebrada y efectos de la compactación diferencial sobre ella. Lutita Lusk, Pennsylvaniano.

PLATE 105 A

Contemporaneous disturbance of bedding. Intraformational flat-pebble conglomerate consisting of flat limestone pebbles (black), partially to completely silicified, embedded in limestone matrix. Conococheague Formation (Cambrian). Approximately 0.25 miles north of Big Spring, Washington County, Maryland, U.S.A.

Synsedimentäre Schichtstörung. Synsedimentäres Flachgeröll-Konglomerat: flache Kalkstein-gerölle (schwarz), zum Teil verkieselt, eingelagert in Kalkstein (hell). Conococheague Formation (Kambrium).

Déformation contemporaine du dépôt. Conglomérat intraformationnel de galets aplatis (noirs); ces galets calcaires sont plus ou moins complètement silicifiés et enrobés dans une pâte calcaire. Formation de Conococheague (Cambrien).

Destrucción mecánica contemporánra a la depositación. Conglomerado intraformacional formado por lajas (negras) de caliza parcial o totalmente silicificada, con mátriz calcárea. Formación Conococheague, Cámbrico.

PLATE 105 B

Pillow structures in loose inverted slabs of limestone. Cincinnatian Series (Ordovician). Generally, pillow has a smooth surface but note tension fractures. Ohio Coordinate System, South Zone; X: 1,635,298; Y: 308,212, along Straight Creek, Pleasant Township, Brown County, Ohio, U.S.A.

Kissenstruktur auf der Unterseite einer losen ,,überkippten" Kalksteinplatte. Cincinnatian Series (Ordovizium). Beachte Dehnungsrisse in der sonst glatten Kissenoberfläche.

Coussins sur des dalles calcaires transportées, retournées. Série de Cincinnati (Ordovicien). Généralement, les coussins ont une surface lisse, mais on peut y observer des fractures de tension.

Estructura almohadillada con superficies generalmente lisas y fracturas tensionales, sobre lajas sueltas e invertidas. Serie Cincinnatiana, Ordovícico.

PLATE 106A

Convoluted beds as seen from above. Upper Priabonien (Tertiary) flysch sedimentation near San Remo, Liguria, Italy (PLESSMANN, 1961, Fig. 30).

Aufsicht auf Wulstbank. Oberes Priabonien (Tertiär).

Couches ondulées vues de dessus. Flysch du Priabonien supérieur (Tertiaire).

Laminación intraplegada (vista en planta) en flysch. Priabonien Superior, Terciario.

PLATE 106B

Convolute lamination. Section at right angles to bedding. Krosno beds (Oligocene), Mymon, central Carpathians, Poland. (Photograph from DZULYNSKI, in press).

Wulstschichtung; Schnitt senkrecht zur Schichtung. Krosno-Schichten (Oligozän).

Feuilletage ondulé. Coupe perpendiculaire à la stratification. Couches de Krosno (Oligocène).

Laminación intraplegada en sección perpendicular a la estratificación. Capas Krosno, Oligoceno.

PLATE 107 A

Disturbed bedding in thin-bedded sandstone. Tar Springs Sandstone (Mississippian). SW 1/4 SW 1/4 NW 1/4 sec. 30, T. 7 S., R. 6 W., Randolph County, Illinois, U.S.A. (Illinois Geological Survey).

Gestörte Schichtung in dünnschichtigem Sandstein. Tar Springs-Sandstein (Mississippian).

Stratification dérangée dans un grès finement lité. Grès de Tar Springs (Mississipien).

Estratificación fina deformada. Arenisca Tar Spring, Mississippiano.

PLATE 107 B

Deformed Pleistocene glaciolacustrine silts and clays. Décollement structure attributed to thrust of overriding glacier. Five miles southwest of Glens Falls, Warren County, New York, U.S.A. (HANSEN et al., 1961, pl. 1, Fig. 2).

Verformte, glaziale, lakustrische Silte und Tone (Pleistozän). Abscherung dem Schub eines überfahrenden Gletschers zugeschrieben.

Limons et argiles glacio-lacustres déformés du Pleistocène. On attribue le décollement des lits à la poussée du glacier sus-jacent.

Deformación de limos y arcillas glacilacustres. Estructura de despegue atribuída al empuje de una masa de hielo. Pleistoceno.

PLATE 108A

Décollement structure (approximately 0.5 m thick). Note undisturbed bedding above and below. Lisan Formation (Pleistocene). 6 km northwest of Sodom (Israel topographic grid, 185/056), Israel (Geological Survey of Israel).

Abscherungsstruktur, ca. 0,5 cm dick. Beachte ungestörte Schichtung im Hangenden und Liegenden. Lisan Formation (Pleistozän).

Exemple de décollement (épaisseur d'environ 0,5 m). Noter les couches indemnes qui encadrent la couche décollée. Formation de Lisan (Pleistocène).

Estructura de despegue (espesor aproximado 0.5 m) limitada entre estratos normales (no deformados). Formación Lisan, Pleistoceno.

PLATE 108B

Contorted bedding, Pleistocene varved clays. Note layers with intricate internal crumplings with undisturbed layers above and below and lack of preferred direction of overturning of folds. Saskatchewan, Canada (Geological Survey of Canada).

Gekröseschichtung, Warventon, Pleistozän. Beachte ungestörte Lagerung im Hangenden und Liegenden von verwickelt gefalteten Schichten und regellose Orientierung der überkippten Falten.

Couches contournées, argiles pleistocènes à varves. Noter les lits à plissotements internes compliqués, intercalés dans des couches indemnes, et le fait que le renversement des plis ne se fait pas dans une direction particulière.

Varves con estratificación deformada. Nótese la laminación plegada sin inclinación preferencial, limitada por capas normales (no deformadas). Pleistoceno.

PLATE 109

Drift boulder of finely laminated argillite of Cobalt Series (Huronian, Precambrian). Note contorted bedding, overturning and flattening of folds against boundary bedding surfaces in the lower of the two disturbed zones. In upper contorted zone, layer is broken into isolated segments. Ontario, Canada (Photograph by W. HILLER).

Driftblock aus feingeschichtetem Argillit der Cobalt Series (Huron, Präkambrium). Beachte Gekröseschichtung, Überkippung und Abplattung der Falten gegen die über- und unterliegenden Schichtflächen in der unteren der beiden gestörten Zonen. In der oberen gestörten Lage ist die Schicht in isolierte Teile zerbrochen.

Bloc erratique d'argilite finement feuilletée, des couches de Cobalt (Huronien, Précambrien). Contournements, avec plis renversés et aplatis contre les surfaces de stratification qui les encadrent, dans la zone dérangée inférieure; dans la zone supérieure, la couche déformée est morcelée en segments isolés.

Rodado glaciario de argillita finamente laminada. Nótese dos niveles deformados: el inferior con pliegues volcados y comprimidos entre dos planos de estratificación; el superior presenta su estructura quebrada y separada en segmentos. Serie Cobalt, Huroniano, Precámbrico.

PLATE 110

Overturned cross-bedding in three units. Note lack of reverse drag in beds above overturned units. Current from left to right. Chinle Formation (Triassic). Sec. 31, T. 34 N., R. 14 E., San Juan County, Utah, U.S.A. (STEWART, 1961, Fig. 54.2).

Überkippte Schrägschichtung in drei Einheiten. Beachte Abwesenheit der umgekehrten Schleppung im Hangenden der überkippten Lagen. Strömungsrichtung von links nach rechts. Chinle Formation (Trias).

Stratification entrecroisée renversée dans trois ensembles. Noter qu'il n'y a pas de retroussement inverse (comme il y en aurait en cas de faille) dans les couches qui recouvrent les ensembles renversés. Le courant allait de gauche à droite. Formation de Chinle (Trias).

Tres unidades con estratificación entrecruzada volcada. Nótese la falta de pliegues de arrastre en los estratos situados sobre las unidades referidas. Corriente de izquierda a derecha. Formación Chinle, Triásico.

PLATE 111 A

Laminated gypsum showing enterolithic folding presumed due to expansion of anhydrite due to partial or complete alteration to gypsum. Castile Formation (Permian), New Mexico, U.S.A.

Feingeschichteter Gips mit Gekrösefaltung, wahrscheinlich durch Quellung des mehr oder weniger vollständig in Gips umgewandelten Anhydrites hervorgerufen. Castile Formation (Perm).

Gypse feuilleté (serpentiforme) avec plissement entérolitique que l'on croit dû à l'expansion de l'anhydrite après l'altération partielle ou complète du gypse. Formation de Castile (Permien).

Yeso laminado mostrando estructura enterolítica, debida presumiblemente a la expansión de anhidrita durante su transformación parcial o completa en yeso. Formación Castile, Pérmico.

PLATE 111 B

Thinly and evenly laminated siltstone from fill of an erosional channel. Note microfaulting of compactional origin; note also grading from coarser (light-colored) to finer (darker) in some layers. Anvil Rock Sandstone, Carbondale Formation (Pennsylvanian). Farmersville Mine, Farmersville, Montgomery County, Illinois, U.S.A. (Illinois Geological Survey).

Fein- und ebengeschichteter Siltstein. Teil der Füllung einer Erosionsrinne. Beachte die durch Sackung entstandenen kleinen Verwerfungen und die in einigen Lagen erkennbare Gradierung von grobem (hell) zu feinem Korn (dunkel). Carbondale Formation (Pennsylvanian).

Microgrès finement et régulièrement feuilleté. Noter les microfailles dues à la compaction, et le classement de grains grossiers (clairs) à fins (plus foncés) dans certains lits. Formation de Carbondale (Pennsylvanien).

Relleno de cauce formado por limolita con laminación regular. Nótese las microfallas compresionales y la gradación de granos gruesos (laminas claras) a finos (láminas oscuras) en algunos niveles. Miembro Anvil Rock, Formación Carbondale, Pennsylvaniano.

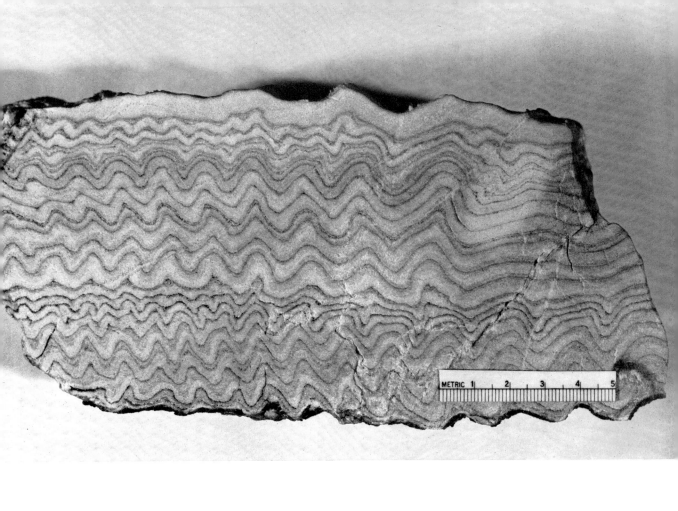

PLATE 112A (left)

Flow structure in sandstone (white) and argillaceous siltstone (dark). Note incipient spiral structure in larger siltstone masses (Pennsylvanian). Natural size. Illinois, U.S.A. (Illinois Geological Survey).

Fließgefüge in hellem Sandstein und dunklem tonigem Siltstein. Beachte das sich entwickelnde Spiralgefüge in größeren Siltmassen (Pennsylvanian). Natürliche Größe.

Pseudo-nodules dans un grès (blanc) et un microgrès argileux (foncé). Noter le début de texture en spirale dans les masses de microgrès de plus grande taille (Pennsylvanien).

Estructura almohadillada en arenisca (clara) y limolita arcillosa (oscura). Nótese la estructura espiralada incipiente en las masas limolíticas mayores. Tamaño natural. Pennsylvaniano.

PLATE 112B (right)

Incipient mottled structure in interbedded sandstone and shale. Natural size (Pennsylvanian). Illinois, U.S.A. (Illinois Geological Survey).

Anfangsstadium gefleckter Schichtung in wechsellagerndem Sandstein und Schieferton. Natürliche Größe (Pennsylvanian).

Début de texture mouchetée dans des couches interstratifiées de grès et d'argile schisteuse. Grandeur nature (Pennsylvanien).

Estructura moteada incipiente en arenisca y lutita interestratificadas. Tamaño natural. Pennsylvaniano.

PLATE 113 A

Clastic dike (dark) of siltstone and sandstone, in metaquartzite of Taxin Group (Precambrian); metaquartzite is vertical and strikes away from observer. Note irregular dike walls. Compare with plate 113 B. Southwest shore of Johnson Island, Lake Athabasca, Saskatchewan, Canada (Geological Survey of Canada).

Klastischer Gang (dunkel) aus Silt- und Sandstein, in Metaquarzit der Taxin Group (Präkambrium). Steilstehender Metaquarzit streicht auf den Beobachter zu. Beachte unregelmäßige Begrenzung des Ganges. Vergleiche mit Tafel 113 B.

Filon clastique (foncé) de microgrès et grès, dans la quartzite métamorphique de Taxin (Précambrien); le pendage de la quartzite est vertical et sa direction vers l'arrière-plan. Noter l'irrégularité des parois du filon. Cf. planche 113 B.

Dique clástico (oscuro) vertical compuesto por materiales limolíticos y areniscosos en metacuarcita. Nótese la irregularidad de sus paredes. Compárese con lámina 113 B. Grupo Taxin, Precámbrico.

PLATE 113 B

Sandstone (quartzite) dike intersecting Espanola formation (Huronian). Note pebbles in center of dike and thin bedding and relief on calcareous silts of Espanola Formation. East end of Griffin Lake, Lot 6, Con. IV, Merritt Township, District of Sudbury, Ontario, Canada.

Sandsteingang (Quarzit), Espanola Grauwacke (Huron) durchsetzend. Beachte Gerölle im Mittelteil des Ganges und dünne Schichtung und Relief der kalkhaltigen Siltsteine der Espanola Formation.

Filon de grès (quartzite) recoupant la grauwacke d'Espanola (Huronien). Noter les cailloux au centre du filon, et le fin litage et le relief du microgrès calcaire de la formation d'Espanola.

Dique cuarcítico, conglomerádico en su parte media, intersectando grauvacas y limolitas calcáreas, estas últimas en realce y finamente estratificadas. Formación Espanola, Huroniano.

PLATE 114

Sand volcano on slump. Sand volcano shows typical form and crater structure. Ross, County Clare, Ireland (GILL and KUENEN, 1958, pl. 35-2).

Sandvulkan auf Rutschungsmasse. Kraterausbildung und Form typisch.

«Volcan de sable» sur un glissement. Le volcan de sable est de forme typique avec un cratère.

Volcán de arena formado sobre depósitos deslizados, con forma y cráter típicos. Actual.

PLATE 115 A

Cylindrical structure in sandstone. Cylinder essentially perpendicular to bedding. Yoshida Formation (Tertiary). Kamigo, Yoshida-machi, Chichibu-gum, Saitama Perfecture, Japan (ARAI, 1959, pl. 1, Fig. 2).

Zylindrische Struktur in Sandstein, praktisch senkrecht zur Schichtfläche. Yoshida Formation (Tertiär).

Filon cylindrique dans un grès. Le cylindre est à peu près perpendiculaire à la stratification. Formation de Yoshida (Tertiaire).

Estructura cilíndrica perpendicular a la estratificación, en una arenisca. Formación Yoshida, Terciario.

PLATE 115 B

Worm borings in sandstone of Galesville Member (Dresbach Sandstone, Upper Cambrian). Sec. 10, T. 11 N., R. 6 E., Sauk County, Wisconsin, U.S.A. (Photograph by DALE FARRIS).

Bohrlöcher im Sandstein des Galesville Member (Dresbach Sandstein, Oberkambrium).

Trous de vers dans le grès de Galesville (Cambrien).

Estructura tubular originada por vermes, en arenisca. Miembro Galesville, Cámbrico.

PLATE 116A

Burrowings seen in horizontal section. Approximately natural scale. Lower Barremien sandstone (Cretaceous). Westphalia, Germany (Koninklijke Shell Exploratie en Produktie Laboratorium; KUENEN, 1961, Fig. 3).

Bauten in horizontalem Schnitt. Ungefähr natürliche Größe. Sandstein, Unteres Barrême (Kreide).

Tubes vus en coupe horizontale, à peu près grandeur nature. Grès Barrémien inférieur (Crétacé).

Estructura tubular vista en sección horizontal. Tamaño aproximadamente natural. Arenisca, Barremiano Inferior, Cretácico.

PLATE 116B

Tubular structure (parallel to knife) in cross-bedded Juniata Formation (Ordovician). On State Highway 74 at Waggoners Gap, Cumberland County, Pennsylvania, U.S.A.

Röhrenstruktur (parallel zum Messer) in schräggeschichteter Juniata Formation (Ordovizium).

Structure tubulaire (parallèle au couteau) dans les grès à stratification entrecroisée de Juniata (Ordovicien).

Estructura tubular (paralela a la navaja) en arenisca con estratificación entrecruzada. Formación Juniata, Ordovícico.

PLATE 117

Armored mud balls left by receding waters of stream. Las Posas barranca between El Rio and Camarillo, California (Soil Conservation Service, U.S. Dept. of Agriculture).

Gespickte Tongerölle, nach dem Rüockzug des Hochwassers gestrandet.

Boules d'argile laissées dans le lit d'un cours d'eau après la décrue.

Rodados de arcilla acorazados, dejados por aguas de inundación en retorno a su cauce. Actual.

Glossary of primary sedimentary structures

The students of sedimentary rocks have been impressed with the abundance and variety of structures and markings—including the hieroglyphs of the early workers. These investigators have exercised little or no restraint in naming these features and as a result there is a multiplicity of terms in the literature. We have attempted to compile these into a glossary.

We have included every term known to us. Some of these are little known, little used, and perhaps best forgotten. Others are synonyms and some choice must ultimately be made and the less desirable terms dropped. The authors, however, do not feel that such decisions can be made wisely at this time. Obviously they have expressed their preference in writing the plate captions. There is need for a glossary which includes obsolete and unusual terms just as there is need to include archaic terms, colloquilisms and the like in a dictionary. The reader will encounter them in the literature and will need a definition.

As the discerning reader will note, the terminology used by geologists is a mixture of terms, some of which are purely descriptive, such as flutes or grooves, whereas others are wholly genetic, such as rain-drop impressions and load casts. Although a purely descriptive, wholly nongenetic nomenclature may seen desirable, such an ideal is unattainable and we shall have to live with our terminological untidiness. Geologists are well aware that a similar mixture of genetic and descriptive terms

confronts him if he examines the names applied to land forms and to rocks.

We have included only terms which apply to or describe primary sedimentary structures. We have excluded post-consolidation structures and all concretionary phenomena. In addition we also exclude those other structures formed by chemical reorganization, such as solution and hydration.

Structures due to organisms have been included only under several general headings such as stromatolites, fucoids, ichnofossils. We have not attempted to include terms denoting varieties of these and related structures. We have omitted entirely self-evident terms such as irregular bedding and horizontal bedding. We omit also terms which are applied to the way in which rocks break, such as flaggy or fissile. The terms included apply to structures found in ancient sediments. There are a few exceptions, such as fulgurite and antidune, the first being a structure, not strictly sedimentary and not certainly known in ancient sediments, and the second being an ephemeral transport form.

Where possible, we have given the reference to the first use of the term; such reference being placed immediately after the term. But inasmuch as some of the structures of sediments, such as ripple marks, were probably recognized in antiquity, it is not always possible to cite an original description. Where such a reference could not be found, we have commonly given a reference to a modern paper in a readily accessible journal in which the structure is described and illustrated. References which we have not seen are marked with an asterisk.

The reader will note that the glossary arrangement is alphabetical. Moreover the adjectival modifier, if there is one, as, for example, tabular cross-bedding, is listed first, not afterwards. This arrangement is simplest to construct; any other raises troublesome problems. For example, it is easier to list and find *flute cast* or *groove cast*, etc., than to list these as *cast, flute* or *cast, groove*.

The term "cast" has been applied to many structures, such as flute and groove casts, in a somewhat loose way. In a strict sense the filling of a primary depression, such as flute, is a mold. Usage, however, has considered the original depression as the mold and the filling as a cast. We defer to usage.

Verzeichnis der Fachausdrücke der primären Sedimentstrukturen

Beim Studium der Sedimente wird man immer wieder von der Vielfalt der Strukturen und Marken, einschließlich der Hieroglyphen älterer Autoren beeindruckt. Allerdings haben sich die Bearbeiter solcher Strukturen nur allzuoft bei der Namensgebung nicht zurückgehalten. Infolgedessen ist die einschlägige Literatur mit Ausdrücken überflutet. Wir haben im folgenden versucht, diese Namen in einem Verzeichnis zusammenzufassen.

Jede uns bekannte Bezeichnung wurde aufgenommen, unbeschadet der Tatsache, daß viele, sei es kaum bekannt, sei es wenig benutzt worden sind und deshalb am besten vergessen werden. Andere sind Synonyme, und wenn man sich auch letztlich für die eine oder andere Bezeichnung entscheiden muß, so haben sich die Verfasser doch gehütet, schon jetzt die Entscheidung zu treffen. Freilich drückt sich ihre Wahl in den Bildunterschriften aus. Es besteht nämlich ein offensichtliches Bedürfnis nach einem Verzeichnis, das veraltete und ungewöhnliche Namen enthält, ebenso wie man altertümliche Begriffe und Ausdrücke der Umgangssprache und dergleichen auch in ein Wörterbuch aufnimmt; denn der Leser, dem solche Ausdrücke in der Literatur begegnen, wird nach ihrer Definition suchen.

Geologische Bezeichnungen sind eine Mischung aus rein beschreibenden Namen, wie Rillen oder Wülste, und anderen, durch und durch genetischen, wie Regentropfen-Eindrücke und Belastungsmarken. Gewiß ist eine rein deskriptive, ganz und gar ungenetische Nomenklatur vorzuziehen und wünschenswert; sie bleibt jedoch ein unerreichbarer Idealfall, und wir werden mit dieser inkonsequenten Terminologie zu leben haben. Eine ähnliche Mischung aus beschreibenden und genetischen Namen wird auf Gesteine und Landformen angewandt, wie der Geologe wohl weiß.

Wir haben nur solche Bezeichnungen aufgenommen, die primäre Sedimentstrukturen beschreiben oder sich doch auf sie beziehen. Konkretionäre Strukturen und solche, die nach der Verfestigung entstehen, haben wir ausgeschlossen. Fernerhin blieben Strukturen unberücksichtigt, die sich bei chemischen Reorganisationen wie Lösung und Hydration bilden.

Von Organismen geschaffene Strukturen finden sich lediglich unter einigen allgemeinen Überschriften, wie Stromatolithen, Fukoiden und Ichnofossilien. Folglich haben wir auch nicht versucht, Namen von Abarten dieser oder ähnlicher Strukturen in das Verzeichnis aufzunehmen. Selbstverständliche Begriffe, wie unregelmäßige und horizontale Schichtung, wurden nicht aufgenommen, und wir haben auch solche Bezeichnungen ausgelassen, die sich auf die Spaltbarkeit eines Gesteins beziehen, wie spaltbar, plattig usw. Die hier vorgelegten Namen stammen von fossilen Sedimenten. Fulgurit und Antidüne sind unter den wenigen Ausnahmen, wobei die erstere nicht unbedingt eine Sedimentstruktur und nicht mit Gewißheit fossil überliefert und die zweite nur eine vergängliche Transportform ist.

Nach Möglichkeit wurde die Veröffentlichung zitiert, in der eine Bezeichnung zum ersten Male gebraucht wird und erscheint unmittelbar nach der Bezeichnung. Indes war es bei schon zu frühen Zeiten bekannten Marken, wie etwa Rippelmarken, nicht immer möglich, Originalquellen zu finden. In solchen Fällen geben wir in leicht zugänglichen Fachzeitschriften veröffentlichte Arbeiten an, in denen jene Strukturen abgebildet und beschrieben sind. Mit einem Sternchen markierte Arbeiten haben wir nicht eingesehen.

Der Leser wird das Verzeichnis alphabetisch angeordnet finden. Auch sind modifizierende Adjektive, sofern vorhanden, z. B. in „tafelige Schrägschichtung", voran-, nicht nachgestellt. Diese Anordnung läßt sich am einfachsten zusammenstellen; jede andere wirft lästige Probleme auf. Strömungswulst oder Rillenmarke ist sicherlich leichter und übersichtlicher anzuführen und zu finden als Wulst, Strömungs-, oder Marke, Rillen-.

In* der englischen Terminologie wird die Bezeichnung „cast" (Ausguß, Ausfüllung) ungenau oder locker angewandt; denn strenggenommen ist das Negativ zu einer primären Eintiefung ein Abdruck („mold"). Es hat sich aber eingebürgert, ursprüngliche Eintiefungen als Abdrücke und ihre Füllungen als Ausgüsse zu bezeichnen. PETTIJOHN und POTTER schließen sich diesem Gebrauch an. In der deutschen Terminologie erscheinen die Namen Abdruck und Ausguß (Ausfüllung) viel seltener, haben jedoch ihren Sinn in ähnlicher Weise verschoben wie ihre englischen Entsprechungen.

* Der folgende Abschnitt ist in der deutschen Übersetzung leicht verändert und ergänzt worden.

278

Glossaire des termes relatifs aux figures sédimentaires primaires

Les spécialistes des roches sédimentaires, frappés par l'abondance et la variété des figures et empreintes qui s'y rencontrent, y compris les hiéroglyphes des anciens auteurs, n'ont pas cherché le moins du monde à se restreindre en nommant ces structures, et il en résulte une multiplicité de termes, que l'on a essayé d'énumerer dans ce glossaire.

Tous les termes dont les auteurs avaient connaissance ont été inclus. Certains sont peu connus, peu usités, et devraient peut-être tomber dans l'oubli. D'autres sont des synonymes, et il faudra éventuellement faire un choix et abandonner les moins appropriés. Les auteurs ne pensent pas, toutefois, que le moment soit mûr pour un choix raisonnable. Ils ont bien évidemment indiqué leur préférence dans la rédaction des légendes des planches. Mais il faut qu'un glossaire contienne les termes anciens ou peu usités, de même qu'on inclut dans un dictionnaire les mots archaïques, familiers et autres. Le lecteur les rencontrera dans la littérature, et aura besoin de définitions.

En utilisant le glossaire avec discernement, on s'apercevra que la terminologie employée par les géologues est un mélange de termes dont certains sont simplement descriptifs, tels que «rainures» ou «cannelures», alors que d'autres impliquent la genèse des phénomènes, par exemple «empreintes de gouttes de pluie» ou «figures de charge». Il serait souhaitable d'arriver à une nomenclature purement descriptive, dénuée d'idée de genèse, mais c'est un idéal irréalisable, et il nous faudra nous accommoder d'une terminologie désordonnée. Les géologues se rendent bien compte qu'ils se trouvent en face d'un mélange comparable de termes descriptifs et génétiques lorsqu'ils étudient de près les noms des formes de terrain ou des roches.

Les auteurs n'ont inclus que les termes qui s'appliquent aux formes sédimentaires primaires, ou qui les décrivent. Ils ont omis les structures formées après consolidation, et les concrétions, et également toutes les figures formées par des transformations chimiques telles que dissolution et hydratation.

Les empreintes dues aux organismes vivants ne sont mentionnées que sous les rubriques générales telles que stromatolithes, fucoïdes ou ichnofossiles. On n'a fait aucun effort pour donner le détail des termes qui se rapportent aux variétés de ces formes. Les termes dont le sens est évident ont été omis, telle «stratification irrégulière», ou «horizontale». Sont omis de même les termes qui s'appliquent aux cassures des roches, par exemple «fissile», ou «en dalles». Les termes énumérés ici s'appliquent aux formes que l'on rencontre dans les sédiments anciens, à quelques exceptions près, telles que fulgurite ou antidune, dont la première est une forme qui n'est pas sédimentaire au sens propre, et qui n'est pas connue avec certitude dans les terrains anciens, et la seconde n'est qu'une forme de transport, très éphemère.

Autant que possible, l'ouvrage où le terme a été proposé est cité immédiatement à la suite de l'en-tête. Mais comme certaines formes sédimentaires, telles que les rides, avaient probablement été reconnues dès l'antiquité, on ne peut pas toujours en citer la description originelle. Dans ce cas on a le plus souvent indiqué un travail récent, d'accès facile, dans lequel la structure est décrite et illustrée. Les ouvrages qui n'ont pas été consultés sont marqués d'un astérisque dans la liste bibliographique.

Le glossaire est alphabétique, et si les substantifs sont modifiés par des adjectifs, ceux-ci restent en position normale (précédant toujours le nom, en anglais). Cet arrangement est le plus facile à établir et cause un minimum de complications; il est plus facile d'écrire, ou de trouver, «fausse stratification» que «stratification, fausse». Les mots composés, eux aussi, gardent leur forme habituelle.

Le terme moulage est utilisé pour désigner bien des formes, parfois de façon peu précise. Nous avons traduit par moule le mot «mold», utilisé couramment pour désigner l'objet creux, et par moulage le terme «cast» utilisé couramment pour désigner son remplissage. (Au sens strict du mot, «mold» devrait s'appliquer au remplissage d'une dépression, mais les auteurs ont respecté l'usage courant.)

Glosario de estructuras sedimentarias primarias

La existencia de una gran variedad de estructuras sedimentarias y el comienzo simultáneo de su estudio por parte de numerosos investigadores, contribuyó a la formación de una terminología frondosa, en muchos casos conceptualmente coincidente, limitando la posibilidad de elaborar una nomenclatura armónica. En base a ello y con fines aclaratorios, hemos considerado conveniente reunir los vocablos en un glosario.

El presente glosario agrupa todos los términos conocidos por nosotros; entre ellos figuran algunos poco divulgados, cuyo empleo podría abandonarse. Además existen muchos sinónimos, de los cuales los difundidos son aquellos aparentemente más apropiados. Sin embargo consideramos inoportuno tomar decisiones, dado el estado actual del desarrollo de esta disciplina. Como es de rigor en todo diccionario, se incluyen términos obsoletos, arcaicos y del lenguaje corriente. El lector hallará sin duda estos vocablos y por lo tanto necesitará una definición.

En la revisión del glosario se advertirá que la terminología empleada por los geólogos resulta una mezcla de términos, de los cuales unos son simplemente descriptivos, tales como *acanaladuras* y *surcos*, mientras que otros implican la génesis del fenómeno, por ejemplo, *marcas de lluvia* y *calcos de carga*. Aunque es obvia la conveniencia de contar con una nomenclatura exclusivamente descriptiva, dicho ideal es de difícil concreción y en consecuencia debe aceptarse su actual estado. Por otra parte, es un hecho conocido por los geólogos la existencia de una mezcla de términos genéticos y descriptivos aplicados a la descripción de formas del paisaje y a las rocas.

Hemos incluido únicamente aquellos términos aplicados a estructuras sedimentarias primarias o que las describen, excluyendo los relacionados con fenómenos posteriores a la consolidación y con las formas de reorganización química originadas por disolución e hidratación; también se omiten todos los fenómenos concrecionales.

Las estructuras originadas por organismos son tratadas bajo la denominación general, tales como *estromatolitas*, *fucoides* e *icnofósiles*, excluyendo las variedades de estas formas. Tampoco se consideran estructuras como *estratificación irregular* y *estratificación horizontal*, en sí mismas evidentes; las relacionadas con la forma en que se hienden ciertas rocas lajosas o con fisilidad. Los términos incluidos se aplican a estructuras halladas en sedimentos antiguos, con algunas excepciones tales como *fulgu*rita y *antiduna;* la primera no es una forma estrictamente sedimentaria y no reconocida con seguridad en terrenos antiguos; la segunda corresponde a una forma efímera de transporte.

En los casos que ha sido posible, se cita la referencia del término original a continuación del mismo. Pero considerando que ciertas estructuras sedimentarias, tales como óndulas, son de antiguo conocidas, no siempre resulta factible indicar la descripción original y por lo tanto se citan trabajos modernos de consulta accesible, donde las estructuras aparecen descriptas e ilustradas. Las fuentes no consultadas se encuentran marcadas con un asterisco.

El glosario presenta un ordenamiento alfabético. Además, si los sustantivos son modificados por los adjetivos, ellos permanecen en posición normal (precediendo siempre al sustantivo en inglés). Esta construcción es la más sencilla puesto que reduce a un mínimo las complicaciones; es menos dificultoso ordenar y ubicar «flute cast» o «groove cast», que «cast, flute» o «cast, groove». El término *calco* ha sido aplicado a muchas estructuras en forma incorrecta. En sentido estricto, el relleno de una depresión primaria, tal como *surco*, determina un *molde*. Sin embargo, el uso corriente ha impuesto el vocablo *molde* para una depresión y *calco* a su relleno. Respetamos el uso corriente.

Glossary of primary sedimentary structures

Abioglyph (VASSOEVICH, 1953, p. 38):
A marking (hieroglyph) of inorganic origin.
Eine Marke (Hieroglyphe) anorganischen Ursprungs. *Abioglyphe.*
Empreinte (hiéroglyphe) d'origine non organique. *Abioglyphe.*
Marca (hieroglifo) de origen inorgánico. *Abioglifo.*

Air heave structure (STEWART, 1956, p. 159):
Small crumplings, which die out downward, found in laminated sands and which are presumed to be formed by rise of air trapped in sand at low tide.
In feinschichtigen Sanden auftretende Verfältelungen, die zum Liegenden hin ausklingen; mutmaßlich durch aufsteigende Luft geformt, die während Ebbe im Sand gefangen war. *Steigmarke, Entgasungs-Kanal.*
Plissotements disparaissant vers le bas, que l'on rencontre dans des sables lités, et que l'on attribue à une montée d'air enfermé dans le sable à marée basse. *Plissotements dûs à un dégagement d'air.*
Arenas laminadas con pequeñas corrugaciones que desaparecen hacia abajo, atribuidas a escape de aire contenido en ellas durante la bajamar. *Corrugaciones por escape de aire.*

Algal balls: See stromatolite.

Algal biscuits (MAWSON, 1929):
Probably a variety of stromatolites, which see.
Wahrscheinlich eine Art von Stromatolithen; s. dort.
Probablement une variété de stromatolithe, q.v.
Probablemente una variedad de estromatolita; véase.

Algal structure: See stromatolite; see also fucoids.

Angular cross-bedding: See plate 34 A.
Cross-bedding in which foreset beds meet underlying surface at sharp, discordant angle.
Schrägschichtung, in der die Leeblätter sich in scharfem Winkel diskordant von der unterliegenden Schichtfläche absetzen. *Winkelige Schrägschichtung* (w).
Stratification oblique, où les couches frontales forment un angle net, une discordance, avec la surface sous-jacente. *Fausse stratification angulaire.*
Estratificación entrecruzada formada por capas frontales con pendiente pronunciada respecto al plano inferior de estratificación. *Estratificación entrecruzada angular.*

Antidune:
(A) An ephemeral or transitory bottom transport form observed at certain velocities (GILBERT, 1914, p. 31) appearing as a sand wave which migrates upcurrent; not believed to be preserved in sediments. (B) Term has been applied to flame structure, which see, by LAMONT (1957).

(A) Eine kurzlebige oder Übergangsform des Bodentransportes, bei bestimmten Transport-Geschwindigkeiten beobachtet (GILBERT, 1914, S. 34). Erscheint als Sandwelle, die stromaufwärts wandert; angeblich nicht in Sedimenten erhalten. (B) Der Ausdruck ist von LAMONT (1957) auf Flammenstruktur angewandt worden; s. dort. *Antidüne* (w).

(A) Dans certaines conditions de transport sur le fond, et à certaines vitesses, une structure éphémère ou transitoire, dont on ne pense pas qu'elle persiste dans les sédiments, et qui prend la forme d'une ride qui remonte le courant. (B) Ce terme a aussi été employé pour désigner la «flame structure», q.v. *Antidune*.

(A) Ondas de arena de dimensiones relativamente amplias, formadas cuando la velocidad de la corriente de agua es grande. El desplazamiento se realiza en sentido opuesto a la corriente. No son preservadas en los sedimentos. (B) Término empleado por LAMONT (1957) como sinónimo de "flame structure", véase. *Antiduna*.

Anti-ripplets (VAN STRAATEN, 1953 b, p. 42):

Small asymmetrical ripples (wave length a few millimeters to 2 cm) which form by wind blowing silt over a moist surface. The silt adheres to the surface and forms ripples with steep face on windward side.

Kleine, unsymmetrische Rippeln (wenige Millimeter bis 2 cm Wellenlänge), die sich formen, wenn Wind Silt über eine feuchte Fläche bläst. Der Silt haftet an der Oberfläche und formt Rippeln mit steiler Luvseite. *Haftrippeln.*

Petites rides asymétriques (longueur d'onde de quelques millimètres à 2 cm) qui se forment quand le vent pousse des poussières sur une surface humide. Les poudres collent à la surface et forment des rides dont la face au vent est la plus raide. *Anti-ripplets.*

Ondula con crestas asimétricas y longitud de onda de hasta 2 cm, formada por partículas finas aventadas sobre una superficie húmeda a la cual se adhieren. Las crestas presentan su mayor pendiente a barlovento. *Anti-óndula.*

Armored mud balls ("pudding balls"): See plate 117.

Subspherical balls of mud, 5 to 30 cm in diameter coated with coarse sand and fine gravel (BELL, 1940).

Kugelige Gebilde aus Schlamm; 5—30 cm in Durchmesser; mit einem Überzug von Grobsand oder feinem Kies (BELL, 1940). *Gespickte Tongerölle.*

Boules de boue presque sphériques, de 5—30 cm de diamètre, et enrobées de sable grossier ou de petit gravier. *Lit.: Boules d'argile (balles argileuses) cuirassées, armées.*

Esferoides arcillosos de 5 a 30 cm de diámetro, revestidos de arena gruesa y grava fina (BELL, 1940). *Rodados de arcilla acorazados.*

Asymmetrical ripple marks: See plate 78 B.

Ripples with asymmetric profile. Steeper lee slope indicates direction of current. In plan view, crests may be either relatively straight or markedly curved. Also called current ripple mark.

Rippeln mit unsymmetrischem Profil. Steilseite stromabwärts gerichtet. Von oben gesehen, erscheinen die Rippelkämme entweder relativ gerade oder ausgesprochen gekrümmt. Auch Strömungsrippeln genannt. *Unsymmetrische Rippelmarken.*

Rides à profil asymétrique, la pente avale, plus raide, indiquant le sens du courant. Vues de dessus, les crêtes peuvent être à peu près rectilignes, ou arquées. *Rides asymétriques, rides de courant.*

Ondula con perfil asimétrico. Los flancos más abruptos están orientados en sentido de la corriente. En planta, las crestas pueden ser rectas o marcadamente curvas. También denominada "current ripple mark". *Ondula asimétrica.*

Backset bedding (DAVIS, 1890, Fig. 3):

Inclined bedding that dips *into* the current. Said to occur at the front of an esker. Also used for the beds deposited on the windward side of a transverse dune.

Geneigte Schichtung, die nach stromaufwärts einfällt. Soll an der Stirnseite von Osern auftreten. Ausdruck auch für Schichten gebräuchlich, die auf der Luvseite von transversalen Dünen abgelagert werden. *Luvblatt-Schichtung.*

Stratification inclinée dont le pendage est vers l'amont du courant, et qui se formerait en avant d'in esker. Désigne parfois aussi les couches déposées sur le côté au vent d'une dune transversale.

Estratificación inclinada con pendiente en sentido contratio a la corriente, formada probablemente en el frente de un esker. También empleado para capas depositadas a barlovento de una duna transversa. *Estratificación inclinada inversa.*

Backwash ripple marks (KUENEN, 1950, p. 292):

Ripple mark on beaches formed by the backwash.

Vom Rückstrom geformte Rippelmarken auf Strandflächen. *Rückstrom-Rippelmarken* (w).

Rides formées par le retrait de vagues. *Rides de retrait.*

Ondula de playa formada por acción de la resaca. *Ondula de resaca.*

Ball structure:

A primary structure characteristic of some limestones and sandstones (see ball-and-pillow structure); a term also applied to the ball structure of coal, called "ball coal" (FOX, 1931, pl. 2). See also armored mud balls, algal balls, lake balls and sea balls.

Primärstruktur, die für gewisse Kalk- und Sandsteine charakteristisch ist (s. auch Ballenstruktur); auch auf knollige Strukturen in Kohleflözen angewandt: die sog. „ball coal" (FOX, 1931, Tafel 2). Siehe auch gespickte Tongerölle, Pflanzenknäuel (lakustrisch und marin).

Structure primaire caractéristique de certains calcaires et grès (cf. ball-and-pillow structure); terme utilisé aussi pour décrire la structure du type de charbon appelé «ball coal». *Lit.: structure en boules.*

Estructura primaria característica en ciertas calizas y areniscas (véase "ball-and-pillow structure"); término también aplicado a ciertos carbones con estructura similar ("ball coal", FOX, 1931, pl. 2). Véase "armored mud balls", "algal balls", "lake balls" y "sea balls".

Ball-and-pillow structure (SMITH, 1916, p. 147) (flow rolls, pseudonodules, etc.): See plates 100A to 104A and 105B.

Structures found in sandstones and calcarenitic limestones, characterized by ball-and-pillow-shaped masses, hemispherical or kidney-shaped, formed by "internal readjustments ... mainly under gravitation" (SMITH, 1916, p. 147—150).

Struktur, die sich in Sandsteinen und Kalksandsteinen findet; halbkugelige oder nierenförmige Massen, die durch „inneren Ausgleich ... als Folge der Schwerkraft" entstanden sind (SMITH, 1916, S. 147—150). *Ballenstruktur* (n), *Rutschungsballen, Sedimentrolle, Wulststruktur, Wickelstruktur, -Falten, Wulstbank, Fließwulst* usw.

Structure particulière à certains grès et calcarénites qui présentent des masses arrondies ou allongées, hémisphériques ou en forme de haricot, qui se forment par «réajustements internes, ... surtout sous l'effet de forces de gravité». *Lit.: structure en boules-et-coussins. Aspect noduleux; grès, calcaire, noduleux.*

Estructura desarrollada en la base de ciertas areniscas y calcarenitas, caracterizada por masas esferoidales, arriñonadas o con aspecto de cojines; originada por reajuste interno, principalmente gravitacional (SMITH, 1916, p. 147—150). *Estructura almohadillada.*

Beach cusps: See plate 98B.

Regularly spaced, crescentic accumulations of sand, pebbles, or cobbles found along beaches. The "horns" of the crescents point seaward (KUENEN, 1948a).

An Stranden vorkommende, bogenförmige Anhäufungen von Sand, Geröllen oder Blöcken in gleichmäßigen Abständen. Die Hörner zeigen meerwärts (KUENEN, 1948). *Strand-hörner, Küstenspitzen, Brandungs-Kiesrücken.*

Accumulations de sable, de cailloux ou de galets, régulièrement espacées et en forme de croissants, que l'on trouve le long des plages. Les pointes («cornes» du croissant) sont tournées vers la mer. *Croissants de plage.*

Arena y grava en acumulaciones semejante a dunas, regularmente espaciadas y alineadas a lo largo de una playa, con sus concavidades hacia el mar (KUENEN, 1948). *Cuspilitos.*

Bed:

"The smallest division of a stratified series, and marked by a more or less well-defined divisional plane from its neighbors above and below" (KEMP, 1922, p. 190). A term loosely employed to denote a unit of less than member or formation rank, composed of two or more strata or laminations, that manifests some degree of lithologic unity (PAYNE, 1942, p. 1724). See also McKEE and WEIR (1953, Table 2).

„Die kleinste Unterteilung einer geschichteten Abfolge, nach oben und unten abgegrenzt durch eine mehr oder weniger gut definierte Trennfläche" (KEMP, 1922, S. 190). Lockere Bezeichnung für eine Einheit aus einer oder mehreren Schichten oder Feinschichten, die eine gewisse lithologische Einheit darstellt (PAYNE, 1942, S. 1724). Siehe auch McKEE und WEIR (1953, Tafel 2). *Schicht.*

«La plus petite division d'une série stratifiée, séparée des couches voisines de dessus et de dessous par un plan de séparation plus ou moins marqué». Dans un sens plus large, moins précis, une unité de rang inférieur au terme et à la formation, formée de 2 (ou plus) strates ou lits, et qui montre une certaine unité lithologique. *Couche, strate.*

Es la división más pequeña de una serie estratificada, limitada arriba y abajo por sendos planos de estratificación más o menos bien definidos (KEMP, 1922, p. 190). Término empleado en sentido amplio, que define a una unidad sedimentaria con rango menor que miembro o formación, compuesto por dos o más subestratos o láminas con cierto grado de uniformidad litológica (PAYNE, 1942, p. 1724). Véase McKEE and WEIR, 1953, Table 2. *Estrato, Capa.*

Bedding:

Denoting the presence of beds, which see. For description and classification of bedding types see ANDRÉE (1915), BIRKENMAJER (1959), and BOTVINKINA (1962).

Bezeichnet die Anwesenheit von Schichten; s. dort. Für eine Beschreibung und Klassifikation von Schichtungstypen s. ANDRÉE (1915), BIRKENMAJER (1959) und BOTVINKINA (1960). *Schichtung.*

Indique la présence de couches (cf. «Bed»). ANDRÉE (1915), BIRKENMAJER (1959) et BOTVINKINA (1960) en décrivent et classifient différents types. *Stratification, litage.*

Término que indica una sucesión de estratos, véase «bed». Para la descripción y clasificación véase ANDRÉE (1915), BIRKENMAJER (1959) y BOTVINKINA (1960). *Estratificación.*

Bioglyph (VASSOEVICH, 1953, p. 38):

A hieroglyph of biologic origin.

Eine Hieroglyphe organischen Ursprungs. *Bioglyphe.*

Hiéroglyphe d'origine biologique. *Bioglyphe, trace vivante, trace d'organisme.*

Hieroglifo de origen orgánico. *Bioglifo.*

286

Bottomset beds:

Term applied to horizontal or gently inclined beds that are deposited in front of a prograding delta.

Bezeichnung für horizontale oder leicht geneigte Schichten, die vor einem sich ausbreitenden Delta abgelagert werden. *Bodenschichten.*

Terme employé pour désigner des couches horizontales ou faiblement inclinées, qui se déposent en avant d'un delta qui avance. *Couches basales, couches de fond.*

Término aplicado a aquellas capas horizontals o ligeramente inclinadas, que se depositan al frente de un delta. *Capas basales.*

Boudinage: See plate 15 A.

Refers to disruption of once continuous layer by stretching and flowage; see "pull-apart" (McCROSSAN, 1958, p. 316).

Betrifft die Unterbrechung vormals zusammenhängender Schichten durch Dehnung und Sedifluktion; s. „Zerrungsstruktur" (McCROSSAN, 1958, S. 316). *Boudinage.*

S'applique à une couche continue à l'origine, mais fragmentée par étirement ou écoulement. *Boudinage.*

Se aplica a un estrato interrumpido en partes por tensión y sedifluxión; véase «pull apart» (McCROSSEN, 1958, p. 316). *Estructura moniliforme.*

Bounce cast (WOOD and SMITH, 1959, p. 168): See plate 68.

Casts of short grooves (up to 5 cm) widest and deepest in middle and fading out at both ends; presumably formed by objects grazing against bottom and rebounding.

Ausgüsse kurzer Rillen (bis zu 5 cm breit), die am weitesten und tiefsten in der Mitte sind und nach beiden Enden hin auslaufen; wahrscheinlich von Gegenständen hervorgerufen, die den Grund schrammen und allmählich wieder absetzen. *Aufprallmarke* (n), *Rückprallmarke, Aufstoßmarke.*

Moulages de petites cannelures (5 cm au plus), dont la largeur et la profondeur sont maxima au centre, et s'atténuent aux deux extrémités; sans doute formées par des objets qui éraflent le fond et rebondissent. *Lit.: Empreinte de rebondissement.*

Calco de surcos cortos (hasta 5 cm), anchos y profundos en su parte media, que se atenúan en sus extremos. Probablemente originados por el roce alternado de cuerpos arrastrados por corrientes, sobre sedimentos que constituyen un fondo. *Calco de roce.*

Brecciated structure:

Characterized by agglomeration of angular fragments. May be a primary sedimentary structure related to desiccation, slump, *etc.*; also produced by tectonic movement and other causes.

Gekennzeichnet durch Anhäufung eckiger Bruchstücke. Kann eine primäre Sedimentstruktur sein, die mit Austrocknung, Rutschung usw. verbunden ist; auch von tektonischen Deformationen und anderen Ursachen hervorgerufen. *Breccienstruktur* (w).

Texture caractérisée par l'agglomération de fragments angulaires. Peut être une structure primaire, due à la dessiccation, au glissement, etc., mais peut aussi résulter d'un mouvement tectonique, ou de causes diverses. *Structure en brèche, s. brèchique.*

Fragmentos angulares aglomerados. Estructura sedimentaria primaria originada por desecación, desplazamientos, movimientos tectónicos u otras causas. *Estructura brechosa.*

Brush mark (DZULYNSKI and SLACZKA, 1958, p. 231) (brush cast): See plates 68 and 69A.

Essentially a bounce cast with a crescentic depression on the down-current end. The depression is interpreted as the cast of a small ridge of mud pushed up by the impinging object.

Im wesentlichen eine Aufprallmarke mit bogenförmiger Vertiefung am stromabwärts gerichteten Ende. Die Eintiefung wird als Abdruck eines kleinen Schlammrückens aufgefaßt, der vom aufstoßenden Objekt aufgeworfen wird. *Quastenmarke* (w).

En essence, une empreinte de rebondissem ent,avec une dépression en croissant en aval. On interprète la dépression comme le moulage d'une petite crête de boue amoncelée par l'objet qui a rebondi. *Lit.: Eraflure.*

Esencialmente es un *calco de roce* ("bounce cast") constituido por depresiones en forma de media luna y alineadas, con sus margenes convexas en sentido de la corriente. Las depresiones son el calco de pequeños lomos producidos por un cuerpo que en su deriva, roza el fondo, empuja y acumula fango alternativamente. *Calco de empuje.*

Bubble impressions:

Small depressions, not marked by raised rim, formed by gas bubbles; superficially resembling raindrop impressions (TWENHOFEL, 1921, p. 359).

Kleine Vertiefungen, nicht durch höheren Randwall markiert; von Gasblasen geformt; oberflächlich sehen sie Regentropfen-Eindrücken ähnlich (TWENHOFEL, 1921, S. 359). *Gas-Trichter.*

Petites dépressions sans rebords élevés, formées par des bulles de gaz, et rappelant les empreintes de gouttes de pluie. *Cratères de bulles.*

Pequeñas depresiones sin bordes en realce, formadas por burbujas gaseosas; superficialmente se asemejan a marcas originadas por gotas de lluvia (TWENHOFEL, 1921, p. 359). *Hoyuelos de burbujeo.*

Burrow: See Plates 17 C, 115 B, 116, A and B.

Tubular openings made by worms and other animals. Usually preserved as fillings; may be vertical, horizontal, or inclined; may be straight or sinuous.

Von Würmern und anderen Tieren geschaffene röhrenförmige Öffnungen; im allgemeinen als Ausfüllung erhalten; sie können senkrecht, waagerecht oder geneigt sein; ebenso gerade oder gewunden. *Bau.*

Tubulure formée par des vers ou autres animaux. En général représentée par le matériau de remplissage (moulage); peut être verticale, horizontale, inclinée, rectiligne ou sinueuse. *Tube, trou, terrier.*

Orificios tubulares excavados por vermes y otras especies animales, generalmente rellenados y así preservados. Pueden presentarse en disposición vertical, horizontal o inclinada y pueden ser rectos o sinuosos. *Estructura tubular.*

"Cabbage leaf" marking (KUENEN, 1957, p. 255): See plate 68. See frondescent cast.

Channel:

(A) Erosional feature that may be meandering and branching and is part of an integrated transport system. See wash-out. (B) Linear erosional feature similar to but larger than a groove. Best developed in turbidite sequences. See channel cast.

(A) Eine Erosionsform, die mäandrieren und sich verzweigen kann und Teil eines integrierten Entwässerungs-Systems ist. Siehe Priel. (B) Eine geradlinige Erosionsform, ähnlich einer Rille, aber größer als diese. Am besten ausgebildet in Ablagerungen von Suspensions-Strömen. Siehe Rinnenmarke. *Erosions-, Strömungs-, Auswaschungs-Rinne* usw.

(A) Forme d'érosion, à méandres ou à plusieurs bras, faisant partie d'un réseau de transport (cf. wash-out). *Chenal.* (B) Petite forme linéaire d'érosion, comparable à un sillon (ou cannelure) mais plus importante. Particulièrement bien développées dans les séquences de turbitite (cf. channel cast). *Rigole.*

(A) Rasgo erosional meandroso y ramificado, integrante de un sistema de transporte. Véase "wash-out". (B) Rasgo lineal erosional de dimensiones superiores a surco ("groove"), con buen desarrollo en turbiditas. Véase "channel cast". *Cauce.*

Channel cast (McIver, 1961, p. 170) (gouge channel, washout, channel fill):
Cast of small groove-like channel or very large groove, 0.5 to 2 m wide and 20—50 cm deep, generally sand-filled and cut in shale.

Ausguß kleiner, rillenähnlicher Rinnen oder sehr großer Rillen, 0,5—2 m breit, und 20—50 cm lang; im allgemeinen in Schieferton eingeschnitten und mit Sand ausgefüllt. *Rinnenmarke, Kolkrinne, Priel-, Rinnenfüllung.*

Moulage d'un petit chenal (comparable à une cannelure) ou d'une très grande cannelure, large de 0,5 à 2 mètres et profond de 20—50 cm, généralement rempli de sable et recoupant un schiste. *Moulage de chenal, structure à chenaux.*

Cauce pequeño o surco muy grande en una pelita, rellenado por arena. El ancho oscila entre 0,50—2,00 m y la profundidad entre 0,20—0,50 m. *Relleno de cauce.*

Chattermarks:
Vibration marks associated with glacial striations and grooves on bedrock surface overridden by ice (Chamberlin, 1888, p. 218; Harris, 1943, p. 250).

Schwingungsmarken, die zusammen mit Kritzen und Rillen auf anstehendem Gestein vorkommen, das vom Eise überfahren ist (Chamberlin, 1888, S. 218; Harris, 1943, S. 250). *Splittermarke* (w).

Empreintes de vibration qu'on rencontre en association avec des stries glaciaires et des rainures à la face supérieure de la roche sur laquelle a passé le glacier. *Empreintes de vibration, marques de percussion, coups de gouge* (ces empreintes ressemblent aussi aux marques laissées par un morceau de craie qui tressaute au lieu de donner une empreinte régulière et continue sur le tableau).

Marcas en forma de media luna (concavidad en sentido del movimiento), asociadas con estriaciones y surcos, originadas por el desplazamiento de una masa de hielo sobre su lecho (Chamberlin, 1888, p. 218). *Marcas de rechinamiento.*

Chevron cross-bedding:
Cross-bedding that dips in different directions in superimposed beds forming a chevron pattern. Carefully described, but not named by White (1881, p. 60). Also called herringbone or zigzag cross-bedding (Shrock, 1948, Fig. 214).

Schrägschichtung, die in übereinanderliegenden Schichten in verschiedene Richtungen einfällt und so ein Grätenmuster hervorruft. Sorgfältig beschrieben, aber nicht benannt von White (1881, S. 60). Auch Gräten- oder Zickzack-Schichtung genannt (Shrock, 1948, Abb. 214). *Kreuzschichtung, Fiederschichtung.*

Stratification entrecroisée dont le pendage varie d'une couche à l'autre pour former des chevrons. Très bien décrite, mais non nommée, par White. Appelée aussi stratification en arêtes de poisson ou en zigzag. *Stratification entrecroisée en arêtes de poisson, en chevrons.*

Unidades sedimentarias superpuestas, formadas por capas frontales con pendiente en direcciones distintas. White (1881, p. 60) describió con sumo detalle esta estructura sin asignarle un término. Esta estructura es también conocida como "herringbone crossbedding" "zigzag cross-bedding" (Shrock, 1948, fig. 214). *Estratificación entrecruzada espigada.*

Chevron mark (Dunbar and Rodgers, 1957, p. 185) (chevron cast): See plates 62 and 63 A. Linear row of chevrons, presumably pointing upstream. See also vibration marks, ruffled groove casts, and herringbone marking.

Eine geradlinige Aneinanderreihung von Pfeilen, die vermutlich stromaufwärts zeigen. Siehe auch Schwingungsmarken und gefiederte Rillenmarke. *Fiedermarke* (n), *Gefiederte Schleifrille*.

Alignement de chevrons dont les sommets sont en principe dirigés vers l'amont. *Empreinte en chevrons*.

Marcas en forma de V alineadas y paralelas a la corriente; se supone que sus vértices están dispuestos en sentido opuesto a la corriente. Vésase "vibration marks", "ruffled groove cast", "herringbone pattern". *Calco espigado*.

"Choppy" cross-lamination (SHROCK, 1948, Fig. 216):
Small-scale trough cross-lamination.

Kleindimensionale, bogige Schrägschichtung. *Wirre, kleindimensionale Schrägschichtung* (w).

«Trough cross-lamination» (q.v.) à petite échelle.

Entrecruzamiento festoneado de pequeñas dimensiones. *Micro-entrecruzamiento festoneado*.

Clastic dikes: See plates 113 A and 113 B.
Dikes resulting from filling of a crevice or fissure. See sandstone dikes.

Durch Füllung von Spalten entstandene Gänge. Siehe Sandsteingänge. *Sedimentgänge*.

Dikes formés par le remplissage d'une fissure. *Filons clastiques (autocicatrisation)*.

Diques formados por relleno de hendiduras. Véase "sandstone dike". *Diques clásticos*.

Clay balls: See mud balls, also armored mud balls.

Clay galls:
Mud curls or cylinders formed by drying and cracking of thin layers of coherent mud; commonly rolled or blown into sand and buried; flattened upon wetting forming a lenticular bleb of clay or shale (BURT, 1930, p. 105).

Gerollte, oft zylindrische Tonhäute, die durch Austrocknen und Schrumpfen dünner Lagen kohärenten Schlammes entstehen; meist in Sandablagerungen verfrachtet und dort eingebettet; in feuchtem Zustand zusammengedrückt und als linsenförmige Körper überliefert (BURT, 1930, S. 105). *Tongallen, eingerollte Schlickgerölle, Tonrollen*.

Pellicules d'argile enroulées en boucles ou en cylindres et formées par dessèchement et fissuration de couches minces d'argile cohésive, le plus souvent roulées ou poussées par le vent et enfouies dans le sable; elles se déroulent sous l'influence de l'eau et forment des lentilles argileuses. *Boule d'argile, balle argileuse, galet mou*.

Láminas delgadas de fango coherente que por desecación y agrietamiento adquieren formas abarquilladas, comunmente arrastradas e incorporadas a sedimentos arenosos; en presencia de humedad ceden, pasando a formas planas vesiculosas. *Vesículas arcillosas*.

Climbing ripples: See plate 39.
Cross-lamination produced by superimposed migrating ripples.

Schrägschichtung, entstanden durch wandernde, sich überlagernde Rippeln. *Kletternde Rippeln* (w).

Stratification entrecroisée formée par la superposition de rides en déplacement. *Lit.: Rides chevauchantes*.

Laminación entrecruzada originada por migración y superposición de óndulas. *Entrecruzamiento de óndulas sobrepuestas*.

Columnar structure:

Columns, 9 to 14 cm in diameter and 1 to 1.4 m in length, found in some calcareous shales or argillaceous limestones; oval to polygonal in section. Columns are perpendicular to bedding. Possibly a desiccation structure. (HARDY and WILLIAMS, 1959, p. 281.)

In gewissen kalkigen Schiefertonen und tonigen Kalksteinen auftretende Säulen, 9 bis 14 cm im Durchmesser und 1 bis 1,4 m lang, oval bis polygonal in der Aufsicht. Die Säulen sind senkrecht zur Schichtung. Möglicherweise eine Austrocknungsstruktur. (HARDY and WILLIAMS, 1959, S. 281.) *Säulenstruktur* (w).

Colonnes de 9 à 14 cm de diamètre et 1 à 1,40 m de longueur, de section ovale ou polygonale, que l'on rencontre dans certains schistes calcaires ou calcaires argileux. Les colonnes sont perpendiculaires à la stratification, et représentent peut-être un accident de dessiccation. *Structure prismée.*

Columnas de sección oval o poligonal, normales a la estratificación, halladas en ciertas lutitas calcáreas o calizas arcillosas. Sus diámetros oscilan entre 0,9—0,14 m y las longitudes entre 1—1,4 m. Estructura probablemente originada por desecación. *Estructura columnar.*

Compound foreset bedding (GILBERT, 1899, Fig. 4):

A cross-bedded unit with tangential foresets and a concave base.

Eine schräggeschichtete Einheit mit tangentialen Leeblättern und einer konkaven Sohlfläche. *Zusammengesetzte Leeblatt-Schichtung* (w).

Ensemble à stratification croisée, avec couches frontales tangentielles sur une base concave.

Unidad con estratificación entrecruzada, formada por capas frontales tangentes a una base cóncava. *Estratificación entrecruzada compuesta.*

Compound ripples (BUCHER, 1919, p. 195):

Type of ripple marks resulting from "simultaneous interference of wave-oscillation with current action."

Eine Art Rippelmarken, die „bei gleichzeitiger Interferenz von Wellen-Oszillation und Strömung" entstehen. *Zusammengesetzte Rippelmarken* (w).

Type de rides qui résultent de «l'interférence simultanée d'oscillations de vagues et d'effets de courants». *Rides complexes, rides d'interférence.*

Es la resultante de la interferencia simultánea de una óndula ácuea y otra ácuea-eólica. *Ondula compuesta.*

Compound ripple marks:

Two or more sets of ripple mark, one superimposed on the other.

Zwei oder mehr sich überlagernde Rippelserien. *Zusammengesetzte Rippelmarken* (w).

Deux ou plusieurs groupes de rides superposés. *Rides complexes, rides superposées.*

Dos o más juegos de óndulas superpuestas. *Ondulas conjugadas.*

Concave cross-bedding (ANDERSEN, 1931, Fig. 38):

Cross-bedding deposited on a lower concave surface. Has also been used to describe cross-bedding with tangential or concave foresets.

Leeblätter auf einer konkaven Sohlfläche angelagert. Auch verwendet, um Schrägschichtung mit tangentialen oder konkaven Leeblättern zu beschreiben. *Bogige Schrägschichtung.*

Stratification entrecroisée formée sur une surface concave. Terme utilisé aussi pour décrire une stratification croisée à couches frontales tangentielles ou concaves. *Stratification entrecroisée arquée, incurvée, ou convexe vers le bas.*

Estratificación entrecruzada depositada sobre una superficie cóncava. También se ha empleado este término para definir a aquella estratificación entrecruzada, con capas frontales cóncavas o tangentes a la base. *Estratificación entrecruzada cóncava.*

Concave inclined-bedding (ANDERSEN, 1931, Fig. 38):
Cross-bedding with concave, generally tangential, foresets.

Schrägschichtung mit konkaven, im allgemeinen tangentialen Leeblättern. *Bogige Schrägschichtung.*

Stratification oblique à couches frontales concaves, généralement tangentielles. *Stratification oblique incurvée, ou arquée.*

Estratificación entrecruzada formada por capas frontales cóncavas, generalmente tangentes al plano principal de depositación. *Estratificación inclinada cóncava.*

Conical flute casts (RÜCKLIN, 1938, p. 97, Fig. 1) (Zapfenwülste): See plate 55 A.
Plain, conical flute casts.

Ebene, konische Strömungswülste. *Zapfenwülste.*

Empreintes de sillons d'érosion simples, coniques.

Turboglifos simples, cónicos. *Turboglifos cónicos.*

Convex incline-bedding (ANDERSEN, 1931, Fig. 38):
Cross-bedding with convex (upwards) foresets.

Schrägschichtung mit nach oben konvexen Leeblättern. *Konvexe Schrägschichtung* (w).

Stratification oblique à couches frontales convexes vers le haut.

Estratificación entrecruzada formada por capas frontales convexas (hacia arriba). *Estratificación inclinada convexa.*

Convolute bedding (KUENEN, 1952, p. 31; 1953, p. 1056) (convolute lamination, curly bedding, slip bedding, etc.): See plates 106A and 106B.

Wavy or contorted laminations that die out both upward and downward within a given sedimentation unit.

Wellige oder verfältelte Feinschichtung, die innerhalb einer gegebenen Sedimentations-Einheit nach oben und unten zu ausklingt. *Wulstschichtung* (n), *endostratische Sedifluktion, Wulstbank, Wulstung, (bankinnere — schichtinterne — synsedimentäre) Einwälz-, Fließ-, Gleit-, Rutsch-, Stauch-, Wickel-, Wulst-Fältelung, -Faltung, -Struktur, -Textur; Fältelungsrutschung, subaquatische Rutschung, Verfältelung, Wickelung usw.*

Se dit d'une structure ondulée ou contournée qui s'atténue et disparaît vers le haut et le bas, au sein d'un ensemble sédimentaire donné. *Stratification ondulée; laminae en volutes, convolutions.*

Unidad sedimentaria con laminación deformada, limitada por estratos normales (no deformados). *Estratificación intraplegada.*

Convolute current-ripple lamination (BOUMA, 1962, p. 137): See plate 38B.
Deformed current-ripple cross-lamination.

Verformte, von Strömungsrippeln erzeugte Schrägschichtung. *Verfältelte Strömungsrippel-Schichtung* (w).

Stratification croisée fine, due à la migration de rides de courant, et déformée.

Laminación entrecruzada formada por migración de óndulas, sometida a deformaciones. *Migro-óndulas entrecruzadas, deformadas.*

Convolute lamination (TEN HAAF, 1956, p. 190):
See convolute bedding.

292

Convolutional balls (DOTT and HOWARD, 1962, pl. 3 A, p. 118):

A comparatively small, concentric "ball" formed in association with convolute bedding. Also called "roll-up structure."

Eine vergleichsweise kleine, konzentrische „Kugel", die sich im Zusammenhang mit Wulst schichtung formt. Auch „*Aufrollstruktur*" genannt. *Wickelstruktur.*

Une «boule» relativement petite, à couches concentriques, associée à une stratification ondulée.

Cuerpos subesféricos concéntricos y relativamente pequeños; asociados con "convolute bedding". También llamada "roll-up structure". *Esferoides intraformacionales.*

Coprolite:

Petrified excrement.

Versteinertes Exkrement. *Koprolith.*

Excrément fossile. *Coprolithe.*

Excremento fósil. *Coprolito.*

Corkscrew flute casts (RÜCKLIN, 1938, p. 97, Fig. 2)(Korkzieherzapfen): See plate 60B.

Flute cast with cork-screw form; with "twisted" beak.

Strömungswülste mit korkenzieherähnlichem, gedrehtem Oberende. *Korkzieher-Zapfen.*

Empreinte en tire-bouchon, à bec tordu. *Empreinte de sillon d'érosion en tire-bouchon.*

Turboglifos en forma de sacacorcho; sus extremos agudos se asemejan a espolones retorcidos. *Turboglifos en sacacorcho.*

Corrosion surface (corrosion zone):

Blackened, pitted, irregular bedding surface found in some limestones; attributed to submarine solution or resorption (SARDESON, 1914; WEISS, 1954).

Geschwärzte, narbige, unregelmäßige Schicht-Oberfläche einiger Kalksteine; wird als submarine Lösungserscheinung oder Resorption aufgefaßt (SARDESON, 1914; WEISS, 1954). *Korrosionsfläche.*

Surface de stratification noircie, irrégulière, à perforations, que l'on rencontre dans certains calcaires, et que l'on attribue à un effet de dissolution et de résorption sous-marines. *Surface durcie, surface limite.*

Superficie irregular de estratificación, ennegrecida y hoyosa, característica en ciertas calizas. Estructura atribuida a solución o resorción submarina. *Superficie de corrosión.*

Creep wrinkles (McIVER, 1961, p. 227) (crinkle marks, pseudo-ripples):

Small microfolds or corrugations on the bedding plane, perpendicular to the direction of movement (slumping or creep).

Runzelung oder Fältelung der Schichtfläche, senkrecht zur Bewegungsrichtung (Rutschung oder Kriechen). *Runzelmarken* (n), *Knittermarken.*

Plissotements, microplis ou petites corrugations sur la surface de stratification, perpendiculaires à la direction du mouvement (glissement ou reptation). *Fronces transversales.*

Pequeños micropliegues o corrugaciones en el plano de estratificación, perpendiculares a la dirección del movimiento (deslizamiento o reptación). *Pseudo-óndula.*

Crescent cast (PEABODY, 1947): See current crescent (cast).

Crescent-type cross-bedding (MOBERLY, 1960, pl. 3, Fig. 1):

Same as trough cross-stratification.

Crescentic fracture:

A curved fracture made by glacial ice that is convex upcurrent (HARRIS, 1943, p. 245).

Von glazialem Eis geschaffener, gekrümmter Bruch, der nach stromaufwärts konvex ist (HARRIS, 1943, S. 245). *Bogiger Bruch* (w).

Fracture arquée, convexe vers l'amont, formée sur la roche de fond par un glacier.

Fractura curva originada por un glaciar durante su desplazamiento sobre una base rocosa; la concavidad está orientada en sentido opuesto al movimiento. *Fractura falciforme.*

Crescentic gouge: See gouge mark.

Crescentic scour mark (RÜCKLIN, 1938, p. 97) (Hufeisenwülste): See current crescent.

Crinkle marks (WILLIAM and PRENTICE, 1957, p. 289, pl. 6):

A series of sub-parallel corrugations of the bedding surface related to very small-scale crumpled internal laminations ascribed to subaqueous solifluction. See also creep wrinkles.

Eine Reihe subparalleler Runzeln der Schichtfläche, die mit kleindimensionaler Verfältelung der Feinschichtung zusammenhängen; im allgemeinen als Folge subaquatischer Sedifluktion gedeutet. Siehe auch Runzelmarken. *Knittermarken* (w).

Séries de corrugations presque parallèles sur la surface de stratification, correspondant à de très fines laminations internes plissotées, et que l'on attribue à la subsolifluxion. *Empreintes froncées, fronces.*

Serie de micropliegues subparalelos en los planos de estratificación, vinculados con la corrugación interna de las láminas que componen un estrato. Su desarrollo es atribuido a solifluxión subácuea. *Pseudo-óndulas.*

Crinkled bedding:

Bedding or laminations displaying minute wrinkles; in carbonate rocks crinkled bedding is believed related to algal mats. Term also used by MIGLIORINI (1950) for convolute bedding.

Schichtung oder Feinschichtung, die kleinformatige Runzeln zeigen; Runzelschichtung in Karbonatgesteinen ist vermutlich Algenrasen verwandt. Von MIGLIORINI (1950) auch im Sinne von Wulstschichtung gebraucht. *Runzelschichtung* (w).

Stratification ou laminations montrant des fronces ou rides très fines; dans les roches carbonatées, on pense que ce phénomène est lié à des feutrages d'algues. Terme utilisé aussi par MIGLIORINI (1950) pour désigner des convolitions («convolute bedding»).

Estratificación o laminación con desarrollo de micropliegues, en rocas carbonáticas; se presume que esta estructura es originada por la actividad biológica de algas. Término empleado por MIGLIORINI (1950) como sinónimo de "convolute bedding". *Estratificación rizada.*

Cross-bedding: See plates 17 A and 17 B, 29 B to 39 and 110.

Loosely used for any inclined bedding but has been restricted to that with foresets thicker than 1 cm by MCKEE and WEIR (1953, Table 2). Defined by POTTER and PETTIJOHN (1963, p. 68) as a structure, confined to a single sedimentation unit (OTTO, 1938, p. 575), consisting of successive systematic, internal bedding, called foreset bedding, inclined to the principal surface of accumulation.

Lose auf jede geneigte Schichtung angewandt; von MCKEE und WEIR jedoch auf Leeblätter beschränkt, die dicker als 1 cm sind (1953, Tabelle 2). Von POTTER und PETTIJOHN definiert als (1963, S. 68) eine Struktur, die auf eine einzelne Sedimentations-Einheit beschränkt ist (OTTO, 1938, S. 575), bestehend aus einer systematischen Aufeinanderfolge innerer Schichten, Leeblatt-Schichtung genannt, die im Winkel zur Hauptanlagerungsfläche steht. *Schrägschichtung, Kreuzschichtung, Winkelschichtung.*

Terme peu précis qui désigne n'importe quel type de stratification inclinée, mais réservé par McKee et Weir à un type où les couches frontales ont plus d'un cm d'épaisseur. Potter et Pettijohn en donnent cette définition: structure confinée à un seul ensemble sédimentaire (Otto), et formée par une stratification interne en superposition normale et systématique (couches frontales) et inclinée par rapport à la surface principale de dépôt. *Stratification oblique, croisée.*

Término generalmente empleado para todo tipo de estratificación inclinada, pero restringido por McKee y Weir (1953, Table 2) a aquella formada por capas frontales mayores de 1 cm de espesor. Definida por Potter y Pettijohn (1963, p. 68) como una unidad sedimentaria (Otto, 1938, p. 575) con estratificación interna (capas frontales) inclinada con respecto a la superficie principal de acumulación. *Estratificación entrecruzada.*

Cross-grooves (Dzulynski and Slaczka, 1958, p. 232): See plate 61.

Two or more intersecting sets of groove casts.

Zwei oder mehr sich überschneidende Reihen von Rillenmarken. *Kreuzrillen* (w).

Groupes de cannelures qui s'entrecroisent (deux ou plus). *Lit.: Cannelures entrecroisées.*

Dos o más juegos de surcos (calcos) que se intersectan. *Surcos intersectos.*

Cross-lamination:

Generally used as a synonym for cross-bedding but restricted by McKee and Weir (1953, Table 2) to cross-stratification with *foresets* less than 1 cm thick.

Im allgemeinen ein Synonym für Schrägschichtung; indes von McKee und Weir (1953, Tabelle 2) auf Schrägschichtung beschränkt, deren Leeblätter 1 cm Dicke nicht übersteigen. *Schräg-Feinschichtung* (w), *feinblättrige Schrägschichtung, Schrägschichtung.*

En général, synonyme de cross-bedding, mais McKee et Weir réservent ce terme pour le cas d'une stratification oblique où les couches frontales ont moins d'un cm d'épaisseur. *Stratification oblique fine.*

Término generalmente empleado como sinónimo de estratificación entrecruzada, pero restingido por McKee y Weir (1953, Table 2) a aquélla formada por capas frontales menores de 1 cm de espesor. *Laminación entrecruzada.*

Cross-ripples (Bucher, 1919, p. 190):

A type of interference ripples.

Eine Art Interferenzrippeln.

Un type de rides interférentes.

Un tipo de óndulas de interferencia.

Cross-stratification:

A general form of cross-bedding.

Eine allgemeine Bezeichnung für Schrägschichtung.

Type général de cross-bedding.

Tipo general de estratificación entrecruzada.

"Crumpled ball" (Kuenen, 1948b, p. 371, pl. 22, Fig. 1):

Highly irregular, crumpled-up masses of laminated sandstone, 5 to 25 cm across, flattened parallel to the bedding as opposed to "slump balls" which have smooth surfaces.

Außerordentlich unregelmäßige, „zerknüllte" Massen feinschichtigen Sandsteins, 5 bis 25 cm im Durchmesser und parallel zur Schichtung abgeflacht; im Gegensatz zu „Rutschungsballen" mit glatter Oberfläche. *Verknäuelter Rutschungsballen* (w), *Verknäuelung.*

Paquets de grès laminé à replis et contours extrêmement irréguliers, de 5 à 25 cm de large, aplatis dans le sens de la stratification. Par contraste, les «slump balls» ont une surface lisse. *Pseudo-nodules irréguliers (lit.: chiffonnés, froissés).*

Masas redondeadas con superficies muy irregulares de arenisca laminada; sus diámetros mayores (0.05 a 0,25 m) se disponen paralelamente a la estratificación. Término empleado para distinguir esta forma de la correspondiente a "slump ball". *Esferoides corrugados.*

Crumpled mud-crack casts (BRADLEY, 1930, p. 136):
Sand fillings (of mud cracks) that display ptygmatic deformation or crumpling produced by adjustment of fillings to compaction of enclosing mud matrix.

Netzleisten aus Sand, die als Ausgleich bei Verfestigung der umgebenden Tonmatrix „ptygmatisch" deformiert oder sonstwie verfältelt werden. *Gefältelte Netzleisten* (w).

Le remplissage sableux des fissures d'une argile montre une déformation ou un plissotement ptygmatiques résultant de l'adaptation du remplissage à la compaction de l'argile encaissante. *Lit.: Moulages de fissures chiffonnés, froissés.*

Grietas de desecación con relleno arenoso, deformadas por ulterior compactación del material pelítico. *Calco de grietas de desecación, deformado.*

Crystal casts (crystal imprints): See plate 98A.
Fillings of cavities left by solution or sublimation of crystals embedded in fine-grained sediment. See salt-crystal casts; also ice-crystal casts (SHROCK, 1948, p. 146).

Ausgüsse von Vertiefungen, die bei der Lösung oder Sublimation von in feinkörnigem Sediment eingebetteten Kristallen entstehen. Siehe Steinsalz-Pseudomorphosen und Eiskristall-Abdrücke (SHROCK, 1948, S. 146). *Kristall-Pseudomorphosen, -Ausgüsse.*

Remplissage des cavités laissées par la dissolution ou la sublimation de cristaux inclus dans un sédiment à grain fin. *Moulages de cristaux.*

Calco de cavidades dejadas por solución o sublimación de cristales en sedimentos de grano fino. Véase "salt crystal casts", "ice crystal casts" (SHROCK, 1948, p. 146). *Calco de cristales.*

Ctenoid cast (BEASLEY, 1914, p. 35; CUMMINS, 1958, p. 36, pl. 2):
Cast with form of an obliquely-cut, longitudinally ribbed cylinder. Probably bounce casts made by equisetiform plant stems. Very rare.

Marken, die von einem schräg-geschnittenen, längs-gerippten zylindrischen Körper hervorgerufen werden. Wahrscheinlich Aufprallmarken von Schachtelhalmen. Sehr selten. *Ctenoiden-Marke* (w), *kammförmige Aufstoßmarke.*

Moulage en forme de cylindre à cannelures longitudinales, coupé obliquement. Probablement formé par le rebondissement de tiges de prêles entraînées par le courant. Très rare. *Empreinte cténoïde.*

Calcos en forma de cilindros cortos seccionados oblícuamente, con finas costillas longitudinales. Probablemente similar a "bounce cast", originados al parecer por la impresión de tallos equisetiformes. *Calco ctenoide.*

Curly bedding (FEARNSIDES, 1910, p. 150) (curled bedding): See convolute bedding.

Current bedding:
A synonym for cross-bedding used principally in Great Britain.
Synonym für Schrägschichtung, vor allem in Großbritannien gebraucht.
Synonyme (Angleterre) de cross-bedding.
Término sinónimo de "cross-bedding" *(Estratificación entrecruzada)*, empleado principalmente en Inglaterra.

296

Current crescent (PEABODY, 1947) (crescentic scour mark, crescent casts): See plates 91 B and 92 B to 93 B.

Small crescentic rounded ridge, commonly with pit in center; crescent convex up-current. A cast of a horse-shoe shaped moat eroded on up-current side of an obstacle such as a pebble, shell, etc.

Kleine, rundliche, halbmondförmige Rücken, gemeinhin mit einer Vertiefung in der Mitte. Der Kamm ist konvex nach stromaufwärts. Ausfüllung eines hufeisenförmigen Grabens, der an der nach stromaufwärts gerichteten Seite eines Gerölls, einer Muschelschale oder eines anderen Hindernisses ausgekolkt ist. *Strömungskamm, Luvgraben, Auskolkung, Kolkmarke, Hufeisen-Wulst.*

Petit tas arrondi, en croissant, avec souvent un creux au milieu, le croissant convexe vers l'amont. Moulage d'un creux en fer-à-cheval, creusé à l'amont d'un obstacle tel qu'un caillou, un coquillage, etc. *Empreinte de cupule en croissant.*

Pequeño lomo redondeado en forma de media luna, con una depresión central; la parte convexa se encuentra orientada en sentido opuesto a la corriente. Calco de marca con aspecto de herradura, formada en la zona de un obstáculo (rodado, conchilla, etc.), opuesta a la corriente. *Calco en herradura.*

Current lineation (STOKES, 1937, p. 52): See parting lineation.

Current mark (TWENHOFEL, 1932, Fig. 89):
"... a term which may be applied to irregular structures on bottoms showing current-erosion effects" (TWENHOFEL, 1932, p. 668); a term earlier used by KINDLE (1917a, p. 36, pl. 17) for what are now called cuspate ripples.

„ ... eine Bezeichnung, die man auf unregelmäßige Strukturen anwenden kann, welche sich auf Flächen finden, die strömungsbedingte Erosionseffekte aufweisen" (TWENHOFEL, 1932, S. 668); früher von KINDLE (1917a, S. 36, Tafel 17) gebraucht für eine Struktur, die man jetzt bogenförmige Rippeln nennt. *Strömungsmarke, Fließmarke.*

«Terme qui peut s'appliquer aux structures irrégulières formées sur des fonds qui montrent un effet d'érosion par les courants.» D'abord utilisé par KINDLE pour désigner ce qu'on appelle maintenant rides de plages en croissant.

Término empleado para formas irregulares debidas a erosión por corrientes (TWENHOFEL, 1932, p. 668); acuñado con anterioridad por KINDLE (1917a, p. 36, pl. 17) y reemplazado en la actualidad por "cuspate ripples". *Ondula linguoide.*

Current ripple mark:
A common synonym for transverse asymmetrical ripple mark.

Häufig benutztes Synonym für unsymmetrische Transversal-Rippeln. *Strömungsrippel, Fließrippel.*

Synonyme fréquent de «transverse (asymmetrical) ripple mark». *Rides de courant.*

Sinónimo de "transverse asymmetrical ripple mark". *Ondula transversa.*

Curved ripple marks:
Ripple marks with crests which appear curved or crescentic in plan view.

Rippelmarken, deren Kämme in der Aufsicht gekrümmt oder bogenförmig erscheinen. *Gekrümmte Rippelmarken (w).*

Rides de plages dont les crêtes, vues de dessus, paraissent arquées ou en croissant. *Rides de plages arquées, en croissant.*

Ondula formada por crestas de diseño curvo en planta. *Ondula curva.*

Cuspate ripple mark: See plate 84 B.

Asymmetric current ripple marks with a somewhat barchan-like shape, the horns pointing into the current. Called "current mark" by KINDLE (1917a, p. 36), linguoid ripples by BUCHER (1919, p. 164), and "cusp ripples" by McKEE (1954, p. 60).

Unsymmetrische Strömungsrippeln von barchan-ähnlicher Form, die Spitzen stromabwärts weisend. Von KINDLE (1917a, S. 36) „Strömungsmarken", von BUCHER (1919, S. 164) zungenförmige Rippeln und von McKEE (1954, S. 60) halbmondförmige Rippeln genannt. *Sichelförmige Rippelmarke* (w).

Rides de courant asymétriques, de forme plus ou moins comparable à celle des barkhanes, les cornes tournées vers l'amont. Synonyme de current mark (KINDLE), linguoid ripples (BUCHER), cusp ripples (McKEE). *Rides arquées en croissant.*

Ondula asimétrica con formas que se asemejan a barjanes, cuyos extremos están orientados en sentido contrario a la corriente. También llamada "current ripple mark" por KINDLE (1917a, p. 36), "linguoid ripple marks" *(Ondula linguoide)* por BUCHER (1919, p. 164) y "cusp ripples" por McKEE (1954, p. 60).

Cut-and-fill:

Small erosional channels subsequently filled. Same as scour-and-fill, wash-out, *etc.*

Kleine, ausgefüllte Erosionsrinnen. Gleichbedeutend mit Kolkrinne, Priel usw.

Petits chenaux d'érosion, remplis ultérieurement. Synonyme de scour-and-fill, wash-out, etc. *Wash-out, chenaux d'érosion.*

Pequeños cauces de erosión rellenados. Semejante a "scour-and-fill", "wash-out", etc. *Relleno de cauce.*

Cylindrical structures (sandstone pipes): See plate 115 A.

Vertical structures in sandstones, a few centimeters to several decimeters in diameter and several or more decimeters in length, with structureless interiors, attributed to rising water columns or spring channels (HAWLEY and HART, 1934, GABELMAN, 1955).

Senkrechte, wenige Zentimeter bis etliche Dezimeter breite und einige bis viele Dezimeter lange Strukturen. Ohne inneres Gefüge. Steigenden Wassersäulen oder Quellkanälen zugeschrieben (HAWLEY & HART, 1934, GABELMAN, 1955). *Zylindrische Strukturen* (w).

Formations verticales dans des grès, de quelques cm à quelques dm de diamètre et de plusieurs dm de long, sans structure interne, et que l'on attribue à la montée d'une colonne d'eau. *Filons clastiques, structures cylindriques.*

Estructura cilíndrica vertical sin estructura interna, en areniscas, con diámetros desde escasos centímetros a varios decímetros y alturas que alcanzan a algunos decímetros; atribuida a columnas ascendentes de agua (HAWLEY and HART, 1934; GABELMAN, 1955). *Estructura cilíndrica.*

Décollement structure: See plates 107 B and 108 A.

A term borrowed from structural geology and applied to folded strata that have slid over underlying, generally undisturbed beds.

Bezeichnung von der Tektonik übernommen und auf synsedimentär gefaltete Schichten angewandt, die über unterliegende, im allgemeinen ungestörte Schichten geglitten sind. *Abscherungs-Struktur.*

Terme employé en géologie structurale pour désigner des couches plissées qui ont glissé sur les couches sous-jacentes, généralement intactes. *Figures de décollement.*

Término geológico estructural empleado para aquella estructura formada por estratos plegados y desplazados sobre otros, generalmente normales (no deformados). *Estructura de despegue.*

Deformed cross-bedding: See plate 110.

Cross-bedding with foresets overturned or buckled in the downcurrent direction usually prior to deposition of the overlying bed. Foreset dip angle may also be altered by subsequent tectonic folding.

Schrägschichtung mit Leeblättern, die gewöhnlich vor Ablagerung der hangenden Schicht überkippt oder nach stromabwärts gebogen sind. Der Einfallswinkel von Leeblättern kann auch von späterer tektonischer Verformung beeinflußt werden. *Verformte Schrägschichtung* (w).

Stratification croisée dans laquelle les couches frontales sont retournées ou pliées vers l'aval, le plus souvent avant le dépôt de la couche sus-jacente. L'angle d'inclinaison des couches frontales peut aussi avoir été changé par plissement tectonique ultérieur. *Stratification oblique dérangée.*

Estratificación entrecruzada con capas frontales volcadas o pandeadas en el sentido de la corriente, a priori a la depositación de la capa suprayacente. La pendiente de las capas frontales puede ser modificada por plegamiento tectónico subsiguiente. *Estratificación entrecruzada deformada.*

Delicate flute casts (KUENEN, 1957, Fig. 9): See furrow flute cast.

Delta bedding:

Refers to inclined bedding presumed to originate as foresets of small deltas.

Geneigte Schichtung, die durch Ablagerung von Leeblättern kleiner Deltas entstehen soll. *Deltaschichtung* (w).

Stratification oblique que l'on suppose formée par les couches frontales de petits deltas. *Stratification deltaïque.*

Estratificación inclinada, presumiblemente originada por capas frontales de pequeños deltas. *Estratificación deltaica.*

Deltoidal cast (BIRKENMAJER, 1958, Figs. 8 and 9, pl. 21, p. 143): See frondescent cast.

Depressed flute casts (RÜCKLIN, 1938, p. 97, Fig. 3) (Flachzapfen):

Depressed, flat, or weakly-developed flute casts.

Flache oder schwach entwickelte Strömungswülste. *Flachzapfen.*

Empreintes de sillons d'érosion aplaties ou peu marquées.

Turboglifos de escaso relieve o débilmente desarrollados. *Turboglifos platiformes.*

Desiccation cracks (desiccation fissures):

Cracks in sediment produced by drying; see mud crack.

In trocknendem Sediment entstandene Risse; s. Trockenriß. *Schrumpfungsrisse, Trockenrisse.*

Fissures produites dans un sédiment par dessiccation. *Fentes de retrait, craquelures de dessiccation.*

Grietas por desecación de sedimentos; véase "mud cracks". *Grietas de desecación.*

Desiccation mark:

Synonym of mud crack, which see.

Synonym für Trockenriß.

Synonyme de mud crack, q.v.

Sinónimo de "mud crack", véase. *Grietas de desecación.*

Diaglyph (VASSOEVICH, 1953, p. 33):

A marking or hieroglyph formed during diagenesis.

Marken oder Hieroglyphen, die sich während der Diagenese formen. *Diaglyphe.*

Empreinte ou hiéroglyphe formé pendant la diagénèse. *Diaglyphe.*

Marca o hieroglifo formado durante la diagénesis. *Diaglifo.*

Diagonal bedding:

Bedding diagonal to the principal surface of accumulation. An obsolete synonym for inclined bedding.

Schichtung diagonal zur Haupt-Anlagerungsfläche. Eine veraltete Bezeichnung für Schrägschichtung. *Diagonalschichtung.*

Stratification en diagonale par rapport à la surface principale de dépôt. Synonyme peu usité de «inclined bedding» *(stratification inclinée). Stratification en diagonale.*

Estratificación oblícua a la superficie principal de acumulación. Término en desuso sinónimo de "incline bedding". *(Estratificación inclinada.)*

Diagonal lamination (HALL, 1843, p. 286—287):

Used as a synonym for cross-bedding.

Synonym für Schrägschichtung.

Synonyme de cross-bedding.

Término empleado como sinónimo de "cross-bedding". *(Estratificación entrecruzada.)*

Diagonal scour marks (DZULYNSKI and SANDERS, 1963, p. 68):

Scour marks formed by concentration of smaller scour marks, generally longitudinal flutes, into distinct rows which alternate with areas where scour marks are less abundant or are absent. Arranged in rows diagonal to the main direction of flow (KUENEN, 1957, Fig. 4).

Kolkmarken, die sich durch Aneinanderreihen kleinerer Kolkmarken, im allgemeinen länglicher Mulden, formen, die mit Flächen abwechseln, auf denen Kolkmarken selten oder abwesend sind. Die Reihen verlaufen diagonal zur Hauptströmungs-Richtung (KUENEN, 1957, Abb. 4). *Diagonale Kolkmarken* (w).

Empreintes d'affouillement formées d'empreintes plus petites, souvent des sillons longitudinaux, concentrées en rangs séparés par des régions où les empreintes sont moins abondantes, ou même absentes. Les rangs sont en diagonale par rapport à la direction principale d'écoulement. *Lit.: empreintes d'affouillement en diagonale.*

Marcas de pequeños cauces formadas por concentración de similares más pequeñas, compuestas de acanaladuras longitudinales y dispuestas en hileras de diferentes direcciones, las que alternan con áreas donde aquéllas son menos abundantes o faltan. La orientación general es diagonal a la dirección principal de la corriente (KUENEN, 1957, Fig. 4). *Marcas de desbaste diagonales.*

Dimpled current mark (JUKES, 1872, p. 163—164, cited in KINDLE, 1917a, p. 37):

An obsolete name used for interference ripple mark.

Ein veralteter Name für Interferenz-Rippeln.

Terme peu usité pour désigner les «interference ripple marks». *Rides en fossettes.*

Sinónimo en desuso de "interference ripple mark". *(Ondula de interferencia.)*

"Dinosaur leather" (CHADWICK, 1948, p. 1315): See plate 59A.

A term locally applied to complex sole markings, probably including both flute and load casts.

Eine lokale Bezeichnung für verwickelte Schichtenflächen-Marken, die vermutlich sowohl Strömungswülste wie Belastungsmarken einschließt.

Terme d'application locale qui désigne des empreintes complexes, comportant des figures de fluxion et de charge. *Lit.: Peau de dinosaure.*

Término localmente empleado para marcas de base complejas, incluyendo probablemente a turboglifos y calcos de carga. *"Cuero de dinosaurio".*

Directional load cast (POTTER and GLASS, 1958, p. 20):

Originally interpreted as flowage cast but same as flute casts, which see.

Synonym mit Strömungswülsten; ursprünglich als Fließmarken interpretiert. *Gerichtete Belastungsmarke* (w).

A l'origine, interprété comme s'appliquant à des figures d'écoulement, mais en fait synonyme de flute casts, q.v.

Estructura interpretada originariamente como "flowage cast"; similar a "flute cast", *(Turboglifo)* véase.

Downslope ripple (McKEE, 1939, p. 72):

A ripple that migrates down a sloping surface.

Rippeln, die hangabwärts wandern.

Une ride qui se déplace en descendant une pente.

Ondula que se desplaza sobre una superficie inclinada en sentido de la pendiente. *Ondula descendente.*

Drag mark (KUENEN, 1957, p. 243; KUENEN and SANDERS, 1956, pl. 38): See plates 62, 63 A and B, 66 and 68.

Cast of long, narrow, even grooves, varying from near microscopic to several centimeters in width and depth. Presumed to be formed by stone or shell pulled along the mud bottom by attached algae. Term proposed by KUENEN to designate grooves formed by dragging objects and to exclude grooves formed by sliding objects (slide marks). See also groove casts.

Ausgüsse dünner, glatter, langgestreckter Rillen, die von nahezu mikroskopischen Dimensionen bis zu mehreren Zentimeter in Weite und Tiefe variieren. Wahrscheinlich durch Steine oder Schalen verursacht, die, an Algen geheftet, über den Meeresboden geschleift werden. Nach dem Vorschlag KUENENs gilt Rillenmarke als Oberbegriff, der Schleifmarke und Gleitmarke einschließt. Siehe auch Rillenmarke. *Schleifmarke, Driftmarke.*

Moulage de cannelures longues, étroites et régulières, microscopiques ou allant jusqu'à quelques centimètres de largeur et de profondeur. On les croit formées par des cailloux ou coquilles traînés sur un fond boueux par des algues qui leur sont attachées. Ce terme a été proposé par KUENEN pour séparer les cannelures formées par des objets traînés, des empreintes de glissement. *Empreinte d'entraînement, de traînage.*

Término propuesto por KUENEN para diferenciar los surcos originados por arrastre de cuerpos y excluir aquéllos producidos por deslizamiento ("slide marks"). Se refiere a calcos uniformes, largos y angostos, cuyas dimensiones en ancho y profundidad, varían desde casi microscópicas a algunos centímetros. Probablemente originadas por rodados, fragmentos rocosos o conchillas arrastrados sobre fondos fangosos, aprisionados por algas. *Calco de arrastre.*

Drag striae (DZULYNSKI and SLACZKA, 1985, p. 234) (striation casts):

Essentially a microgroove cast and presumed to form in same manner as the large groove casts.

Im wesentlichen eine sehr kleine Rillenmarke, und vermutlich gleicher Entstehung wie die größeren Rillenmarken. *Schleifriefen* (w).

En fait des cannelures très fines, que l'on suppose formées de la même façon que les grandes. *Stries d'entraînement, de traînage.*

Calco de surcos muy delgados originados probablemente en forma semejante a "groove cast". *Calco de microsurco.*

Drift bedding (SORBY, 1857, p. 279):

A term proposed to replace false bedding. An obsolete term for inclined bedding.

Ausdruck, der Pseudoschichtung ersetzen soll. Eine veraltete Bezeichnung für geneigte Schichtung.

Terme proposé pour remplacer false bedding; terme peu usité pour stratification inclinée.

Término propuesto para reemplazar a "false bedding"; actualmente en desuso y sinónimo de "inclined bedding". *(Estratificación inclinada.)*

Dune:

An elongate mound or ridge of granular material, usually sand, that occurs at the sediment interface, commonly having a triangular profile, and usually transverse to direction of flow. Generally dunes occur in systems.

Ein länglicher Kamm oder Rücken aus körnigem Material — gewöhnlich Sand —, der an der Grenzfläche Transportmedium-Sediment auftritt. Im allgemeinen mit dreieckigem Profil und gewöhnlich quer zur Strömungsrichtung. Dünen treten in der Regel gruppenweise auf. *Düne.*

Monticules (ou crêtes) allongés, de matériau granulaire, le plus souvent de sable, qui se rencontrent à la surface de séparation de sédiments; en général à profil triangulaire, et en général perpendiculaires à la direction d'écoulement. Le plus souvent, les dunes forment des complexes dunaires. *Dune.*

Colina elongada generalmente arenosa, de perfil triangular, formada durante la fase de transporte de material y transversa a la dirección del flujo. Comunmente componen sistemas. *Duna.*

Eddy markings (RIGBY, 1959, Figs. 1 and 2):

Circular or semi-circular markings on bedding planes that may either be concentric or overlap.

Kreis- oder halbkreisförmige Marken auf Schichtflächen; konzentrisch oder sich überlappend. *Wirbelmarken* (w).

Marques circulaires ou semi-circulaires, que l'on rencontre sur les plans de stratification et qui peuvent être concentriques, ou se recouvrir. *Lit.: Empreintes en tourbillons.*

Plano de estratificación con marcas circulares, semicirculares o curvas; pueden ser concéntricas o estar sobrepuestas. *Marcas de remolino.*

Edgewise structure: See plate 105 A.

An arrangement of more or less tabular pebbles set at varying and steep angles to the bedding; some such arrangements have been attributed to sliding (HADDING, 1931, p. 383).

Eine Anordnung mehr oder weniger tafeliger Gerölle, die in allen möglichen, steilen Winkeln zur Schichtung stehen; in einigen Fällen sind derartige Anordnungen auf Gleitung zurückgeführt worden (HADDING, 1931, S. 383). *Kantengestelltes Konglomerat* (w), *Primärbreccie* (?).

Pierres plus ou moins plates, arrangées sous une inclinaison variable, mais toujours raide par rapport à la stratification; cette disposition est parfois interprétée comme résultant d'un glissement. *Structure de champ.*

Rodados más o menos tabulares dispuestos de tal manera que sus elongaciones forman con el plano de estratificación, ángulos de marcada pendiente. Estructura atribuida a deslizamientos (HADDING, 1931, p. 383). *Sesgoconglomerado.*

Elongate irregular marks (DZULYNSKI and SANDERS, 1963, p. 66):

Elongate, somewhat irregular scour marks intermediate in size between channels and flutes. Probably the Hauptwülste of RÜCKLIN (1938, p. 100, Fig. 6).

Längliche, etwas unregelmäßige Kolkmarken; in Größe ungefähr zwischen Erosionsrinnen und Erosionsmulden. Wahrscheinlich identisch mit RÜCKLIN's Hauptwülsten (1938, S. 100, Abb. 6). *Hauptwülste (?)*.

Empreintes de creusement allongées, de forme irrégulière, intermédiaires comme taille entre les chenaux et les cannelures. Probablement équivalentes des Hauptwülste de RÜCKLIN.

Micro causes elongados algo irregulares, de tamaño intermedio entre "channel" y "flute". Probablemente similar a "Hauptwülste" (RÜCKLIN, 1938, p. 100, fig. 6). *Micro cauces elongados*.

Endoglyph (VASSOEVICH, 1953, 37):

A hieroglyph found within a single layer.

Eine Hieroglyphe, die in einer einzelnen Lage auftritt. *Endoglyphe*.

Hiéroglyphe confiné au sein d'une seule couche. *Endoglyphe*.

Hieroglifo localizado en el seno de un estrato. *Endoglifo*.

Epiglyphs (VASSOEVICH, 1953, p. 37):

A hieroglyph on top of a bed.

Hieroglyphen auf Schicht-Oberflächen. *Epiglyphe*.

Hiéroglyphe situé au toit d'une couche. *Epiglyphe*.

Hieroglifo en el techo de un estrato. *Epiglifo*.

Erosion grooves (DZULYNSKI and SANDERS, 1963, p. 66):

Appear to be the "longitudinal ripple marks" of VAN STRAATEN (1951). Closely-spaced lines of straight-sided scour marks.

Wahrscheinlich die „longitudinalen Rippelmarken" VAN STRAATEN's (1951). Eng gescharte Reihen von eben begrenzten Kolkmarken. *Erosionsrillen* (w).

Probablement équivalent des «longitudinal ripple marks» de VAN STRAATEN. Empreintes de creusement rectilignes en faisceaux.

Marcas de desbaste estrechamente agrupadas y de flancos rectos. Se asemejan a las — óndulas longitudinales — de VAN STRAATEN (1951). *Surcos de erosión*.

Erpoglyph (VASSOEVICH, 1953, p. 74):

A term applied to worm castings.

Eine auf Wurmspuren angewandte Bezeichnung. *Erpoglyphe*.

Terme qui désigne les tortillons de vers. *Erpoglyphe*.

Término aplicado a rastros de vermes. *Erpoglifo*.

Exoglyph (VASSOEVICH, 1953, p. 37):

A hieroglyph on surface of bed as opposed to internal hieroglyph (endoglyph).

Eine Hieroglyphe auf Schicht-Oberflächen; im Gegensatz zu internen Hieroglyphen (Endoglyphen). *Exoglyphe*.

Hiéroglyphe que l'on rencontre à la surface d'une couche, par opposition aux hiéroglyphes internes (endoglyphes). *Exoglyphe*.

Hieroglifo localizado en la superficie de un estrato, en oposición a hieroglifos internos (Endoglifos). *Exoglifo*.

False bedding:

An old term for cross-bedding, which see. See also pseudo-cross-stratification.

Eine alte Bezeichnung für Schrägschichtung. Siehe dort. Siehe auch Pseudo-Schrägschichtung.

Terme ancien, équivalent de cross-bedding, q.v. *Fausse stratification.*

Término actualmente en desuso sinonimo de "cross-bedding". *Estratificación entrecruzada.*

False mud cracks:

Some polygonal patterns, such as those formed in soils, and some fucoidal networks (plate 70B) resemble those produced by drying and hence have been called false mud cracks; see also parting cast.

Gewisse polygonale Muster, die sich in Böden formen, und gewisse „fukoidale" Netzwerke (Tafel 70B) ähneln den polygonalen Trockenrissen und sind deshalb falsche Trockenrisse genannt worden. Siehe auch Dehnungsmarke. *Falsche Trockenrisse* (w).

Certains arrangement polygonaux, tels que ceux qui se forment dans des sols, et certains réseaux de fucoïdes qui ressemblent aux effets de la dessiccation, et qu'on appelle donc fausses fentes de dessiccation. *Pseudo-fentes de dessiccation.*

Diseño poligonal con reticulado fucoidal semejante al formado en suelos (lám. 70B). Esta estructura se asemeja a la producida por desecación ("mud crack"); véase "parting cast". *Pseudogrietas de desecación.*

False-stratification:

An early term (LYELL, 1839, Fig. 3), for cross-bedding. Rarely used in modern times in America.

Eine alte Bezeichnung für Schrägschichtung (LYELL, 1839, Abb. 3). In Amerika heutzutage kaum noch gebraucht.

Terme ancien, rarement usité actuellement aux Etats-Unis, équivalent de cross-bedding. *Fausse stratification.*

Término propuesto por LYELL (1839, fig. 3), sinónimo de "cross-bedding" *(Estratificación entrecruzada)*; actualmente en desuso.

Feather-like flow markings (KSIAZKIEWICZ, 1958, pl. 5, Fig. 1):
See frondescent casts and ruffled groove casts.

Festoon cross-bedding:
Same as festoon cross-lamination.

Festoon cross-lamination (KNIGHT, 1929, Fig. 21): See plates 32 and 33A.

Elongate, semi-ellipsoidal, cross-cutting erosional troughs filled in with conformally laminated strata. Same as trough cross-stratification. Cross-bedding deposited on a concave surface.

Längliche, halb-ellipsoidale, diskordante Erosionströge, mit konkordanten Schichtblättern gefüllt. Synonym für trogförmige Schrägschichtung. Schrägschichtung an einer konkaven Fläche angelagert. *Bogige Schrägschichtung, Muldenschichtung, synklinale Schrägschichtung, trogförmige Schrägschichtung, Kreuzschichtung.*

Lits minces qui remplissent, en en suivant les contours, des auges d'érosion allongées et de section semi-elliptique. Equivalent de «trough cross-stratification». Stratification croisée reposant sur une surface concave. *Lit.: Stratification croisée en festons.*

Superposición imbricada de cubetas elongadas y semi-elípticas, con relleno laminado concordante a sus bases cóncavas. Sinónimo de "trough cross-lamination". Estratificación entrecruzada depositada en una superficie cóncava. *Entrecruzamiento festoneado.*

304

Flame structure (Walton, 1956, p. 267) (load wave, "antidunes" of Lamont, 1957):
See plate 53 B.

The mud plumes separating the downward bulging load pockets or load casts of sand at sand-shale interface. Also described as "streaked-out ripples".

Fahnenartige Tongebilde, welche die sich nach unten hin ausbauchenden Belastungstaschen und Belastungsmarken aus Sand an der Sand-Schieferton Grenzfläche voneinander trennen. Auch als „ausgewalzte Rippeln" beschrieben. *Flammenstruktur* (w).

Désigne les plumes ou panaches de boue séparant les poches formées par le sable qui s'enfoncent vers le bas par suite de charge (empreintes de charge), au passage d'une couche sableuse à une couche schisteuse. *Lit.: Structure en flamme.*

Crestas de fango que separan depresiones arenosas (originadas por carga), en una interface arena-fango. También denominada „streaked-out ripples". *Estructura flamiforme.*

Flaser structure: See plates 17 A and 17 B.

Lenticles of fine sand or silt, commonly aligned and usually cross-bedded, which superficially resemble the flaser structure of some mylonites and other sheared metamorphic rocks.

Kleine Linsen aus Feinsand oder Silt, im allgemeinen ausgerichtet und schräggeschichtet, die oberflächlich der Flaserstruktur einiger Mylonite oder anderer zerscherter metamorpher Gesteine ähneln. *Flaserschichtung.*

Petites lentilles de sable fin ou de limon, alignées et généralement en stratification croisée, qui ont une ressemblance superficielle avec la texture filamenteuse de certaines mylonites et autre roches métamorphiques écrasées.

Lentes pequeños compuestos por arena fina o limo, generalmente alineados y con estratificación entrecruzada. Se asemeja a aquella estructura hallada en ciertas milonitas y otras rocas metamórficas cizalladas. *Estratificación flaser.*

Flat-topped ripple mark (Wegner, 1932, Fig. 3):

Ripples with flat, wide crests separated by narrow troughs (Tanner, 1958).

Rippeln mit flachen, breiten Kämmen, die von engen Tälern getrennt sind (Tanner, 1958). *Flachkämmige Rippelmarken* (w).

Rides à crêtes larges et plates, séparées par des sillons étroits.

Ondulas con crestas amplias y planas, separadas por depresiones estrechas (Tanner, 1958). *Ondula de crestas planas.*

Flow-and-plunge structure:

A term which has been applied to cross-lamination by Shrock (1948, p. 242).

Ein seltener Name für Schrägschichtung (Shrock, 1948, S. 242).

Terme appliqué à la stratification croisée par Shrock.

Término empleado por Shrock (1948, p. 242), sinónimo de "cross-lamination". *(Laminación entrecruzada).*

Flow cast (Shrock, 1948, p. 156):

"The 'rolls', lobate ridges, and other raised features produced and preserved in the overlying sandstone are ... given the designation *flow cast* because they represent the filling of the negative features produced by the flowage of the soft underlying sediment" (Shrock, 1948, p. 156). The features illustrated by Shrock seem to be largely what are now termed "load casts", which see. The term was redefined by Prentice (1956) for load casts modified by horizontal flowage of the burden during or after emplacement. See "flowage casts".

„Den Rollen, Wülsten und anderen erhabenen Gebilden, die im hangenden Sandstein entstanden sind, wird die Bezeichnung Fließmarke gegeben, weil sie Eintiefungen ausfüllen, die durch Fließbewegungen im weichen unterliegenden Sediment entstanden" (SHROCK, 1948, S. 156). Die von SHROCK illustrierten Gebilde scheinen weitgehend den neuerdings sog. „Belastungsmarken" zu entsprechen. Siehe dort. Die Bezeichnung wurde neudefiniert von PRENTICE (1956) für Belastungsmarken, die durch horizontales Fließen der Belastungsmasse während oder nach der Platznahme hervorgerufen wurden. *Gefließ-Marke.*

«Les rouleaux, renflements, lobes, et autres accidents en relief, produits et préservés dans le grès sus-jacent, sont désignés par le terme *flow cast*, parcequ'ils représentent le remplissage d'accidents négatifs formés par l'écoulement du sédiment mou sous-jacent» (SHROCK). Les formes illustrées par SHROCK paraissent en général être ce qu'on appelle maintenant «load casts» (empreintes de charge). Le terme a été re-défini par PRENTICE pour des empreintes de charge modifiées par écoulement horizontal de la charge pendant ou après la mise en place.

Estructura en forma de rollos, lóbulos y otros rasgos positivos, preservados en la base de areniscas. Representan el relleno de depresiones originadas por la acción de corrientes sobre fondos fangosos. La ilustración de SHROCK se asemeja más a "load casts", véase. El término fué definido nuevamente por PRENTICE (1956) y aplicado para "load cast" modificados por flujo durante o posteriormente al emplazamiento; véase "flowage cast". *Calco de flujo.*

Flow mark (RICH, 1950, p. 72): See flute cast.

Flow roll (PEPPER, *et al.*, 1954, p. 88) (flow structures, slump balls, pseudonodules, storm rollers):
Pillow-sized and pillow-shaped bodies of sandstone which characterize certain beds. Presumed to form by deformation, perhaps a product of large-scale load-casting or of subaqueous slump. See ball-and-pillow structure.

Ballenartige Gebilde, in Form und Größe Kissen ähnlich; meistens aus Sandstein und auf bestimmte Schichten beschränkt. Vermutlich durch Verformung entstanden, wahrscheinlich als Folge gewaltiger Belastung oder subaquatischer Rutschungen. Siehe Ballenstruktur. *Sedimentrolle, Rutschungsballen, -körper, Wickelfalten, -strukturen, Sedimentwalze* usw.

Masses de grès, de la taille et de la forme d'un coussin, caractéristiques de certains bancs. On pense qu'elles sont dues à une déformation, résultant peut-être d'un effet de charge à grande échelle, ou de glissements sous-aquatiques.

Masas areniscosas de forma y tamaño semejantes a los de una almohada, que caracterizan a ciertos estratos. Originadas probablemente por deformación de carga en gran escala o por desplazamiento subácueo. Véase "ball-and-pillow structure" *(Estructura almohadillada).*

"Flow structure" (COOPER, 1943, p. 190):
See ball-and-pillow structure. Also referred to as "flow layer", "flow-fold", "sandstone flow".

Flowage cast (BIRKENMAJER, 1958, p. 141, Figs. 1—5) (flow cast of PRENTICE, 1956):
Structures thought to be formed by the flowage of mobile, hydroplastic sand over the uneven bottom in the direction of slope. May be *transverse, longitudinal*, or *multi-directional*. Those, which seem to be produced by a combination of load-casting and current-oriented flow, have been termed flow cast by PRENTICE (1960, p. 220). They are related to "flame structures" (WALTON, 1956, p. 267).

Strukturen, die man sich als durch Fließen beweglichen, hydroplastischen Sandes über unregelmäßigen geneigten Boden entstanden denkt. Sie mögen *transversal, longitudinal* oder in *vielfacher Ausrichtung* vorkommen. Diejenigen unter ihnen, die durch eine Kombination von Belastung und strömungsorientiertem Fließen geschaffen sind, wurden von PRENTICE (1960, S. 220) Fließmarken genannt. Sie stehen in Beziehung zu den „Flammenstrukturen" (WALTON, 1956, S. 267). *Gefließ-Marke.*

Accidents qu'on attribue à l'écoulement de sable mobile, hydroplastique, qui coule dans la direction de la pente sur un fond irrégulier. Peuvent être transversaux, longitudinaux ou à directions multiples. Ceux qui paraissent résulter d'une combinaison d'effet de charge et d'écoulement orienté par le courant ont été appelés flow casts par PRENTICE et s'apparentent aux «flame structures». *Figures de fluxion; fronces d'écoulement.*

Estructura originada por el flujo de arena hidroplástica sobre un fondo irregular con pendiente. Puede ser transversa, longitudinal o multidireccional. Se supone que su desarrollo se debe a efectos combinados de carga y flujo orientado (PRENTICE, 1960, p. 220); está relacionada con "flame structures".

Flute (MAXON and CAMPBELL, 1935):

Discontinuous grooves and pockets, 2 to 10 or more cm long, formed on bedrock by action of turbulent flow of water. A loosely-used synonym for flute cast.

Unzusammenhängende Rillen und Kolke, 2 bis 10 cm oder mehr lang, die im anstehenden Gestein durch die Tätigkeit von turbulentem fließendem Wasser entstanden sind. Eine lose Bezeichnung für Strömungswülste. *Kolkmarke, Kolk, Fließrinne, Auskolkung, Erosionsrinne, -furche.*

Poches et sillons discontinus, de 2 à 10 cm de long, formés par l'action d'un courant d'eau turbulent sur la roche sous-jacente. Terme utilisé aussi comme synonyme de «flute cast».

Surcos y bolsillos discontínuos, de 2 a 10 o más cm de longitud, formados por la acción de una corriente turbulenta de agua sobre la roca subyacente. Sinónimo general de "flute cast".

Flute cast (CROWELL, 1955, p. 1359) (flow mark, scour cast, scour finger, vortex cast, lobate rill mark, turboglyph): See plates 55 A to 59 B and 60 B.

A sole-mark, a raised sub-conical structure, the up-current end of which is rounded or bulbous, the other end flaring out and merging with the bedding plane. Formed by filling of an erosional scour or flute. Other terms that have been used include "Kegelwülste" (FUGGER, 1899, p. 298) and "Zapfenwülste" (simple cones), "Korkzieher-Zapfen" (cork screw), "Flachzapfen" (depressed cone) and "Hufeisenwülste" (horse-shoe) by (RÜCKLIN, 1938). TEN HAAF (1959, p. 28, Fig. 13) described fan-shaped, linguiform, bulbous and voluted forms.

Eine fast konische, erhabene Schichtflächen-Marke, deren stromaufwärts gerichtetes Ende rundlich abgestumpft oder gewulstet ist, während das andere Ende sich verbreitert und allmählich in die Schichtfläche übergeht. Entsteht durch Ausfüllen einer muldenförmigen Erosions- oder Kolkrinne. Unter weiteren gleichsinnigen Bezeichnungen seien u. a. „Kegelwülste" (FUGGER, 1899, S. 298), „Zapfenwülste", „Korkzieher-Zapfen", „Flachzapfen" und „Hufeisen-Wülste" (RÜCKLIN, 1938) genannt. TEN HAAF (1959, S. 28, Abb. 13) beschrieb fächerförmige, zungenförmige, bauchige und gewundene Ausbildungen. *Strömungswulst, Fließwulst, -marke, -rille, Kolk-, Strömungsmarke, Kolkausguß-Sandstein-Wulst.*

Figures que l'on observe à la face inférieure de certaines couches. Accidents subconiques en relief, dont l'extrémité amont est arrondie ou bulbeuse, l'autre extrémité s'étalant et passant graduellement au plan de stratification. Formés par le remplissage d'un creux ou sillon affouillé par l'érosion. Parmi les autres termes qui s'appliquent aux mêmes accidents on peut citer: «Kegelwülste» (FUGGER) et les «Zapfenwülste» (cônes simples), «Korkzieher-Zapfen» (tire-bouchon), «Flachzapfen» (cône aplati) et «Hufeisenwülste» (en fer à cheval) de RÜCKLIN. TEN HAAF en a décrit des formes en éventail, en langues, bulbeuses ou contournées. *Empreinte de sillon d'érosion, moulage d'échancrure.*

Estructura formada por realces hemicónicos hallados en la base de ciertos estratos. Estas formas presentan sus extremos bulbosos orientados en sentido contrario a la corriente y los opuestos pierden relieve y pasan gradualmente al plano de estratificación. Se originan por relleno de pequeñas depresiones excavadas por flujo. Sinónimos: "Kegelwülste" (FUGGER, 1899, p. 298), "Zapfenwülste" ("simple cones"), "Korkzieher-Zapfen" ("cork screw"), "Flachzapfen" ("depressed cone"), "Hufeisenwülste" ("horse-shoe") (RÜCKLIN, 1938). TEN HAAF (1959, p. 28, fig. 13) describió formas en abanico, linguoides, bulbosas y en espiral. *Turboglifo.*

Force-aparts (ARAI, 1960, p. 70):
A structure similar to ball-and-pillow structure, which see.
Ähnlich Ballenstruktur; s. dort.
Accident comparable à «ball-and-pillow structure», q.v.
Estructura similar a "ball-and-pillow structure", véase.

Foreset bed: See plates 29 B to 37.
One of the inclined, internal, systematically arranged layers of a cross-bedded unit.
Eine der geneigten, inneren, systematisch angeordneten Lagen einer schräggeschichteten Einheit. *Leeblatt, Vorschüttungsblatt, Leeschicht, Stirn-Absatz, Mittelschicht, Schrägschicht.*
Une des couches internes inclinées, arrangées en système, d'un ensemble à stratification croisée. *Couche frontale, couche de front.*
Una de las capas inclinadas internas y regularmente ordenadas de una unidad sedimentaria con estratificación entrecruzada. *Capa frontal.*

Foreset bedding:
A synonym for cross-bedding.

Frondescent cast (TEN HAAF, 1959, p. 30, Fig. 5) ("cabbage leaf" marking, deltoidal cast): See plate 68.
"The flat casts are usually several decimeters long, resembling certain shrubs or large cabbage leaves ... The spreading 'foliage' is always directed down-current" (TEN HAAF, 1959, p. 30).
,,Diese flachen Marken sind im allgemeinen mehrere Dezimeter lang und Büschen oder Kohlblättern ähnlich. Das Netzwerk der Blattfächer breitet sich immer nach stromabwärts aus" (TEN HAAF, 1959, S. 30). *Fächermarke, fächerförmige Fließmarke.*
«Moulages aplatis, généralement longs de plusieurs décimètres, et qui ressemblent à certaines plantes arbustives, ou à une grande feuille de chou ... Le ‹feuillage› s'étale toujours dans la direction du courant» (vers l'aval) (TEN HAAF). *Empreinte arborescente, en feuille de chou.*
Calco de relieve plano, generalmente de varios decímetros de longitud, que se asemeja a ciertos arbustos o largas hojas de col. El "follaje" está orientado en sentido de la corriente (TEN HAAF, 1959, p. 30). *Calco caulifoliado.*

Fucoid: See plate 70 B.
An old term for a structure, once thought to be algal in origin, now used in a loose way for burrows, roots, trails, etc. Distinguished from hieroglyphs by being *within* layers and formed by material more or less unlike the matrix (VASSOEVICH, 1953, p. 21).
Alte Bezeichnung für eine Struktur, die man einst für Algen hielt. Neuerdings in lockerer Weise auf Bauten, Wurzeln, Spuren usw. angewandt. Tritt im Gegensatz zu Hiero-

glyphen innerhalb einer Lage auf und besteht aus Material, das sich mehr oder weniger von dem der Matrix unterscheidet (WASSOJEWITSCH, 1953, S. 21). *Fukoiden.*

Terme ancien qui désigne des empreintes dont on croyait autrefois qu'elles étaient d'origine végétale (algues); utilisé actuellement, avec assez peu de précision, pour désigner des terriers, empreintes de racines, traces, etc. Ces empreintes se distinguent des hiéroglyphes par le fait qu'elles se rencontrent au sein d'une couche, et sont formées d'un matériau différant plus ou moins de la roche encaissante (VASSOEVICH). *Fucoïdes.*

Término antiguo empleado para marcas de supuesto origen vegetal (algas). Empleado actualmente en sentido general para rastros y huecos tubulares originados por especies animales excavadoras y raíces. Estas marcas se distinguen de los hieroglifos por hallarse en el seno de un estrato formado por un material más o menos diferente al de su relleno. *Fucoide.*

Fulgurite:

An irregular, glassy, tube-like structure formed by fusion of sand by lightning.

Eine unregelmäßige, glasige, röhrenartige Struktur, die sich bei Blitzschlag durch Aufschmelzen von Sand formt. *Fulgurit.*

Tubulure irrégulière, vitreuse, formée dans un sable par fusion au passage de la foudre. *Fulgurite.*

Estructura tubular, vítrea e irregular, debida a fusión de arena por caída de un rayo. *Fulgarita.*

"Furious" cross-lamination (REICHE, 1938, p. 926):

Foreset beds which are themselves cross-bedded. Also called doubly cross-laminated by REICHE (POTTER and PETTIJOHN, 1963, Figs. 4—6).

Schräggeschichtete Leeblätter. Auch doppelte Schrägschichtung genannt (POTTER und PETTIJOHN, 1963, Abb. 4—6). *Zusammengesetzte Schrägschichtung* (n).

Se dit de couches frontales qui sont elles-même entrecroisées. Appelées aussi «doubly cross-laminated» (doublement croisées).

Capas frontales con estratificación entracruzada. También denominada "doubly cross-laminated" *(Estratificación entrecruzada doble)* por el mismo autor (POTTER and PETTIJOHN, 1963, figs. 4—6).

Furrow cast (McBRIDE, 1962, p. 58) ("rill mark" and "load-casted langitudinal ripplemark" of some authors): See furrow flute cast.

Furrow flute cast (McBRIDE, 1962, p. 57) ("delicate flute cast," "sludge cast," "rill mark"):

Cast of furrow-like depressions, with up-current noses similar to flute casts, but differ in being much longer and separated from parallel, adjacent furrows by narrow septa which appear as grooves in the cast. If the up-current terminations are missing, the structure is called a furrow cast (POTTER and PETTIJOHN, 1963, pl. 21 A and B).

Ausguß einer furchenähnlichen Vertiefung, deren nach stromaufwärts gerichtete „Nasen" denen der Strömungswülste ähnlich sind, sich jedoch von diesen darin unterscheiden, daß sie länger und von parallelen benachbarten Furchen durch enge Septen getrennt sind, die im Ausguß als Rillen erscheinen. Fehlt das obere Ende, dann nennt man die Marke „Furchenstruktur" (POTTER und PETTIJOHN, 1963, Tafel 21 A und B). *Furchenmarke* (n).

Moulage de dépressions en forme de sillons, dont les terminaisons amont sont semblables à celles des «flute casts», mais qui diffèrent de ces dernières en ce qu'elles sont beaucoup plus longues et séparées des sillons parallèles et adjacents par des septes étroits, qui forment des cannelures dans le moulage. Si les terminaisons amont manquent, cette figure s'appelle «furrow cast». *Sillons d'érosion allongés en faisceaux parallèles.*

Calcos de depresiones con extremos bulbosos similares a "flute cast", pero difieren por ser alargados y estar separados entre sí por tabiques angostos y paralelos (surcos en el calco). Si los extremos orientados en sentido contrario a la corriente disminuyen gradualmente en relieve, la estructura se denomina "furrow cast" (POTTER and PETTIJOHN, 1963, pl. 21 A and B). *Turboglifos interseptos.*

Gas pits (MAXON, 1940, p. 142):
Circular pits found in mud, from 2.5 cm to 30 cm in diameter and from less than 3 cm to 30 cm or more deep. Produced by escaping gas.

Runde Schlammtrichter, von 2,5 cm bis 30 cm oder mehr im Durchmesser und von weniger als 3 cm bis 30 cm oder mehr tief. Von aufsteigenden Gasen hervorgerufen. *Gas-Trichter, Entgasungs-Marken, -Trichter.*

Dépressions circulaires de la boue, de 2,5 à 30 cm de diamètre ou plus, et de moins de 3 cm à 30 cm ou plus de profondeur. Formées par l'échappement de gaz. *Cratères d'échappement de gaz.*

Hoyos circulares con diámetros que oscilan aproximadamente entre 2,5—30 cm y profundidades entre 3—30 cm, formados en fangos por escape de gas. *Hoyos por escape gaseoso.*

Gnarly bedding:
A synonym for disturbed bedding; see curly bedding, convolute bedding, etc.

Synonym für gestörte Schichtung. Siehe Kräuselschichtung, Wulstschichtung usw.

Synonyme de disturbed bedding.

Sinónimo de "curly bedding", "convolute bedding", *etc.*

Gouge channel (KUENEN, 1957, p. 142): See channel cast.

Gouge marks (crescentic gouge, lunoid furrow):
Crescentic marks concave up-current formed by glacial plucking on bedrock surface (GILBERT, 1905, pl. 37; HARRIS, 1943, p. 245).

Nach stromaufwärts zu konkave Marken, die bei der hobelnden Einwirkung von Gletschereis auf anstehendes Gestein entstehen (GILBERT, 1905, Tafel 37; HARRIS, 1943, S. 245). *Schürfmarken* (n).

Empreintes en croissant, à concavité tournée vers l'amont, et formées par le burinage saccadé produit par un glacier sur la roche de fond. *Coups de gouge.*

Marcas en forma de media luna, originadas por fracturación y desprendimiento de fragmentos rocosos del lecho de un glaciar; la concavidad de estas marcas están orientadas en sentido opuesto al avance (GILBERT, 1905, pl. 37; HARRIS, 1943, p. 245). *Marcas lunadas.*

Graded bedding: See plates 1 B, 42 A and 42 B.
Bedding characterized by gradual change in grain size from coarse at base of bed to fine at top; generally such beds have a sharp base and indefinite top. Several types of grading are possible including reversed grading (KUENEN, 1953, Fig. 1). See also sorted bedding.

Durch einen allmählichen Korngrößen-Wechsel von grob (unten) nach fein (oben) gekennzeichnete Schichtung. Im allgemeinen haben solche Lagen eine scharf abgesetzte Basis, während ihr Übergang zur Dachfläche allmählich ist. Zu den mannigfachen Typen von gradierter Schichtung gehört u. a. „umgekehrt gradierte Schichtung" (KUENEN, 1953, Abb. 1). Siehe auch sortierte Schichtung. *Gradierte Schichtung, — Bettung, Saigerungs-Schichtung.*

Stratification caractérisée par le changement progressif de la taille des éléments, qui passe de grossière à la base à fine au sommet de la couche (granodécroissance). En général, les couches ainsi formées ont une limite nette au mur et mal définie au toit. Il existe divers types de granoclassement, y compris un type inverse *(granocroissance)*. *Granoclassement, couches granoclassées, granulo-classées, granulo-classement vertical, sédimentation dégradée (vers le haut)*.

Estratificación caracterizada por el gradual decrecimiento en el tamaño de los granos desde la base hacia el techo. Generalmente las capas presentan una base bien definida y un techo indefinido. La gradación puede realizarse en varios tipos, incluyendo la gradación invertida (KUENEN, 1953, fig. 1). Véase "sorted bedding". *Estratificación gradada*.

Grooves:

A term applied to any straight linear depression which, unlike most channels, has a uniform cross-section and depth and great length. See also groove casts. Grooves include drag marks and slide marks, which see. Applied also to glacial grooves in bedrock.

Eine auf geradlinige Vertiefungen jeglicher Art angewandte Bezeichnung, die im Gegensatz zu den meisten Rinnen ein gleichmäßiges Profil, gleichmäßige Tiefe und eine große Länge aufweist. Siehe auch Rillenmarke. Rillen schließen Schleif- und Gleitmarken ein. Siehe dort. Auch auf glaziale Schrammen angewandt. *Rillen* (n), *Furchen, Rinnen*.

Ce terme s'applique à toute dépression rectiligne très longue dont le profil transversal et la profondeur, contrairement au cas de la plupart des chenaux, sont uniformes. Les «grooves» comprennent des «drag marks» et «slide marks», q.v. Ce terme s'applique aussi aux rainures glaciaires sur la roche de fond. *Cannelures, rainures*.

Término aplicado a depresiones lineales y rectas, las cuales a diferencia de la mayoría de los cauces, presentan secciones y profundidades uniformes; además sus longitudes son considerables. También empleado para surcos originados por glaciares sobre la roca de base. Los "grooves" incluyen a "drag marks" y "slide marks", véase. *Surcos*.

Groove cast (SHROCK, 1948, p. 162) (drag mark, or drag cast): See plates 57A, 59A, 60A, 61 to 68.

Rounded or sharp-crested rectilinear ridges produced by filling of grooves. Called "mud furrow" by HALL (1843, p. 234). See also striation cast.

Durch Ausfüllen von Rillen entstandene, abgerundete oder scharfkantige, geradlinige Rücken. Von HALL (1848, S. 234) „Schlammfurchen" genannt. Siehe auch Riefenmarke. *Rillenmarke* (n), *Schleifmarke, Rinnenausfüllung, Driftmarke*.

Arêtes rectilignes, arrondies ou à crête aigue, formées par le remplissage de cannelures. Appelées "mud furrow" (sillon de boue) par HALL. *Moulage de cannelure, de rainure; cannelures, rainures*.

Lomos rectilíneos, redondeados o con crestas agudas, originados por relleno de surcos. Denominada "mud furrow" por HALL (1843, p. 234). Véase "striation cast". *Calco de surco*.

Groove ruffling (TEN HAAF, 1959, p. 32, Fig. 18): See ruffled groove casts.

Hail (hailstone) imprints (hail pits):

Larger but otherwise similar to raindrop impressions (TWENHOFEL, 1921, p. 359).

Größer als Regentropfen-Eindrücke, aber diesen ähnlich (TWENHOFEL, 1921, S. 359). *Hagel-Eindrücke* (w).

Empreintes laissées par des grêlons; plus grandes, mais par ailleurs semblables à celles de gouttes de pluie. *Empreintes de grêlons, de grêle*.

Marcas similares a aquéllas originadas por lluvia, pero de mayor tamaño (TWENHOFEL, 1921, p. 359). *Marcas de granizo*.

Hassock bedding, hassock structure: See convolute bedding: see also pillow-form structure.

Herringbone cross-bedding:

A term found in SHROCK (1948, Fig. 214). See chevron cross-bedding.

Von SHROCK gebrauchte Bezeichnung (1948, Abb. 214). Siehe Fiederschichtung.

Terme employé par SHROCK.

Término empleado por SHROCK (1948, Fig. 214). Véase "chevron cross-bedding". *(Estratificación entrecruzada espigada)*.

Herringbone marking (KUENEN, 1957, p. 256, pl. 2B) (vibration mark and ruffled groove cast): See chevron mark.

Hexagonal interference ripples (BUCHER, 1919, Fig. 8):

Interference ripples with hexagonal pattern.

Interferenzrippeln mit hexagonalem Muster. *Hexagonale Interferenzrippeln* (w).

Rides interférentes formant un réseau hexagonal. *Rides hexagonales*.

Ondulas que al interferirse originan formas hexagonales. *Ondulas hexagonales*.

Hieroglyph (FUCHS, 1895):

A term adopted and applied to any markings found on bedding planes, generally applied to sole markings regardless of origin. VASSOEVICH (1953) divided hieroglyphs into *hypoglyphs* (on the bottom of the bed), and *epiglyphs* (on the top of the bed). He also divided hieroglyphs according to the time of origin: *synglyphs* (contemporaneous with sedimentation), *diaglyphs* (during diagenesis), *kataglyphs* (during katagenesis, i.e. alteration after deep burial), *metaglyphs* (during metamorphism), and *hyperglyphs* (during weathering). Hieroglyphs may be considered also as *bioglyphs* (organic in origin) or *mechanoglyphs* (or mechanical origin). VASSOEVICH also designated many special types such as *turboglyphs* (flute casts) and *ksimoglyphs* (groove casts).

Eine umfassende Bezeichnung, die auf Schichtflächen-Marken jeglicher Art angewendet wird. WASSOJEWITSCH (1953) unterteilt Hieroglyphen in Hypoglyphen (auf der Schicht-Unterseite) und Epiglyphen (auf der Schicht-Oberseite). Weiterhin teilt er Hieroglyphen nach dem Zeitpunkt ihres Entstehens ein: Synglyphen (synsedimentär), Diaglyphen (während der Diagenese), Kataglyphen (während der Katagenese, d. h. Veränderung nach tiefer Versenkung), Metaglyphen (während der Metamorphose) und Hyperglyphen (während der Verwitterung). Hieroglyphen kann man ferner aufteilen in Bioglyphen, sind sie organischen, und Mechanoglyphen, sind sie mechanischen Ursprungs. Schließlich führte WASSOJEWITSCH Spezial-Bezeichnungen ein, wie Turboglyphen (Strömungswülste) und Ksimoglyphen (Rillenmarken). *Hieroglyphe*.

Ce terme s'applique à toutes les empreintes qu'on rencontre sur les surfaces de stratification, et surtout à la face inférieure des couches, quelle que soit leur origine. VASSOEVITCH distingue les *hypoglyphes* (face inférieure) et *épiglyphes* (sommet des couches). Il les sépare aussi suivant le moment de leur formation en *synglyphes* (contemporains de la sédimentation), *diaglyphes* (formés pendant la diagénèse), *cataglyphes* (formés pendant la catagénèse, c.à.d. transformation après recouvrement par des couches épaisses), *métaglyphes* (pendant le métamorphisme), et *hyperglyphes* (pendant l'altération). On peut aussi séparer les *bioglyphes* (d'origine organique) et les *mécanoglyphes* (d'origine mécanique). VASSOEVITCH a nommé aussi des types particuliers tels que *turboglyphes* (flute casts) et *ksimoglyphes* (groove casts). *Hiéroglyphe*.

Término aplicado a cualquier tipo de marcas ubicadas en los planos de estratificación; generalmente empleado para marcas de base (de un estrato), sin tener en cuenta el origen. VASSOEVICH (1953) dividió los hieroglifos en:

"*hypoglyphs*" *hipoglifos* (en la base de un estrato),

"*epiglyphs*" *epiglifos* (en el techo de un estrato),

"*synglyphs*" *singlifos* (contemporáneos con la sedimentación),

"*diaglyphs*" *diaglifos* (durante la diagénesis),
"*kataglyphs*" *cataglifos* (durante la catagénesis),
"*metaglyphs*" *metaglifos* (durante el metamorfismo),
"*hyperglyphs*" *hiperglifos* (durante la meteorización),
"*bioglyphs*" *bioglifos* (de origen orgánico),
"*mechanoglyphs*" *mecanoglifos* (de origen mecánico).
Entre las estructuras menos generalizadas figuran:
"*turboglyphs*" *turboglifos* ("flute casts"),
"*ksimoglyphs*" *ksimoglifos* ("groove casts").

Horse-shoe flute casts (RÜCKLIN, 1938, p. 97, Fig. 4) (Hufeisenwülste): See plate 93B.
Horse-shoe shaped flute cast; same as current crescent.
Hufeisenförmige [Strömungswülste, Synonym für Strömungskämme. *Hufeisen-Wülste*.
Sillons d'érosion en fer à cheval. Synonyme de «current crescent».
Realce con aspecto de herradura; sinónimo de "current crescent". *Calco en herradura.*

Hyperglyph (VASSOEVICH, 1953, p. 33):
A hieroglyph formed during weathering.
Eine Hieroglyphe, die während der Verwitterung entsteht. *Hyperglyphe.*
Hiéroglyphe formé pendant l'altération de la roche. *Hyperglyphe.*
Hieroglifo formados durante la meteorización. *Hiperglifo.*

Hypoglyphs (VASSOEVICH, 1953, p. 37):
A hieroglyph on base of bed.
Eine Hieroglyphe auf einer Schichtunterseite. *Hypoglyphe.*
Hiéroglyphe formé à la face inférieure des bancs. *Hypoglyphe.*
Hieroglifo en la base de un estrato. *Hipoglifo.*

Ice crystal marks (casts): See plate 96A.
Cracks left by sublimation of ice crystals. Commonly sand-filled and appearing as straight, slightly-raised ridges on base of sandstone beds (UDDEN, 1918).
Oft mit Sand ausgefüllte Eintiefungen, die nach der Sublimation von Eiskristallen entstehen. Sie treten als geradlinige, ein wenig erhabene Kanten an der Unterseite von Sandstein-Schichten auf (UDDEN, 1918). *Eiskristall-: -Abdrücke, -Marken, -Pseudomorphosen, -Spuren.*
Fissures qui se forment après sublimation de cristaux de glace; elles sont généralement remplies de sable et se présentent comme des arêtes rectilignes, peu élevées, à la face inférieure de grès. *Moulages de cristaux de glace.*
Grietas originadas por sublimación de cristales de hielo, posteriormente rellenadas por arena; aparecen como lomos rectos de escaso relieve en la base de ciertas areniscas (UDDEN, 1918). *Calco de cristales de hielo.*

Ichnofossil (organic hieroglyph):
General term for fossil trails, tracks, and burrows; Lebensspuren. See ABEL (1935), SEILACHER (1953) and CASTER (1957) for more details.
Allgemeine Bezeichnung für fossile Spuren, Fährten, Bauten; Lebensspuren. Siehe ABEL (1935), SEILACHER (1953) und CASTER (1957) für eingehendere Beschreibungen. *Ichnofossil.*
Terme général qui désigne les pistes fossiles, traces de pas et terriers. Pour plus de détails, voir ABEL (1935), SEILACHER (1953) et CASTER (1957). *Vestiges fossiles de vie, traces d'organismes.*

Término general empleado para huellas, rastros y huecos tubulares fósiles; sinónimo "Lebensspuren". Véase ABEL (1935), SEILACHER (1953) y CASTER (1957) para mayores detalles. *Icnofósiles*.

Imbricate structure (shingle structure): See plate 43 A.

A structure characterized by overlapping of tabular fragments or pebbles which display up-current shingling or dip. Described by JAMIESON (1860, p. 349), BECKER (1893, p. 54) and others.

Eine Struktur, bei der plattenförmige Gesteinsfragmente oder Gerölle wie Dachziegel sich überlagern und nach stromaufwärts einfallen. Von JAMIESON (1860, S. 349), BECKER (1893, S. 54) und anderen beschrieben. *Dachziegel-Lagerung, -Stellung, dachziegelartige —*.

Structure caractérisée par des fragments aplatis ou des cailloux, inclinés vers l'amont. *Structure imbriquée, en écailles*.

Estructura caracterizada por la disposición imbricada de fragmentos tabulares, cuyas pendientes están orientadas en sentido opuesto a la corriente. Descripta por JAMIESON (1860, p. 349), BECKER (1893, p. 54) y otros. *Estructura imbricada*.

Impact cast (RADOMSKI, 1958, p. 403):

Cast of marking produced by object striking the mud bottom. The steeply raised end of impact casts are always oriented down-current. Essentially same as prod cast. See also bounce cast and skip cast.

Ausguß einer Vertiefung, die entsteht, wenn ein Objekt in schlammigen Grund einschlägt. Das steil aufgeworfene Ende der Einschlagmarken zeigt immer stromabwärts. Im Grunde das gleiche wie Stoßmarke. Siehe auch Aufprallmarke und Hüpfmarke. *Einschlagmarke* (n), *Stoßmarke, -eindruck*.

Moulage de l'empreinte laissée par un objet qui heurte le fond boueux. L'extrémite relevée et raide de tels moulages est toujours vers l'aval. En gros, synonyme de «prod cast».

Calco de marcas producidas por cuerpos que al ser arrastrados por corrientes, golpean sobre fondos fangosos; la pendiente más abrupta está orientada en sentido de la corriente. Similar a "prod cast"; véase "bounce cast" y "skip cast". *Calco de impacto*.

Inclined bedding: See plates 40 A and B, 41 A and B.

An inclusive term describing bedding inclined to principal surface of accumulation.

Ein weitgefaßter Ausdruck, der jegliche Schichtung einschließt, die in einem Winkel zur Haupt-Anlagerungsfläche steht. *Geneigte Schichtung* (w), *Diagonalschichtung*.

Terme général qui désigne une stratification inclinée par rapport à la surface principale de dépôt. *Stratification inclinée*.

Término general que indica estratificación inclinada con respecto a la superficie principal de acumulación. *Estratificación inclinada*.

Incomplete ripples (SHROCK, 1948, p. 122):

A term used to describe isolated crests of ripple mark. Also called starved ripples.

Eine Bezeichnung für isolierte Rippeln. Auch „ausgehungerte Rippeln" genannt. *Einzelrücken*.

Terme qui décrit des crêtes isolées de rides de vagues. Appelées aussi «starved ripples».

Crestas aisladas de una óndula; también denominada "starved ripples". *Ondula incompleta*.

Interference ripple marks (KINDLE, 1914, p. 158—161): See plate 87A.

A special form of compound rippling consisting of polygonal pits arranged side by side forming a cell-like structure. Called "tadpole nests" by HITCHCOCK (1858, p. 121—123), "dimpled current mark" by JUKES (1872, p. 163—164, cited in KINDLE, 1917a, p. 37) and "cross-ripple" by BUCHER (1919, p. 190).

Eine besondere Form zusammengesetzter Rippeln, bei der polygonale Vertiefungen in zellenartigem Muster aneinandergereiht sind. „Kaulquappen-Nester" von HITCHCOCK (1858, S. 121—123) genannt, „grubenförmige Strömungsmarken" von JUKES (1872, S. 163—164, zitiert in KINDLE, 1917a, S. 37) und „Kreuzrippeln" von BUCHER (1919, S. 190). *Interferenzrippeln.*

Type particulier de rides complexes, qui forment un réseau de dépressions polygonales comme des cellules. Appelé «tadpole nests» (nids de têtards) par HITCHCOCK, «dimpled current marks» (empreintes de courant à fossettes) par JUKES et «cross ripples» (rides entrecroisées) par BUCHER. *Rides interférentes, d'interférence; rides en fossettes.*

Forma especial de óndulas compuestas que consiste en depresiones poligonales con contornos communes a modo de celdas. Denominada "tad-pole nests" por HITCHCOCK (1858, p. 121—123), "dimpled current mark" por JUKES (1872, p. 163—164, KINDLE, 1918a, p. 37) y "cross ripple" por BUCHER (1919, p. 190.) *Ondulas de interferencia.*

Intraformational corrugation:

A term sometimes applied to small-scale intraformational folding (SHROCK, 1948, p. 271).

Manchmal für synsedimentäre Faltung gebraucht (SHROCK, 1948, S. 271). *Synsedimentäre Runzelung (w), endostratische Sedifluktion.*

Terme parfois appliqué à des plissements intraformationnels à petite échelle. *Plissotements intraformationnels.*

Término empleado para describir plegamientos intraformacionales de pequeñas dimensiones (SHROCK, 1948, p. 271). *Corrugacion intraformacional.*

Intraformational folds: See plates 107 A to 109.

Folding confined to a stratum lying between undeformed beds which is attributed to processes syngenetic with those responsible for the bed itself; may be due to sliding or slump but also produced by other means.

Faltung, die sich auf einzelne Lagen beschränkt, die zwischen unverformten Schichten liegen. Wird auf Ursachen zurückgeführt, mit denen die Entstehung der Schicht selbst zusammenhängt, kann aber durch Gleitungen, Rutschungen oder andere Prozesse bedingt sein. *Synsedimentäre Falten, endostratische Sedifluktion.*

Plissement confiné à une strate dont les couches encaissantes ne sont pas déformées, et qu'on attribue à un processus de même origine que celui qui a formé la strate elle-même. Peut résulter de glissements ou d'écoulements, ou avoir d'autres causes encore. *Plissements intraformationnels.*

Pliegues en un estrato limitado entre otros dos no plegados; su formación se debe a procesos simultáneos con aquéllos que originaron la unidad sedimentaria que los contiene. Se forman por deslizamientos, entre otros mecanismos. *Pliegues intraformacionales.*

Intra-stratal contortions (RICH, 1950, p. 729, Fig. 9): See convolute bedding.

Intra-stratal flowage (RICH, 1950, Fig. 15): See ball-and-pillow structure.

Intrastratal flow structure (WILLIAMS, 1960):

A type of convolute bedding, which see.

Eine Art Wulstschichtung; s. dort. *Endostratische Sedifluktion.*

Type particulier de convolute bedding, q.v.

Variedad de "convolute bedding", véase. *Estructura de flujo intraestratal.*

Kataglyph (VASSOEVICH, 1953, p. 33):

A hieroglyph formed during katagenesis; i.e. under a cover set of beds.

Eine Hieroglyphe, die sich während der Katagenese, d.h. unter der Belastung von Schichtpaketen formt. *Kataglyphe.*

Hiéroglyphe formé pendant la catagenèse, c.a.d. sous des couches sus-jacentes.

Hieroglifo formado durante la catagénesis (bajo una considerable cubierta de estratos). *Cataglifo.*

Keazoglyph (VASSOEVICH, 1953, p. 64):

Small transverse displacements along cracks; the parting cast of BIRKENMAJER (1959, p. 111), which see.

Geringe transversale Verschiebungen entlang Spalten; die Dehnungsmarken BIRKEN-MAJER (1959, S. 111). *Keazoglyphe.*

Petits déplacements transversaux le long de fissures. Même phénomène que les parting casts de BIRKENMAJER, q.v.

Pequeños desplazamientos transversos a lo largo de grietas; "Parting cast" (grietas tensionales) de BIRKENMAJER (1959, p. 111). *Keazoglifo.*

Klizoglyph (VASSOEVICH, 1953, p. 64):

A term employed to designate desiccation cracks.

Eine Bezeichnung für Trockenrisse. *Klizoglyphe.*

Terme qui désigne les fentes de dessiccation.

Término empleado para grietas de desecación. *Klizoglifo.*

Kneaded sandstone (COOPER, 1943, p. 198):

A vague structure of some sandstone beds resembling kneaded dough, probably a variation of ball-and-pillow structure.

In einigen Sandsteinen auftretende, unausgeprägte Struktur, die geknetetem Teig ähnelt. Wahrscheinlich eine Art Ballenstruktur. *Gekneteter Sandstein* (w).

Structure peu nette de certains grès, ressemblant à une pâte pétrie, malaxée; probablement une variante de la «ball-and-pillow structure». *Lit.: grès pétris.*

Estructura indefinida que se presenta en ciertas areniscas, a modo de masas malaxadas; es probable que sea una variedad de "ball-and-pillow structure". *Arenisca malaxada.*

Knob-and-trail:

A structure found on glacial pavements characterized by protruding resistant "knob" and lee-side ridge of weaker rock (CHAMBERLIN, 1888, p. 245).

Eine Struktur, die sich auf glazial geschrammtem Untergrund findet. Vorstehende Buckel, an die sich „stromabwärts" ein Rücken weicheren Gesteins anschließt (CHAMBER-LIN, 1888, S. 245). *Geschwänzte Buckelstruktur* (w).

Accident de certains pavages glaciaires, caractérisé par un pointement d'une roche plus résistante en aval duquel la roche sous-jacente plus tendre, protégée contre le rabotage, forme un petit remblai.

Estructura desarrollada sobre lechos glaciarios, originada por obstáculos resistentes que proyectan en sentido del movimiento del hielo, lomos hemicónicos de roca menos resistente (CHAMBERLIN, 1888, p. 245). *Criosombra de obstáculo.*

Ksimoglyph (VASSOEVICH, 1953, p. 61, 72): Drag mark, which see.

Lake balls:

Natural felt balls found in shallow waters of some lakes (OHLSON, 1962, p. 377). Also called "hair balls", "water-rolled weed balls" and "burr balls". See also sea balls.

Auf natürliche Weise „verfilzte" Ballen, die sich in seichten Teilen einiger Seen finden (OHLSON, 1962, S. 377). Auch „Haarballen", „wassergerollte Häckselknäuel" usw. genannt. Siehe auch *marine Pflanzenknäuel. Lakustrische Pflanzenknäuel* (n).

Boules feutrées naturelles, que l'on rencontre dans les parties peu profondes de certains lacs. Appelées aussi «hair balls» (pelotes de poils). *Pelotes végétales.*

Esferoides fibrosos hallados en aguas poco profundas de ciertos lagos (OHLSON, 1962, p. 377); también llamada "hair balls", "water-rolled weed balls" y "burr balls". Véase "sea balls". *Esferoides lacustres.*

Lamination:
Layer or bed less than 1 cm in thickness (PAYNE, 1942, p. 1724); loosely applied to the thinnest units of stratification differing from one another in grain size and/or composition.

Weniger als 1 cm mächtige Lagen oder Schichten (PAYNE, 1942, S. 1724); auf feinste Einheiten der Schichtung verwandt, die sich voneinander in Korngröße und/oder Zusammensetzung unterscheiden. *Feinschichtung, Lamellargefüge, Lamination.*

Lit de moins de 1 cm d'épaisseur (PAYNE); moins précisement, les unités les plus minces d'un ensemble stratifié, diffèrant les unes des autres par la taille du grain ou par la composition (ou les deux). *Straticules, laminations, laminae.*

Estrato menor de 1 cm de espesor (PAYNE, 1942, p. 1724); empleado en sentido amplio para las unidades más delgadas de estratificación, que difieren unas de otras en granometria y/o composición. *Laminación.*

Laminar corrugations: See intraformational corrugation.

Lebensspuren:
A general term for sundry tracks, trails, and burrows or casts derived therefrom. See ichnofossil and biohieroglyph.

Eine allgemeine Bezeichnung für alle möglichen organischen Spuren, Fährten, Bauten und deren Ausgüsse. Siehe Ichnofossil und Biohieroglyphe. *Lebensspuren.*

Terme qui dénote divers types de traces, pistes et terriers, et leurs moulages.

Término general empleado para huellas, rastros y huecos tubulares, o sus respectivos calcos. de especies animales. Véase "ichnofossil" y "bioglyph".

Lee-side concentration:
Small-scale cross-bedding formed by deposition on lee-side of ripples (McKEE, 1938, Fig. 4).

Kleinformatige Schrägschichtung, die sich an der Leeseite von Rippeln bildet. *Rippelschichtung.*

Stratification croisée à petite échelle, formée par dépôt à l'abri des rides, sur leur pente avale. *Lit.: Concentration en aval des rides, stratification de rides.*

Estratificación entrecruzada de pequeñas dimensiones, formada por depositación a sotavento de las crestas que componen una óndula (McKEE, Fig. 4). *Entrecruzamiento de crestas.*

Lenticular bedding:
Strata which wedge or pinch out within the confines of a given outcrop, hand specimen or core. See also flaser structure.

Schichten, die innerhalb der Aufschluß-, Handstück- oder Bohrkerngrenzen auskeilen. Siehe auch Flaserschichtung. *Linsenschichtung, linsige Schichtung.*

Couches qui disparaissent en coin au sein d'un affleurement, d'un échantillon de roche, ou d'une carotte de sondage. *Lit.: Stratification lenticulaire.*

Acuñamiento de un estrato observable en un trecho limitado de un afloramiento, en una muestra o en un testigo de perforación. Véase "flaser bedding". *Estratificación lenticular.*

Lenticular cross-bedding:

Cross-bedding whose outline, in vertical section, is lenslike (SHROCK, 1948, Fig. 208).

Schrägschichtung, deren Umrisse in senkrechtem Schnitt linsenähnlich sind (SHROCK, 1948, Abb. 208). *Linsige Schrägschichtung.*

Ensemble à stratification croisée dont le contour, en coupe verticale, a l'aspect d'une lentille. *Lit.: Stratification croisée en lentille.*

Estratificación entrecruzada cuyos contornos, en sección vertical, tienen la forma de una lente (SHROCK, 1948, Fig. 208). *Estratificación entrecruzada lenticular.*

Level-surface ripple (McKEE, 1939, p. 72):

A ripple that migrates along a level surface.

Rippeln, die auf einer Horizontalfläche entlang wandern. *Horizontal-Rippeln* (w).

Ride qui se déplace sur une surface plane.

Ondula que se desplaza sobre una superficie horizontal. *Ondula horizontal.*

Lineation:

A descriptive term for any kind of linear structure on or within a rock. As applied to sedimentary rocks it usually refers to a linear sole mark, or a linear fabric of aligned clasts or fossils, or any other linear structure of megascopic or microscopic nature.

Ein allgemeiner Ausdruck für lineare Strukturen jeglicher Art an der Oberfläche oder innerhalb eines Gesteins. Auf Sedimente angewandt meint Lineation im allgemeinen lineare Schichtflächen-Marken, lineares Gefüge ausgerichteter Partikel oder Fossilien oder andere lineare Strukturen makroskopischer oder mikroskopischer Art. *Lineation, Striemung, Strömungsriefen.*

Terme employé pour décrire toutes sortes de structures linéaires à la surface ou au sein d'une roche. Dans le cas des roches sédimentaires, s'applique surtout à des empreintes rectilignes, à l'arrangement linéaire de dépôts fragmentaires ou de fossiles, et à toute autre structure macroscopique ou microscopique. *Linéation.*

Término empleado para toda clase de estructura lineal, desarrollada en la superficie de una roca o dentro de ella. En rocas sedimentarias se refiere a marcas de base, clastos y fósiles alineados o cualquier otra estructura lineal de naturaleza mega o microscópica. *Lineación.*

Linguoid ripple marks (BUCHER, 1919, p. 164): See plates 84A and 84B.

A name given to asymmetric current ripples that have a barchan-like shape, the horns pointing into the current. Also called cuspate ripples.

Eine Bezeichnung für unsymmetrische Strömungsrippeln, die Barchanen ähneln und deren Spitzen stromabwärts weisen. Auch sichelförmige Rippeln genannt. *Linguoidrippeln.*

Rides de courant asymétriques, qui ressemblent à des barkhanes, les cornes tournées vers l'amont. Appelées aussi «cuspate ripples». *Rides arquées en croissant.*

Ondula con crestas que se asemejan a barjanes con sus extremos orientados en sentido contrario a la corriente. Sinónimo de "cuspate ripple mark". *Ondula linguoide.*

Load cast (KUENEN, 1953, p. 1048 and 1058) (load pocket): See plates 52A to 53B.

The bulbous, mammillary or papilliform downward protrusions of sand produced by load deformation of underlying hydroplastic mud; due to yielding under unequal load. Called "flow cast" by SHROCK (1948, p. 156).

Die grobknolligen, wulst- bis warzenförmigen Ausbauchungen aus Sand, die in von Sand belastetem und verformtem, hydroplastischem Schlamm entstehen, der bei ungleicher Belastung nachgibt. Von SHROCK Fließmarke genannt (1948, S. 156). *Belastungsmarke, Fließwulst, Sandstein-Wulst, Unterseiten-Wulst.*

Protubérances bulbeuses, mammelonnées ou en papilles, formées par du sable qui s'enfonce, par déformation due à la charge, dans la boue hydroplastique sous-jacente, laquelle ne cède pas de manière uniforme (la boue peut même remonter dans le sable sus-jacent en formant la «flame structure»). Appelées «flow casts» par SHROCK. *Empreintes de charge, moulages de surcharge.*

Protuberancias arenosas desarrolladas hacia abajo, de formas bulbosas, mamilares o papiliformes, originadas por efecto de carga diferencial con la consiguiente deformación de la capa de fango infrayacente. Denominada "flow cast" por SHROCK (1948, p. 156). *Calco de carga.*

Load-cast lineation (CROWELL, 1955, p. 1358):

Small-scale, rather poorly defined, irregular, linear structures appearing as casts on base of sandstones attributed to dense sluggish turbidity current moving over soft mud.

Ziemlich schlecht definierte, kleindimensionale, unregelmäßige, lineare Strukturen, die als Ausfüllungen an der Unterseite von Sandstein-Bänken auftreten. Sie sollen entstehen, wenn dichte, träge Suspensionsströme über weichen Schlamm fließen. *Belastungslineation* (w).

Accidents linéaires de petite taille, assez peu marqués, irréguliers, qui se présentent comme des moulages à la face inférieure de grès et qu'on attribue à un courant de turbidité, dense et très lent, se déplaçant sur de la boue molle.

Calcos pequeños mal definidos, irregulares y alineados, hallados en la base de ciertas areniscas, atribuidos al desplazamiento de corrientes de turbidez sobre fondos de fango plástico. *Lineación de calcos de carga.*

Load-cast striations (TEN HAAF, 1959, p. 46, Fig. 31; KUENEN, 1957, p. 255, Fig. 22):

A rill-like pattern of uncertain origin.

Riefenmuster ungewissen Ursprungs. *Belastungsriefen* (w).

Réseau comparable à des rigoles de plage, d'origine indéterminée.

Estructura de aspecto semejante a — marcas de escurrimiento —, de origen dudoso. *Calco de estriaciones de carga.*

Load-casted current markings (KUENEN, 1957, p. 248):

Flutes, grooves and other current marks swollen and misshapen due to load-casting.

Erosionsmulden, Rillen und andere Strömungsmarken, durch Belastung ausgeweitet und umgestaltet. *Belastete Strömungsmarken* (w).

Sillons, cannelures et autres empreintes de courant grossies et déformées par effet de charge.

Acanaladuras, surcos y otras marcas originadas por corrientes, engrosadas y deformadas por carga. *Marcas de corriente agravadas.*

Load-casted sole marks: See load-casted current markings.

Load fold (SULLWOLD, 1959, p. 1247):

Load folds are plications of the underlying stratum which are believed to be the result of unequal pressure from the overlying load pockets and waves.

Belastungsfalten sind Fältelungen im Liegenden von Belastungsmarken und vermutlich durch deren ungleichmäßige Auflastung entstanden. *Belastungsfalte* (w).

Plissements d'une couche sous-jacente résultant probablement de la pression inégale exercée par les poches et vagues de charge qui la surmontent. *Lit.: Plissement (ou pli) de charge.*

Pliegues en un estrato originados probablemente por carga diferencial (bolsillos y ondas de carga) del estrato suprayacente. *Pliegue de carga.*

Load mold (SULLWOLD, 1960, p. 635):

The depression in the underlying bed occupied by the load pocket.

Eintiefung im Liegenden von Belastungstaschen. *Belastungsmulde* (w).

Dépression de la couche sous-jacente occupée par une poche de charge. *Lit.: Moule de charge.*

Depresiones en el estrato infrayacente originadas por carga. *Moldeado de carga.*

Load pocket (SULLWOLD, 1959, p. 1247):

A load pocket is a sole mark characterized by a "bulge of sand pressing into the underlying stratum." Same as load cast.

Eine Belastungstasche ist eine Schichtflächen-Marke, die eine «Ausbauchung von Sand in die unterliegende Schicht hinein» darstellt. Dasselbe wie Belastungsmarke. *Belastungstasche* (w).

Empreinte formée par «une protubérance de sable qui s'enfonce dans la couche sous-jacente». Synonyme de load cast. *Lit.: Poche de surcharge.*

Realce bulboso en la base de un estrato arenoso, originado por carga sobre otro infrayacente. Similar a "load cast". *Bolsillo de carga.*

Load wave (SULLWOLD, 1959, p. 1247):

A load wave (wisp or plume) " ... is the salient curved unevenness of underlying material which appears to have been squirted up into the superjacent turbidity-current deposit." See also flame structure.

Eine Belastungsfahne ist eine «scharf ausgeprägte, flammenartige Struktur, bei der unterliegendes Material in hangende Turbidite eingepreßt worden zu sein scheint». Siehe auch Flammenstruktur. *Belastungsfahne* (w).

Dans le cas d'un effet de charge, «la protubérance arquée, irrégulière, de matériel sous-jacent, qui semble avoir été injectée vers le haut dans le dépôt sus-jacent dû à un courant de turbidité». *Lit.: vague de charge.*

Crestas irregulares compuestas por material del estrato infrayacente, que aparecen como inyectadas en la turbidita suprayacente. Véase "flame structure". *(Estructura flamiforme).*

Lobate plunge structure (HESSLAND, 1955, pl. 1, Fig. 1): See flute cast.

Lobate rill mark (SHROCK, 1948, p. 131):

The lobate rill mark figured by SHROCK (Fig. 92) is apparently the same as flute cast, which see.

Die bei SHROCK (Abb. 92) abgebildete Struktur ist allem Anschein nach mit Strömungswülsten identisch. Siehe dort.

La figure de SHROCK semble indiquer que ce terme équivaut à «flute cast», q.v.

Esta estructura descripta por SHROCK (fig. 92) es, aparentemente, semejante a "flute cast", véase. *Turboglifo.*

Longitudinal ripple marks (VAN STRAATEN, 1951):

Relatively straight crests parallel to current direction. May be symmetrical or asymmetrical.

Relativ gerade, gestreckte, symmetrische oder asymmetrische Rippelkämme, die parallel zur Strömung liegen. *Longitudinale Rippelmarken* (w).

Crêtes assez rectilignes et parallèles à la direction du courant. Elles peuvent être symétriques ou asymétriques. *Rides longitudinales.*

Ondula con crestas relativamente rectas, paralelas a la dirección de la corriente; las crestas pueden ser simétricas o asimétricas. *Ondula longitudinal.*

Loop-bedding (BRADLEY, 1931, p. 29):

Small groups of laminae that are otherwise quite regular are sharply constricted or even end abruptly at intervals, giving the effect of long, thin loops or links of a chain. Thought to be a desiccation feature. Observed in some fine calcareous sediments and also in oil shale.

In sonst regelmäßigen Feinschichten auftretende Unterteilung der Schichtung in gekrümmte Pakete mit mehr oder weniger scharfen Grenzflächen, im ganzen den Gliedern einer Kette vergleichbar. Wird als eine Form von Austrocknung aufgefaßt und findet sich in feinkörnigen, kalkhaltigen Sedimenten und Ölschiefern. *Girlandenschichtung* (w).

Petits paquets de laminae, très régulières ailleurs, mais nettement resserrées, ou même brusquement interrompues à intervalles, ce qui donne un effet d'anneaux longs et minces, ou de maillons d'une chaîne. On pense que c'est une figure de dessication. Se rencontre dans des sédiments calcaires fins et dans des schistes bitumineux.

Pequeños grupos de láminas generalmente muy regulares, estrechamente agrupados y de bordes abruptos, con aspecto de lazos o eslabones; se supone originados por desecación. Observados en ciertos sedimentos finos calcáreos y en lutitas bituminosas. *Estratificación eslabonada.*

Mammillary structure: See pillow-structure.

Massive bedding: See plates 24 A and B, 25, 26, 50 A and B, 51.

A term commonly applied to very thick homogeneous beds. Also used to describe absence of structure in rocks regardless of thickness.

Auf sehr dicke, homogene Schichten angewandte Bezeichnung. Auch auf strukturlose Schichten unterschiedlicher Mächtigkeit angewandt. *Massige, massive Schichtung, Bankung.*

Se dit communément de bancs très épais et homogènes. S'emploie aussi pour décrire l'absence de structure dans des roches quelle que soit leur épaisseur. *Série massive, banc massif.*

Término comunmente empleado para estratos homogéneos y de considerable espesor. También empleado para describir la falta de estructura en las rocas, prescindiendo del espesor. *Estratificación masiva.*

Mechanoglyph (VASSOEVICH, 1953, p. 38):

A hieroglyph of mechanical origin.

Eine Hieroglyphe mechanischen Ursprungs. *Mechanoglyphe.*

Hiéroglyphe d'origine mécanique. *Mécanoglyphe.*

Hieroglifo de origen mecánico. *Mecanoglifo.*

Megaripple (VAN STRAATEN, 1953, p. 1—2):

Any type of ripple with a wave length greater than 1 m.

Rippeln jeglicher Art mit einer Wellenlänge größer als 1 m. *Großrippel.*

Rides de tous types, de longueur d'onde supérieure à 1 m. *Rides de plage géantes, mégarides.*

Cualquier tipo de óndula con una longitud de onda mayor de 1 metro. *Mega-óndula.*

Metaglyph (VASSOEVICH, 1953, p. 33):

A hieroglyph formed during metamorphism.

Während der Metamorphose geformte Hieroglyphen. *Metaglyphe.*

Hiéroglyphe formé pendant le métamorphisme. *Métaglyphe.*

Hieroglifo formado durante el metamorfismo. *Metaglifo.*

Meta-ripples (Bucher, 1919, Fig. 4):

A term used to describe large asymmetrical sand waves. See also Shrock (1948, p. 102).

Eine Bezeichnung für große, unsymmetrische Sandwellen. Siehe auch Shrock (1948, S. 102). *Großrippeln.*

Désigne de grandes vagues de sable asymétriques. *Grandes rides, vagues de sable, métarides.*

Término empleado para describir amplias ondas asimétricas de arena. Véase Shrock (1948, p. 102). *Meta-óndula.*

Microgroove cast (McBride, 1962, p. 56):

Cast of striations less than 2.5 cm long. Thought to be a cast of a groove cut by coarse sand grains.

Ausguß von Riefen, die weniger als 2,5 cm lang sind. Als Ausfüllung einer Rille aufgefaßt, die von groben Sandkörnern erodiert ist. *Kleine Rillenmarke* (w).

Moulage de stries de moins de 2,5 cm de long. Probablement moulage de stries formées par des grains de sable grossier. *Lit.: Moulage de stries.*

Calco de surcos muy delgados con longitudes menores de 2,5 cm, probablemente formados por granos gruesos de arena. *Calco de microsurco.*

Micro-cross-lamination (Hamblin, 1961, p. 390—401): See plate 38A.

Small distinctive cross-lamination similar to trough cross-lamination. Also called rib-and-furrow (Stokes, 1953, p. 17). Originally called "Schrägschichtungsbögen" by Gürich (1933, Figs. 1, 4, and 8).

Kleindimensionale, aber ausgeprägte Schrägschichtung, ähnlich bogiger Schrägschichtung. Von Stokes (1953, S. 17) auch „Rippen- und Furchenstruktur" genannt. Ursprünglich als „Schrägschichtungsbögen" bezeichnet (Gürich, 1933, Abb. 1, 4 und 8). *Kleindimensionale Schrägschichtung.*

Petit ensemble individualisé à stratification croisée, semblable au «trough cross-lamination». Appelé aussi «rib-and-furrow». *Lit.: Microstratification croisée.*

Laminación entrecruzada de pequeñas dimensiones, similar a "trough cross-bedding". También denominada "rib-and-furrow" (Stokes, 1953, p. 17). El término original "Schrägschichtungsbögen" fué empleado por Gürich (1933, figs. 1, 4, 8). *Laminación micro-entrecruzada.*

Mottled structure:

Primary mottling consists of discontinuous lumps, tubes, pods, and pockets of a sediment randomly enclosed in a matrix of contrasting textures; usually formed by filling of animal borings and burrows (Moore and Scruton, 1957, p. 2727).

Primäre Fleckung aus allerlei rundlichen, knolligen, röhrenförmigen und unregelmäßigen Gebilden aus Sediment, die von einer Grundmasse kontrastierenden Gefüges umgeben sind. Sie sind im allgemeinen auf organische Bohrlöcher und Bauten zurückzuführen (Moore und Scruton, 1957, S. 2727). *Gefleckte Schichtung, Wühlgefüge, Fossitextur.*

Mouchetage primaire, formé de masses, tubes, lentilles et poches discontinus d'un sédiment, répartis au hasard dans une matrice de structure différente; résultent généralement du remplissage de tubes et terriers d'animaux. *Texture mouchetée, tachetée.*

Aspecto que presenta un sedimento cuando ha sido parcialmente removido por la actividad biológica de especies animales excavadoras. La estructura consiste en huecos y orificios tubulares rellenados, cuya textura contrasta con aquélla del sedimento no removido (Moore and Scruton, 1957, p. 2727). *Estructura moteada.*

Mud-buried ripple mark (SHROCK, 1948, p. 109): See plate 82 A.

Ripple mark with troughs filled with mud.

Rippelmarken, deren Mulden mit Schlamm gefüllt sind. *Schlammbedeckte Rippelmarken* (w).

Rides dont les creux sont remplis de boue.

Ondula con sus depresiones cubiertas por fango. *Cobertura de óndula*.

Mud crack (cast): See plates 94 and 95, 96 B and 97.

Filling of desiccation cracks in mud, generally by sandstone; generally preserved as raised ridges (casts) arranged in polygonal patterns on underside of sandstone bed (see KINDLE, 1926).

Ausfüllung von in Schlamm entstandenen Trockenrissen, im allgemeinen durch Sand. Im Normalfall als erhabene Leisten in polygonalem Muster auf der Unterseite von Sandstein-Bänken erhalten (s. KINDLE, 1926). *Trockenriß, Netzleiste, Schwundriß, Schrumpfungsriß, Luftriß.*

Remplissage des fentes de dessiccation d'une boue, généralement par un grès, et le plus souvent conservé sous forme de moulages en relief, formant un réseau polygonal à la face inférieure de bancs de grès. *Moulages de fentes de retrait.*

Grietas en fango generalmente rellenadas por arena y preservadas en la base de areniscas, como calcos en forma de lomos con diseño poligonal (KINDLE, 1926). *Grietas de desecación*.

Mud furrows (casts) (HALL, 1843, p. 234):

An early, now obsolete term for groove casts.

Eine alte, inzwischen aufgegebene Bezeichnung für Rillenmarken.

Terme ancien peu usité équivalent de «groove casts».

Término en desuso sinónimo de "groove cast". *(Calco de surco).*

Mud lumps:

Upwelling of mud from buried clays forming islands or mounds which rise 2 to 4 meters above sea level; known from delta of Mississippi River (SHAW, 1913, p. 19; MORGAN, 1961).

Aus abgelagertem Ton aufsteigende Schlammklumpen, die Inseln oder Hügel formen, und von 2 bis 4 m über den Meeresspiegel aufsteigen; vom Mississippi-Delta bekannt (SHAW, 1913, S. 19; MORGAN, 1961). *Mudlump.*

Remontées de boue venant d'argile profonde, et formant des îlots ou des monticules qui s'élèvent de 2 à 4 m au-dessus du niveau de la mer; observés dans le delta du Mississipi. *Lit.: Paquets de boue.*

Efluencia fangosa de depósitos arcillosos cubiertos, con la subsiguiente formación de islas o promontorios que sobresalen entre 2 y 4 m sobre el nivel del mar; observados en el delta del Río Mississippi (SHAW, 1913, p. 19; MORGAN, 1961). *Promontorios fangosos.*

Mud volcanoes:

Conical accumulation about an orifice formed by eruption of sulfurous and bitumen-laden mud.

Konische Anhäufungen um einen Krater, der von einer Eruption schwefelhaltigen und hochbituminösen Schlammes geformt ist. *Schlammvulkan.*

Accumulation conique autour d'un orifice formé par l'éruption de boue sulfureuse et riche en bitume. *Volcans de boue.*

Acumulación cónica alrededor de un orificio formado por efluencia de fango sulfuroso y bituminoso. *Volcán de fango.*

Nailhead striation (nailhead scratch):

A glacial scratch or marking with blunt end, generally, but not always, down-current.

Glaziale Schrammen, deren stumpfes Ende im allgemeinen, aber nicht immer, stromabwärts zeigt. *Nagelkopf-Schrammen* (w).

Strie ou empreinte glaciaire dont la terminaison abrupte se trouve généralement (mais pas toujours) vers l'aval. *Lit.: strie en tête de clou.*

Marca originada por atrición de un glaciar, orientada, aunque no siempre, con su extremo romo en sentido del movimiento. *Estría roma.*

Neptunian dikes:

Dikes filled by sediment, generally sand, in contrast to plutonic dikes filled by volcanic materials.

Vom Sediment — im allgemeinen Sand — gefüllte Gänge; im Gegensatz zu plutonischen Gängen, die mit vulkanischem Material gefüllt sind. *Neptunische Gänge* (w).

Filons dont le remplissage est sédimentaire (en général sable) par contraste avec les «filons plutoniens» à remplissage volcanique. *Lit.: Filons, dikes, neptuniens.*

Diques rellenados por sedimentos, generalmente arena, en contraste con aquéllos rellenados por materiales volcánicos. *Diques clásticos.*

Nodular bedding: See plates 15 B, 16 A, B, C.

Layers which consist of scattered to loosely packed nodular bodies of rock in matrix of like or unlike character, also called lumpy bedding.

Lagen aus verstreuten bis lose gepackten knolligen Gesteinskörpern, deren umgebende Grundmasse von gleicher oder verschiedener Zusammensetzung sein kann. Auch „klumpige Schichtung" genannt. *Knollige Schichtung.*

Couches formées de morceaux noduleux d'une roche, disséminés ou au contraire plus ou moins tassés, dans une roche de caractère analogue ou différent. S'appelle aussi «lumpy bedding». *Couches à nodules, noduleuses.*

Estratos con cuerpos nodulares dispersos o estrechamente agrupados, en una mátriz de naturaleza distinta. También denominada "lumpy bedding". *Estratificación nodular.*

Nodule:

A term generally applied to non-regular bodies of contrasting composition from the matrix in which they are embedded. Most nodules seem to be secondary, such as chert nodules, but some investigators claim others have a primary origin, such as the manganese nodules on present deep-sea floor.

Eine allgemeine Bezeichnung für unregelmäßige Körper, deren Zusammensetzung von der der sie umschließenden Matrix verschieden ist. Die meisten Knollen scheinen sekundär zu sein, wie die Hornstein-Knollen, doch nimmt man für andere, wie die Manganknollen, die sich auf dem heutigen Tiefsee-Boden formen, eine primäre Entstehung an. *Knolle, Knauer.*

Désigne d'une manière générale des corps irréguliers, de composition différente de celle de la roche dans laquelle ils sont inclus. La plupart semblent être d'origine secondaire, tels les silex, mais d'après certains auteurs, il en existerait d'origine primaire, tels que les nodules de manganèse qu'on trouve sur les grands fonds océaniques. *Nodules, concrétions.*

Término general aplicado a todo cuerpo irregular de composición diferente a la mátriz que lo circunda. La mayoría son de origen secundario, como los de ftanita. Ciertos investigadores sostienen el origen primario de algunos; es el caso de aquéllos de manganeso, hallados en fondos marinos actuales. *Nódulo.*

Normal ripple marks (SHROCK, 1948, p. 101):

Simple asymmetrical ridges with various ground plans.

Einfache, unsymmetrische Kämme mit verschiedener Grundanordnung. *Normale Rippelmarken* (w).

Rides simples, asymétriques, de formes et d'arrangements divers. *Rides normales.*

Ondula con crestas asimétricas simples y de trazado variado en planta. *Ondula normal.*

Oblique bedding:

An archaic term meaning inclined bedding.

Eine veraltete Bezeichnung für geneigte Schichtung.

Terme archaïque pour «inclined bedding».

Término en desuso sinónimo de "incline bedding".

Oblique lamination:

An early term (PHILLIPS, 1836, p. 91) for cross-bedding.

Ein sehr alter Ausdruck für Schrägschichtung (PHILLIPS, 1836, S. 91).

Terme ancien pour «cross-bedding».

Término en desuso (PHILLIPS, 1836, p. 91) sinónimo de "cross-bedding" *(Estratificación entrecruzada).*

Oblique stratification: See oblique bedding.

Olistoglyph (VASSOEVICH, 1953, p. 61):

A slide mark or mark due to interlaminar gliding.

Eine Gleitmarke oder Marke, die sich bei internem Gleiten formt. *Olistoglyphe.*

D'après DZULYNSKI et al., une empreinte de glissement, ou une empreinte due au déplacement d'une couche par rapport à une autre. *Olistoglyphe.*

Marca de deslizamiento o marca originada por deslizamiento interlaminar. *Olistoglifo.*

Olistostrome (FLORES, in discussion of BENEO, 1955, p. 122):

A bed or layer accumulated as result of sliding. Generally without internal bedding and composed of intimately mixed heterogeneous materials.

Eine Lage aus gerutschtem Sediment. Im allgemeinen eine strukturlose Masse aus durcheinander gemischtem heterogenem Material. *Olistostrom.*

Couche formée à la suite d'un glissement. En général, dénuée de litage interne, et formée de matériaux hétérogènes bien mélangés. *Olisthostrome.*

Capa acumulada por deslizamiento. Generalmente sin estratificación interna como consecuencia de una mezcla íntima de materiales heterogéneos. *Olistostroma.*

Ordinary rolling strata (ANDERSEN, 1931, Fig. 38):

Relatively uniform asymmetrical wavy bedding with conformable crests and troughs; called "uniform deposition" or ripples "superimposed in rhythm" by McKEE (1939, p. 72).

Relativ einförmige, unsymmetrisch wellige Schichtung mit konkordanten Kämmen und Tälern; von McKEE (1939, S. 72) „gleichförmige Ablagerung" oder „sich rhythmisch überlagernde Rippeln" genannt.

Lits à ondulations régulières, asymétriques, les crêtes et les creux étant superposés d'un lit à l'autre; McKEE appelle ce type de stratification, «uniform deposition» ou rides «superimposed in rhythm» (dépôt uniforme ou rides superposées rhythmiquement). *Lit.: Couches ondulées ordinaires.*

Estratificación ondulada relativamente uniforme y asimétrica, compuesta por crestas y depresiones concordantes; denominada "uniform deposition" o "ripples superimposed in rhythm" por McKee (1939, p. 72). *Estratificación corrugada.*

Oscillation ripple marks:

Ripple marks with relatively straight crests and symmetrical profile (Bucher, 1919, p. 183—184).

Rippeln mit relativ geraden Kämmen und symmetrischem Querschnitt (Bucher, 1919, S. 183—184). *Oszillationsrippeln, Seegangsrippeln.*

Rides à crêtes relativement rectilignes, et à profil symétrique. *Rides d'oscillation.*

Ondula con crestas de perfil simétrico y relativamente rectas en planta (Bucher, 1919, p. 183—184). *Ondula de oscilación.*

Parallel ripple mark (McKee, 1954, p. 57—58): See plate 77.

Ripple mark with relatively straight crests. Crests may be asymmetrical in cross-section.

Rippelmarken mit relativ geraden Kämmen, die im Querschnitt unsymmetrisch sein können. *Parallel-Rippelmarke* (w).

Rides à crêtes relativement rectilignes. Les crêtes peuvent avoir un profil asymétrique. *Rides parallèles.*

Ondula con crestas de perfil generalmente asimétrico y relativamente rectas en planta. *Ondula paralela.*

Para-ripples (Bucher, 1919, p. 262—263): See plate 89A.

A term, introduced by Bucher, applied to large symmetrical and asymmetrical ripples in limestone.

Von Bucher eingeführter und auf große, symmetrische und unsymmetrische Rippeln in Kalkstein angewandt. *Pararippeln* (w).

Terme proposé par Bucher pour décrire de grandes rides symétriques ou asymétriques dans des calcaires. *Rides de grande taille.*

Término propuesto por este autor para describir óndulas formadas en arenas calcáreas; las crestas pueden ser simétricas o asimétricas. *Para-óndula.*

Parting cast (Birkenmajer, 1959, p. 111):

The pseudo-mud cracks of Ksiazkiewicz (1958, pl. 16, Fig. 2). Sand-filled tension cracks due to creep on sea bottom; possibly related to pull-apart structure.

Die Pseudo-Trockenrisse von Ksiazkiewicz (1958, Tafel 16, Abb. 2). Mit Sand gefüllte Dehnungsrisse, die durch Kriechen des Meeresbodens entstehen; möglicherweise den Zerrungs-Strukturen verwandt. *Dehnungsmarke* (w).

Equivalent des «pseudo-mud cracks» de Ksiazkiewicz. Fissures de tension, dues à un effet de rampement sur le fond de la mer, et remplies de sable; peut-être apparenté à la «pull-apart structure».

"Pseudo-mud cracks" de Ksiazkiewicz (1958, pl. 16, fig. 2). Grietas tensionales rellenadas de arena, originadas por desplazamiento de sedimentos en fondos marinos. Estructura probablemente relacionada con "pull-apart structure". *Grietas tensionales.*

Parting lineation (Crowell, 1955, p. 1361): See plates 76A and 92B.

A structure characteristically found on bedding planes of horizontally laminated sandstones. Divided by McBride and Yeakel (1963, p. 780) into (a) parting-plane lineation, "subparallel linear shallow grooves and ridges of low relief (generally less than 1 mm) on lamination surfaces," and (b) parting-step lineation, characterized by subparallel, step-like ridges where the parting surface cuts across several adjacent laminations. Called "current lineation" by Stokes (1947), "primäre Richtung" by Cloos (1938, Fig. 2,

p. 358), described by Sorby (1856, p. 114) as, "a peculiar linear graining on the surface of horizontal beds."

Eine Struktur, die sich häufig auf Schichtflächen parallel feingeschichteter Sandsteine findet. Von McBride und Yeakel unterteilt in (1963, S. 780) a) ebene Strömungs-Streifung, „subparallele, lineare, flache Rillen und Kämme niedrigen Reliefs (im allgemeinen unter 1 mm) auf der Schichtfläche", und b) stufenförmige Strömungs-Streifung, die sich „durch subparallele Stufen auszeichnet, wo die Trennfläche mehrere Feinschichten durchsetzt". Von Stokes (1945) Strömungsriefung genannt, und „primäre Richtung" von Cloos (1938, Abb. 2, S. 358); von Sorby (1856, S. 114) beschrieben als „eine eigenartige, lineare Körnung auf der Oberfläche horizontaler Schichten". *Strömungs-Streifung, Strömungsriefen, feine Parallelstriemung, geradlinige Riefung, Stromstreifung.*

Figure qui se rencontre typiquement sur les plans de stratification de grès lités horizontalement. McBride and Yeakel les classent en: (a) «parting-plane lineation» *(linéation dans le plan de délit)*, «rainures peu profondes et crêtes très peu marquées (en général moins d'un mm de haut) linéaires et presque parallèles, sur le plan de lamination», et (b) «parting-step lineation» *(linéation de délit en gradins)* avec des «crêtes presque parallèles, comme des gradins, le long desquelles le délit recoupe plusieur lits». Phénomène appelé «current lineation» par Stokes, «primäre Richtung» par Cloos, et décrit par Sorby comme «un arrangement linéaire particulier du grain, à la surface de couches horizontales». *Lit.: Linéation de délit.*

Estructura desarrollada en los planos de estratificación de ciertas areniscas laminadas. Dividida por McBride y Yeakel (1963, p. 780) en: a) "parting-plane lineation", surcos y lomos subparalelos de muy escaso relieve (generalmente menor de 1 mm), formados en los planos de estratificación de sedimentos laminados; b) "parting-step lineation", superficies escalonadas subparalelas que cortan varias láminas superpuestas. Denominada "current lineation" por Stokes (1947) y "primäre Richtung" por Cloos (1938, p. 358). Descripta por Sorby (1856, p. 114) como una disposición lineal muy particular de granos en los planos de estratificación. *Lineación por corriente.*

Peiroglyph (Vasseovich, 1953, p. 37):

A cross-cutting structure, such as a sandstone dike.

Eine durchgreifende Struktur wie z. B. Sandstein-Gang. *Peiroglyphe.*

Structure qui recoupe les couches, comme par exemple filon clastique. *Peiroglyphe.*

Estructura que corta a otra, como por ejemplo los diques clásticos. *Peiroglifo.*

Pericline ripple mark (ten Haaf, 1959, Fig. 10b):

Usually arranged in an orthogonal pattern parallel and transverse to current direction with wave length up to 80 cm and up to 30 cm high (ten Haaf, 1959, p. 22).

Gewöhnlich in rechtwinkeligem Muster, parallel und quer zur Strömungsrichtung angeordnet; mit Wellenlängen bis zu 80 cm und Amplituden bis zu 30 cm (ten Haaf, 1959, S. 22). *Perikline Rippelmarke* (w).

«Généralement disposées en un réseau orthogonal de rides parallèles et perpendiculaires à la direction du courant, jusqu'à 80 cm de longueur d'onde et jusqu'à 30 cm de hauteur.» *Lit.: Rides périclinales.*

Ondula con diseño ortogonal en planta, paralela y transversa a la dirección de la corriente, con una longitud de onda de hasta 80 cm y una altura de hasta 30 cm (ten Haaf, 1959, p. 22). *Ondula periclinal.*

Pillow-structure (Smith, 1916, p. 147) (pseudo-nodules, slump balls, flow structure, etc.):
See plates 100—103, incl.

A structure assumed by some sandstones resembling balls or pillows; most characteristic of lower portions of sandstone resting on shales; see ball-and-pillow structure.

In einigen Sandsteinen auftretende ballen- oder kissenähnliche Struktur. Häufig im Unterteil von Sandstein-Bänken, die Schieferton überlagern. Siehe Ballenstruktur. *Kissenstruktur* (w).

Structure que prennent certains grès, et qui ressemble à des boules ou des coussins; particulièrement caractéristique de la base d'un banc de grès reposant sur des argiles schisteuses. *Lit.: Structure en coussins.*

Estructura hallada en ciertas areniscas, con formas de esferoides o cojines; su desarrollo es más completo en la base de ciertas areniscas que yacen sobre pelitas; yéase "ball-and-pillow structure" *(Estructura almohadillada).*

Pit-and-mound structure (KINDLE, 1916, p. 542):

A term introduced by KINDLE to describe "small circular pits and mounds which sometimes mark the surfaces of sedimentary strata" (KINDLE, 1916, p. 542). Pit-and-mound structure may resemble some rain-drop impressions but it tends to have a small pit surrounded by a raised mound (SHROCK, 1948, p. 132—136).

Ein von KINDLE eingeführter Name, um „kleine, runde Trichter und Kegel zu beschreiben, die auf manchen Schicht-Oberflächen zu finden sind" (KINDLE, 1916, S. 542). Trichter- und Kegelstrukturen können Regentropfen-Eindrücken ähneln; sie bestehen jedoch aus Trichter und umgebendem Wall (SHROCK, 1948, S. 132—136). *Trichter- und Kegelstruktur* (w), *Sandkegel.*

Terme proposé par KINDLE pour désigner de «petites cuvettes et monticules circulaires qui marquent parfois la surface de couches sédimentaires». Peuvent ressembler à des empreintes de gouttes de pluie, mais tendent à avoir une petite cuvette entourée d'un monticule.

Término empleado por este autor para describir diminutos montículos con hoyuelos en las cúspides; suelen definir planos de estratificación (KINDLE, 1916, p. 542). Estas formas se asemejan a aquellas originadas por lluvia, pero se distinguen por presentar hoyuelos rodeados por montículos (SHROCK, 1948, p. 132—136). *Estructura de hoyuelo y montículo.*

Planar cross-bedding: See plate 34 B.
Cross-bedding characterized by planar foresets (POTTER and PETTIJOHN, 1963, Fig. 4-1).
Durch ebene Leeblätter ausgezeichnete Schrägschichtung (POTTER und PETTIJOHN, 1963, Abb. 4-1). *Ebene Schrägschichtung* (w), *Diagonalschichtung, gerade gestreckte Schrägschichtung.*
Stratification oblique caractérisée par des couches frontales planes.
Estratificatión entrecruzada con capas frontales planas (POTTER and PETTIJOHN, 1963, fig. 4-1). *Estratificación entrecruzada plana.*

Planar cross-stratification (McKEE and WEIR, 1953, Fig. 2):
Lower bounding surface of cross-stratified unit has a planar surface of erosion.
Schrägschichtungs-Einheit, deren Sohlfläche eine ebene Erosionsfläche ist. *Ebenflächig begrenzte Schrägschichtung* (w).
La face limite inférieure d'un terme à stratification oblique est une surface d'érosion aplanie. *Groupe planaire de laminae.*
Estratificación entrecruzada depositada sobre una superficie erosional plana. *Estratificación entrecruzada basiplana.*

Prod mark (DZULYNSKI and SLACZKA, 1958, p. 232) (prod cast): See plates 66 and 68.
A short ridge, parallel to the current, which unlike flute casts, rises down-current and ends abruptly.
Ein parallel zur Strömung verlaufender kurzer Kamm, der im Gegensatz zu Strömungswülsten stromabwärts zu ansteigt und unvermittelt absetzt. *Stoßmarke* (n), *Stechmarke.*

Petite protubérance parallèle au courant, mais qui, au contraire des «flute casts» va en s'élevant vers l'aval et se termine brusquement.

Lomos cortos que aumentan de relieve progresivemente en sentido de la corriente y disminyen con pendiente abrupta (semejantes a "flute casts" pero de orientación inversa). *Calco de punzamiento.*

Progressive sand waves (Bucher, 1919, p. 168):

Sand waves that migrate down-current.

Sandwellen, die stromabwärts wandern. *Progressive Sandwellen* (w).

Vagues de sable qui se déplacent dans le sens du courant.

Ondulación arenosa que se desplaza en sentido de la corriente. *Onda de arena progresiva.*

Proglyph (Vassoevich, 1953, p. 36):

Term applied to a *cast* of markings such as grooves.

Ein Name für Ausgüsse von Marken, wie z. B. Rillen. *Proglyphe.*

Terme qui désigne le moulage, le remplissage d'empreintes telles que des cannelures.

Término aplicado a calcos de marcas, tales como surcos. *Proglifo.*

Prolapsed bedding (Wood and Smith, 1959, p. 172, pl. 8, Fig. 3):

A series of flat folds with near-horizontal axial planes contained entirely within a bed with undisturbed boundaries.

Eine Reihe flacher Falten mit nahezu horizontalen Achsenflächen; auf eine einzige Schicht beschränkt, die von ungestörten Schichten über- und unterlagert ist. *Zusammengefaltete Schichtung* (w).

Série de plis plats, dont les plans axiaux sont presque horizontaux, entièrement contenue au sein d'une couche dont les surfaces limites ne sont pas dérangées. *Lit.: Stratification oblique couchée.*

Pliegues achatados cuyos planos axiales están dispuestos horizontalmente. Se encuentran limitados entre estratos normales (no deformados). *Pliegues volcados intraformacionales.*

Pseudo cross-stratification:

Inclined bedding which by unusual sorting, commonly in response to ripple mark migration, produces foreset beds that appear to dip into the current (Spurr, 1894, p. 45—46).

Geneigte Schichtung mit ungewöhnlicher Sortierung, die im allgemeinen auf wandernde Rippeln zurückzuführen ist. Die Leeblätter scheinen nach stromaufwärts einzufallen (Spurr, 1894, S. 45—46). *Pseudo-Schrägschichtung* (w).

Stratification inclinée dans laquelle un granoclassement particulier, le plus souvent dû au déplacement de rides de courant, produit des couches frontales qui semblent pendre vers l'amont. *Lit.: Pseudo-stratification oblique.*

Estratificación inclinada que presenta selección excepcional debido frecuentemente a migración de óndulas, formando capas frontales inclinadas que parecen buzar en sentido opuesto a la corriente. *Pseudo-estratificación entrecruzada.*

Pseudo mud cracks (Ksiazkiewicz, 1958, pl. 16, Fig. 2): See parting cast.

Pseudo-nodules (Macar, 1948, p. 48):
See ball-and-pillow structure.

Pseudo ripple marks (Kuenen, 1948b, p. 372):
A type of deformation attributed to lateral pressure, characterized by regularly spaced corrugations or small-scale similar folds; occurs in beds immediately beneath slump sheet. Term also applied to tectonic corrugations (Ingerson, 1940).

Eine auf seitlichen Druck zurückgeführte Verformung, die sich durch in gleichen Abständen laufende Runzeln oder kleine „ähnliche Falten" auszeichnet. Tritt unmittelbar unter Rutschhorizonten auf. Bezeichnung wird auch auf tektonische Runzeln angewandt (INGERSON, 1940). *Scheinrippeln.*

Type de déformation que l'on attribue à une pression latérele, et qui est caractérisée par des ondulations régulièrement espacées, ou autres microplis semblables; se rencontre dans les couches situées immédiatement sous la nappe de glissement. Ce terme s'applique aussi à des ondulations d'origine tectonique. *Lit.: Pseudo-rides, fausses rides.*

Tipo de deformación atribuida a presiones laterales y caracterizada por corrugaciones regularmente espaciadas o pliegues pequeños homogéneos. Se presenta en ciertos estratos cuando el suprayacente ha sido desplazado lateralmente. Término también aplicado a corrugaciones de origen tectónico (INGERSON, 1940). *Pseudo-óndulas.*

Pull-apart structure (NATLAND and KUENEN, 1951, p. 89):
Beds which have been stretched and torn apart into shapes similar to those of boudinage.
Durch Dehnung in boudinähnliche Formen gestreckte und zerrissene Schichten. *Zerrungs-Struktur* (w).
Couches qui ont été étirées et déchirées, et qui finissent par avoir un aspect comparable au boudinage. *Couches étirées, phénomène d'arrachage.*
Capas desgarradas por fuerzas tensionales, con aspecto similar a la estructura "boudinage" *(Estructura moniliforme).*

Quake sheet (KUENEN, 1958, p. 20):
A product of seismic shock quake ("quake load-casting") without horizontal slip. Probably transitional to true slump sheets.
Eine Lage, deren Verformung als das Ergebnis von Erdbeben angesehen wird. Seitliche Verschiebungen (Gleitungen) fehlen. Übergänge zu wahren „Rutschungs-Schichten" sind wahrscheinlich. *Erschütterungs-Schicht* (w).
Paquet de sédiment dont l'enfoncement a été déclanché par un choc séismique ou un tremblement de terre (par «quake load-casting»). Il peut exister des stades de transition entre ce type et les nappes de glissement à déplacement latéral.
Estrato deformado por carga (sin desplazamiento horizontal) debido a movimientos sísmicos. Pueden existir estados de transición entre este tipo y aquellos estratos deslizados. *Estrato sísmico.*

Radiate mud cracks:
Mud cracks that display an incomplete radiate pattern and lack normal polygonal development (KINDLE, 1926, p. 73, pl. 3).
Trockenrisse, die ein unvollständig radiales Muster zeigen und denen normale polygonale Entwicklung fehlt (KINDLE, 1926, S. 73, Tafel 3). *Radiale Trockenrisse* (w).
Fentes de dessiccation qui forment un réseau radiaire incomplet, et n'atteignent pas le stade polygonal normal. *Fentes de dessiccation radiaires.*
Grietas de desecación en diseño radiado incompleto sin desarrollo poligonal (KINDLE, 1926, p. 73, pl. 3). *Grietas radiadas de desecación.*

Rain prints (casts) (rain-drop impressions): See plate 94.
Small circular pits formed in soft sediment, usually mud, generally preserved as casts on underside of superjacent sandstone bed (TWENHOFEL, 1921).
Kleine, rundliche Vertiefungen, die in weichem Sediment (meistens Schlamm) entstehen, und im allgemeinen als Ausgüsse auf der Unterseite des hangenden Sandsteins erhalten sind (TWENHOFEL, 1921). *Regentropfen-Eindrücke, -Marken.*

Petites cuvettes circulaires formées dans les sédiments mous, en général dans la boue, et conservées le plus souvent sous forme de moulages à la face inférieure du grès sus-jacent. *Empreintes de pluie, de gouttes de pluie.*

Hoyuelos circulares formados sobre sedimentos pelíticos, preservados como calcos en la base de ciertas areniscas (TWENHOFEL, 1921). *Marcas de lluvia.*

Rectangular interference ripples (BUCHER, 1919, p. 190):

Interference ripples with rectangular pattern (KINDLE, 1917a, pl. 23).

Interferenzrippeln mit rechteckiger Anordnung (KINDLE, 1917a, Tafel 23). *Rechteckige Interferenzrippeln* (w).

Rides d'interférence formant un réseau orthogonal.

Ondulas de interferencia con diseño rectangular (KINDLE, 1917a, pl. 23). *Ondulas rectangulares.*

Reed casts (OSBORNE, 1958, p. vii):

A term used for vertical cylindrical sand-casts presumably of reed-roots. See root cast.

Vertikale, zylindrische Sandstrukturen, die vermutlich Ausgüsse von Schilfwurzeln darstellen. Siehe Wurzelstruktur.

Moulages sableux verticaux et cylindriques attribués à des racines de roseaux.

Término empleado para huecos cilíndricos y verticales, rellenados por arena en reemplazo de presumibles raices de juncáceas. Véase "root cast". *Calco junco-radiciforme.*

Regressive ripples (JOPLING, 1961, p. D-15):

A series of asymmetric current ripples with steep sides pointing into the current.

Eine Reihe unsymmetrischer Strömungsrippeln, deren Steilseiten nach stromaufwärts zeigt. *Regressive Rippeln* (w).

Une série de rides de courant asymétriques, dont les faces abruptes sont tournées vers l'amont. *Rides de courant régressives.*

Ondula con crestas asimétricas, cuyos flancos más abruptos están orientados en sentido opuesto a la corriente. *Ondula regresiva.*

Regressive sand wave (BUCHER, 1919, p. 165):

A sand wave that migrates up-current; see antidune.

Eine Sandwelle, die stromaufwärts wandert; s. Antidüne. *Regressive Sandwelle* (w).

Vague de sable qui se déplace vers l'amont, qui recule.

Ondulación arenosa que se desplaza en sentido opuesto a la corriente. Véase "antidune" *(Antiduna).*

Rheoglyph (VASSOEVICH, 1953, p. 55):

A hieroglyph produced by syngenetic deformation, slump and related processes, as opposed to turboglyphs which are current produced.

Eine Hieroglyphe, die durch synsedimentäre Deformation, Rutschung und ähnliche Prozesse hervorgerufen wird; im Gegensatz zu den strömungserzeugten Turboglyphen. *Rheoglyphe.*

Hiéroglyphe produit par déformation syngénétique, par un glissement ou coulée ou autres processus, par opposition à turboglyphe, produit par un courant. *Rheoglyphe.*

Hieroglifo originado por deformación singenética, debido a desplazamientos y procesos afines, en oposición a los turboglifos que se originan por corrientes. *Reoglifo.*

Rhomboid ripple mark (BUCHER, 1919, p. 153):

Ripple mark with a reticulate, rhomboid pattern (KINDLE, 1917a, pl. 19b).

Rippelmarken mit gitterartiger, rhomboider Anordnung (KINDLE, 1917a, Tafel 19b). *Rhomboidrippel, rautenförmige Rippel.*

Rides de plage formant un réseau rhomboïdal. *Rides rhomboïdales, en losange, losangiques.*

Ondula con diseño reticulado romboidal (KINDLE, 1917a, pl. 19b). *Ondula romboide.*

Rib-and-furrow (STOKES, 1953, p. 17—21); See plate 38A.

The bedding plane expression of micro-cross-stratification. Originally called *Schräg-schichtungsbögen* by GÜRICH (1933, Figs. 1, 2, 4 and 6).

Kleindimensionale Schrägschichtung in der Aufsicht. Ursprünglich von GÜRICH (1933, Abb. 1, 2, 4 und 6) Schrägschichtungsbögen genannt. *Schrägschichtungsbögen.*

Microstratification croisée vue dans le plan de stratification. A l'origine appelée Schräg-schichtungsbögen par GÜRICH. *Lit.: Structure en côtes et sillons.*

Aspecto del plano de estratificación (en planta) de un estrato con micro estratificación entrecruzada; véase "micro-cross-lamination". Término original "Schrägschichtungs-bögen" (GÜRICH, 1933, figs. 1, 2, 4 and 6). *Estructura de costillas y surcos.*

Rill-cast (DZULYNSKI and SLACZKA, 1958, p. 230, pl. 26, Fig. 3):

Probably same as furrow flute casts, which see.

Entsprechen wahrscheinlich den gefurchten Strömungswülsten; s. dort.

Probablement équivalent de «furrow flute cast», q.v.

Probablemente equivalente a "furrow flute cast", véase.

Rill mark: See plate 92A.

Dendritic, bifurcating-upstream, rivulet commonly found on subaerial portion of beaches, sand bars, and sand flats formed by flow of thin sheet of water (SHROCK, 1948, p. 128). Some rill marks show a distributary pattern of down-current bifurcation.

Dendritische, sich nach stromaufwärts gabelnde Rinnsale, die man häufig auf den der Luft ausgesetzten Partien von Stranden, Sandbänken und -flächen findet. Sie formen sich beim Abfluß einer dünnen Wasserschicht (SHROCK, 1948, S. 128). Das Verteilungssystem mancher Rieselmarken gabelt sich nach stromabwärts. *Rieselmarke.*

Rigoles dendritiques, divergentes vers l'amont, qu'on rencontre souvent sur la partie sub-aérienne des plages, bancs de sable, ou étendues sableuses, et qui sont formées par l'écoulement d'une mince couche d'eau. Dans certains cas, les rigoles forment un réseau divergent vers l'aval. *Rigoles de plage.*

Canalículos con distribución dendrítica, bifurcados en sentido opuesto a la corriente, formados en la zona subaereal de playas, barras y planicies arenosas; su desarrollo se debe al flujo de una delgada lámina de agua (SHROCK, 1948, p. 128). En algunos casos se bifurcan en sentido de la corriente. *Marcas de escurrimiento.*

Ring-mark (DZULYNSKI and SLACZKA, 1958, p. 236): See plate 68.

Ring-like ridges commonly appearing in a line, the higher side is upcurrent. Incomplete rings, forming semicircles with concavities downcurrent are common. Considered a species of saltation mark produced by fish vertebrae.

Ringförmige Wälle, deren höhere Seite stromaufwärts zeigt. Oft aneinandergereiht. Un-vollständige, halbkreisförmige Ringe, deren konkave Seite stromabwärts zeigt, sind häufig. Sie werden als eine Art Springmarke angesehen, die von Fischwirbeln hervorge-rufen werden. *Ringmarke* (w).

Crêtes en anneau, plus élevées vers l'amont, et en général alignées. Il y a souvent des anneaux incomplets, formant des demi-cercles à concavité vers l'aval. On les considère comme un type de figure de culbutage, produit par des vertèbres de poisson qui se déplacent par bonds.

Realces semicirculares generalmente orientados, con sus sectores más sobresalidos orientados en sentido opuesto a la corriente; las concavidades están orientadas en sentido de la corriente. Se atribuye su formación a vertebras de peces transportadas por saltación. *Marcas anulares.*

Ripple bedding:
Bedding surface characterized by ripple marks.
Schicht-Oberfläche von Rippeln besetzt. *Rippelschichtung.*
Plan de stratification caractérisé par la présence de rides.
Plano de estratificación caracterizado por la presencia de una óndula. *Estratificación ondulítica.*

Ripple cross-lamination (McKee, 1938, p. 80): See plates 38 B and 39.
Cross-lamination of small thickness, usually less than 2 cm, formed by migrating ripple mark.
Schrägschichtung geringer Dicke, gewöhnlich weniger als 2 cm, die von wandernden Rippeln geformt wird. *Rippelschichtung, Rippelflaserung, Kleinrippel-Schrägschichtungs-Flaserung.*
Ensemble à stratification croisée, de moins de 2 cm d'épaisseur en général, et formé par la migration d'une ride. *Stratification croisée de rides.*
Laminación entrecruzada de escaso espesor (generalmente menor de 2 cm) formada por migración de una óndula. *Migro-óndulas entrecruzadas.*

Ripple drift (Sorby, 1857, p. 278):
Small-scale cross-bedding formed by migrating ripples.
Kleindimensionale Schrägschichtung, die von wandernden Rippeln hervorgerufen wird. *Rippelschichtung.*
Stratification croisée à petite échelle, formée par des rides qui se déplacent.
Estratificación entrecruzada en pequeña escala formada por migración de óndulas. *Migro-óndulas entrecruzadas.*

Ripple mark: See plates 77 to 91 A.
Periodic undulations of primary origin at interface between a fluid and granular material; usually on a small scale; many varieties. See sand wave.
Periodische Undulationen primären Ursprungs an der Grenzfläche zwischen fluidalen und körnigen Medien; im allgemeinen kleindimensional; große Vielfalt. Siehe Sandwellen. *Rippelmarke, Wellenfurche.*
Ondulations périodiques, d'origine primaire, dans le plan de séparation entre un fluide et un matériau granulaire; en général de petite taille; il en existe de nombreux types. Cf. sand wave. *Rides de plage.*
Ondulación periódica primaria formada en interfaces de materiales granulares y flúidos, generalmente de pequeñas dimensiones; abundan las variedades. Véase "sand wave". *Ondula.* Nota: el término *ondulita* se emplea para aquellas óndulas fósiles.

Ripple scour (Potter and Glass, 1958, pl. 5): See plate 91 A.
A shallow, linear trough with transverse ripple mark.
Ein seichter, langgestreckter Trog mit Transversalrippeln. *Gerippelte Erosionsmulde* (w).

Rides transversales dans un creux allongé, peu profond.

Cauce superficial y elongado con óndula de crestas transversas. *Ondula de cauce.*

Roll-mark (Dzulynski and Slaczka, 1958, p. 234):

A series of similar marks appearing in line in direction of flow made by a rolling object. Rolling is commonly combined with saltation. See Pavoni (1959, p. 945, Fig. 2).

Eine parallel zur Strömung verlaufende Reihe einander ähnlicher Marken, die von rollenden Gegenständen erzeugt werden. Einherrollen ist häufig verbunden mit sprungartigen Bewegungen. Siehe Pavoni (1959, S. 945, Abb. 2). *Rollmarke.*

Une série d'empreintes semblables, alignées dans la direction du courant et formées par un objet qui roule. Ce roulement est souvent associé à des bonds ou culbutes. *Lit.: Empreintes de roulement.*

Serie de marcas similares y alineadas, paralelas al flujo, originadas por rolido de cuerpos, frecuentemente combinadas con marcas de saltación. Véase Pavoni (1959, p. 945, fig. 2). *Marcas de rolido.*

Roll-Spuren (Krejci-Graf, 1932, p. 27): See roll-mark.

Rolling incline-bedding (Andersen, 1931, Fig. 38):

Small scale cross-bedding related to ripple-mark migration; asymmetrical with more pronounced lee-side accumulation than in unilateral rolling strata, which see.

Kleindimensionale Schrägschichtung, die mit wandernden Rippeln in enger Beziehung steht. Unsymmetrisch; mit stärker ausgeprägter Akkumulation an der Leeseite als in „unilateral rolling strata"; s. dort.

Stratification croisée de petite taille, liée à la migration de rides; asymétrique, et avec une accumulation avale (lee side accumulation) plus fréquente et plus prononcée que dans «unilateral rolling strata», q.v. *Stratification onduleuse asymétrique.*

Estratificación entrecruzada de pequeñas dimensiones, originada por desplazamiento de óndulas. Su formación se debe a la mayor acumulación de material a sotavento de las crestas que en el caso de "unilateral rolling strata", véase. *Estratificación pseudo-inclinada.*

Rolling strata (Andersen, 1931, Fig. 38):

A little-used term for ripple cross-lamination; also called wavy bedding.

Eine selten gebrauchte Bezeichnung für Rippel-Schrägschichtung; auch wellige Schichtung genannt. *Rippelschichtung.*

Terme peu usité pour «ripple cross-lamination»; appelé aussi «wavy bedding». *Couches ondulées.*

Término poco empleado, sinónimo de "ripple cross-lamination"; también denominada "wavy bedding". *Migro-óndulas entrecruzadas.*

Roll-up structure (Dott and Howard, 1962, p. 119, pl. 3 a):

Same as convolutional ball, which see.

Dasselbe wie Wulstballen; s. dort. *Wickelstruktur.*

Synonyme de «convolutional ball», q.v.

Similar a "convolutional balls" *(Esferoides intraformacionales),* véase.

Root cast:

Slender, tubular, near-vertical and commonly downward-branching structure formed by filling of tubular openings left by roots. Many resemble some organic borings.

Schmale, röhrenförmige, nahezu vertikale und gewöhnlich sich nach unten verzweigende Struktur, die sich durch Ausfüllen bei von Wurzeln hinterlassenen, röhrenförmigen Öffnungen bildet. In vielen Fällen ähneln sie organischen Bohrlöchern. *Wurzelstruktur* (w).

Accidents tubulaires, minces, presque verticaux et qui se ramifient souvent vers le bas, formés par le remplissage de canalicules laissés par des racines. Ressemblent souvent à des perforations d'origine animale. *Lit.: moulages de racines.*

Estructura formada por relleno de oquedades tubulares, ramificadas y aproximadamente verticales, dejadas por raices. En muchos casos se asemejan a aquellas originadas por especies animales. *Calco de raices.*

Ruffled groove cast (TEN HAAF, 1959, p. 32, Fig. 18):
A groove cast with a feather pattern; i.e. a groove with lateral wrinkles that join the main cast in the down-current direction at an acute angle (POTTER and PETTIJOHN, 1963, pl. 17B).

Eine Rillenmarke mit Fiedermuster, d.h. eine Hauptrille mit geraden Seitenrillen, die nach stromabwärts in spitzem Winkel auf die erstere treffen (POTTER und PETTIJOHN, 1963, Tafel 17B). *Gefiederte Rillenmarke.*

Moulage d'une cannelure avec un dessin en plume, i.e. cannelure avec des micro-rides latérales qui rejoignent la dépression principale dans la direction avale en formant un angle très aigu. *Lit. cannelure tuyautée.*

Calco de surco con el aspecto de una pluma de ave. Las "barbas" formadas por crestas y surcos alternados, se unen al "raquis" principal en ángulo agudo orientado en sentido de la corriente (POTTER and PETITJOHN, 1963, pl. 17B). *Calco de surco espigado.*

Sag structure (TEN HAAF, 1959, p. 43):
A general term for load-casts and related structures.

Eine allgemeine Bezeichnung für Belastungsmarken und ähnliche Strukturen. *Sackungs-Struktur* (w), *Belastungs-Setzung.*

Terme général pour les empreintes de charge et accidents semblables appelés parfois *figures,* ou *poches d'enfoncement.*

Término general empleado para calcos de carga y estructuras emparentadas. *Estructura de carga.*

Saltation marks (DZULYNSKI and SLACZKA, 1958, p. 235): See plate 68.
Marks and the casts thereof, made by an object proceeding along a saltatory path. Related to roll-mark and includes ring-mark.

Ausgüsse von Eindrücken eines Gegenstandes, der sich sprungweise vorwärtsbewegt. Ähnlich den Rollmarken, und Ringmarken einschließend. *Springmarken* (w).

Empreintes (et leurs moulages) formées quand un objet se déplace par bonds. Apparentées aux «roll-marks», et comprenant les «ring-marks». *Lit.: figures de bondissement, de saltation.*

Marcas (o su calco) originadas por saltación de un cuerpo transportado. Estructura vinculada con "roll-mark" y "ring-mark". *Marcas de saltación.*

Sand shadow (DZULYNSKI and SLACZKA, 1958, p. 237, Fig. 13):
A lee-side accumulation of sand behind a fixed object.

Leeseitige Anhäufung von Sand hinter einem festen Objekt. *Sedimentfahne.*

Accumulation de sable dans le creux aval, derrière un objet immobile.

Acumulación de arena a sotavento de un cuerpo fijo. *Sombra arenosa.*

Sand streak:
Linear, parallel, low ridges with symmetrical cross-section that form at interface of sand and air or water and are parallel to direction of flow.

Gerade, niedrige, parallele Kämme mit symmetrischem Profil, die sich an der Grenzfläche von Wasser oder Luft formen und parallel zur Strömung laufen. *Strömungsriefen.*

Crêtes basses, linéaires et parallèles, à profil symétrique, qui se forment à la surface de séparation du sable et de l'air ou de l'eau, et qui sont parallèles à la direction d'écoulement. *Lit.: Filets de sable.*

Lomos alargados, bajos, de perfil simétrico, formados en una interface de aire o agua, paralelos a la dirección del flujo. *Trazas arenosas.*

Sand volcanoes (GILL and KUENEN, 1958, p. 442): See plate 114.

Miniature volcano-like accumulations of sand generally situated on top of slump sheets.

Miniaturvulkanen ähnliche Anhäufungen aus Sand, die sich häufig an der Oberfläche von Rutschungsmassen finden. *Sandvulkane, Sandkegel.*

Accumulations de sable comme des volcans en miniature, situées en général à la surface d'une nappe de glissement. *Lit.: volcans de sable.*

Acumulaciones de arena con aspecto de volcanes diminutos, formadas sobre mantos deslizados. *Volcanes de arena.*

Sand wave: See plates 27, 28, 29 A, 77 to 83 B.

"A ridge on the bed of a stream formed by the movement of the bed material, which is usually approximately normal to the direction of flow, and has a shape somewhat resembling a wave" (LANE, 1947, p. 938). Sand waves are usually periodic and are also common in other fluid systems such as those in aeolian and marine environments.

„In Flußbetten auftretende Rücken (Barren), die sich durch Fortbewegung des Materials formen, aus dem das Flußbett besteht. Sie verlaufen häufig normal zur Strömungsrichtung und besitzen eine wellenartige Form" (LANE, 1947, S. 938). Sandwellen sind im allgemeinen periodisch. Sie treten häufig auch in den Ablagerungen anderer bewegter Medien auf, wie in äolischen und marinen Ablagerungen. *Sandwelle* (w), *Strombank, Sandrücken.*

«Crête formée sur le lit d'un cours d'eau par le mouvement des matériaux du lit, en général perpendiculaire à la direction du courant, et dont la forme ressemble à celle d'une vague" (LANE). Les vagues de sable sont en général périodiques et se rencontrent fréquemment dans d'autres milieux fluides, tels que les milieux éolien ou marin. *Lit.: vague de sable. Ride géante.*

Onda formada por desplazamiento del material de un río, aproximadamente perpendicular a la dirección de la corriente (LANE, 1947, p. 938). Generalmente son periódicas y se forman también por transporte eólico y en ambientes marinos. *Onda de arena.*

Sandstone balls: See plates 104 A and B. See pillow-structure.

Sandstone dike (Clastic dike): See plates 113 A and B.

A body of sandstone contained in a cross-cutting fissure; related to sandstone sills, a product of quicksand intrusion (DILLER, 1890).

Spaltenfüllender Sandstein-Körper. Steht in Beziehung zu Sandstein-Lagergängen, die Quicksand-Intrusionen darstellen (DILLER, 1890). *Sandstein-Gang, Sedimentgang.*

Masse de grès contenue dans une fissure qui recoupe les sédiments; apparentée aux «sandstone sills», lesquels résultent de l'intrusion de sable mouvant, fluent. *Filon de grès, filon clastique.*

Cuerpo de arena controlado por una fisura que corta a una estructura; relacionada con filones capas areniscosos originados por la intrusión de arenas móviles (DILLER, 1890). *Dique areniscoso.*

Sandstone pipes: See cylindrical structures.

Scalloping (Shrock, 1948, p. 121):

A name given to a structure of unknown origin found only in slates and superficially resembling ripple marks.

Eine Struktur unbekannter Herkunft, die nur in Tonschiefer auftritt und oberflächlich Rippelmarken ähnelt. *Wellung.*

Accident d'origine inconnue, qui ne se rencontre que dans des ardoises, et qui offre une ressemblance superficielle avec les rides de plages. *Lit.: Festons.*

Término asignado a una estructura de origen desconocido, hallada únicamente en pizarras. Su aspecto es semejante a una óndula. *"Ondulita" en pizarra.*

Scour-and-fill (cut-and-fill, washout, channel):

Small-scale bottom scour or channels that are subsequently filled.

Kleindimensionale Kolke oder Erosionsrinnen, die später ausgefüllt werden. *Kolkförmige Erosion, kleindimensionale Erosionsrinne.*

Creusement du fond du lit ou chenaux qui seront ultérieurement comblés, de petites dimensions. *Creusement et remblaiement.*

Pequeño cauce rellenado. *Relleno de cauce.*

Scour cast (KINGMA, 1958, Figs. 1 and 2): See flute cast.

Scour channel: See channel.

Scour finger (BOKMAN, 1953, p. 159, pl. 1 C): See flute cast.

Scour lineation (McIVER, 1961, p. 178):

Smooth ridges, 2 to 5 cm wide of very low relief, characterized by symmetrical ends. Gives line of current movement but not direction. Believed to be due to scour.

Glatte Rücken, 2—5 cm weit, mit sehr niedrigem Relief und ausgezeichnet durch symmetrische Enden. Sie geben den Sinn, aber nicht die Richtung der Strömung an. Man hält sie für Auskolkungen. *Kolkriefen.*

Petits cordons réguliers, de 2 à 5 cm de largeur et très bas, à terminaisons symétriques. Indiquent la direction, mais pas le sens, du courant. Attribués à un creusement.

Lomos de 2,5 a 5 cm de ancho, de escaso relieve y extremos simétricos; estructura debida a desbaste por agua. Indican dirección de la corriente pero no sentido. *Lineación de cauces.*

Scour marks (DZULYNSKI and SANDERS, 1963, p. 65):

Sole marks, generally preserved as casts, of depressions produced by current erosion or scour.

Im allgemeinen als Ausgüsse erhaltene Vertiefungen. Durch Strömungserosion oder Auskolkung geformt. *Kolkmarke, Auskolkung, Erosionsmarke, Fließrille, Strömungsrinne.*

Empreintes, généralement conservées par moulage, de dépressions résultant de l'érosion ou du creusement par un courant. *Empreintes de creusement.*

Marcas de base generalmente preservadas como calcos de depressiones, debidas a erosión o desbaste por corrientes. *Marcas de desbaste.*

Sea-balls:

Ball-like masses of somewhat fibrous material of organic origin mechanically collected by wave movement in shallow waters (CRONEIS and GRUBBS, 1939, p. 598). See also lake balls.

Aus faserigem Material organischer Herkunft bestehende Ballen, die mechanisch durch Wellen-Tätigkeit in seichtem Wasser zusammengespült werden (CRONEIS und GRUBBS, 1939, S. 598). Siehe auch lakustrische Pflanzenknäuel. *Marine Pflanzenknäuel, Pflanzenballen*.

Masses en boules de matériaux assez fibreux d'origine organique, assemblés mécaniquement par l'effet des vagues en eau peu profonde. *Pelotes de mer*.

Cuerpos esferoidales constituidos por materiales algo fibrosos, de origen orgánico, congregados mecánicamente en aguas poco profundas (CRONEIS and GRUBBS, 1939, p. 598). Véase "lake balls". *Esferoides marinos*.

"Sealing-wax flow" (FAIRBRIDGE, 1946, p. 8): See convolute bedding.

Sedimentation unit (OTTO, 1938, p. 574):

A layer or deposit formed under conditions of essentially constant flow and sediment discharge; distinguished from like units by changes in grain size and/or fabric indicating changes in velocity and/or direction of flow.

Eine Lage, die sich unter im wesentlichen konstanter Strömung und Sedimentzulieferung formt; die einzelnen Einheiten unterscheiden sich voneinander durch Wechsel in Korngröße und/oder Gefüge, die Veränderungen in Geschwindigkeit und/oder Richtung der Strömung anzeigen. *Sedimentations-Einheit* (w).

Couche ou dépôt formé dans des conditions inchangées de courant et d'apport de sédiment; séparé des unités voisines par des changements granulométriques ou de texture (ou les deux), qui indiquent un changement de vitesse ou de direction d'écoulement (ou les deux). *Ensemble sédimentaire*.

Capa depositada bajo condiciones constantes de velocidad de flujo y descarga de material. Se distingue de otra similar por cambios en la granometría y/o fábrica, que indican a su vez cambios en la velocidad y/o sentido de la corriente. *Unidad sedimentaria*.

Shale crescents (SHROCK, 1948, Fig. 86): See plate 82A.

Crescents formed by shale filling the troughs of ripple mark.

Bogige Kämme aus Schieferton, der die Tröge von Rippelmarken füllt. *Schieferton-Schmitzen* (w).

Croissants formés par une argile schisteuse qui remplit les creux entre des rides.

Pequeños lentes arcillosos formados por relleno parcial de las depresiones de una óndula. *Lentecillos arcillosos*.

Shingle structure: See imbricate structure.

Shooting flow cast (WOOD and SMITH, 1959, p. 169):

Strong parallel ridges, up to 30 cm wide, 10 cm deep, and up to 2 m long.

Streng parallele Rücken, die bis zu 30 cm breit, 10 cm tief und 2 m lang werden können. *Gescharte große Fließmarken* (n).

Crêtes parallèles, bien marquées, ayant jusqu'à 30 cm de largeur et 10 de hauteur, et jusqu'à 2 m de long.

Lomos paralelos bien marcados, de hasta 30 cm de ancho, 10 cm de altura y hasta 2 m de largo. *Mega calcos de flujo*.

Shrinkage cracks:

See desiccation cracks; see also syneresis cracks.

Simple cross-stratification (McKEE and WEIR, 1953, Fig. 2):

Lower bounding surface of cross-stratified unit is a nonerosional surface.

Die die Schrägschichtungs-Einheit unterliegende Fläche ist nicht durch Erosion entstanden. *Einfache Schrägschichtung* (w).

La surface à la base d'un ensemble à stratification croisée n'est pas une surface d'érosion.

Unidad sedimentaria con estratificación entrecruzada, depositada sobre una superficie no erosional. *Estratificación entrecruzada simple.*

Skip casts: See skip marks.

Skip marks (DZULYNSKI and SLACZKA, 1958, p. 231) (skip casts): See plate 68.

Regularly spaced marks (or casts) in a straight line produced by regular intermittent impingement on bottom. Essentially a regular series of bounce or brush casts, which see.

In regelmäßigen Abständen aufeinanderfolgende Abdrücke (oder Ausgüsse), die durch gleichmäßig unterbrochenes Aufprallen auf den Grund entstehen. Im Grunde eine regelmäßige Reihe von Aufprall- oder Quastenmarken. Siehe dort. *Hüpfmarke.*

Empreintes (ou moulages) régulièrement espacées et en ligne droite, produites par des heurts intermittents et réguliers contre le fond. En fait, une série régulière d'empreintes de rebondissement ou d'éraflures.

Marcas (o su calco) regularmente espaciadas y alineadas, producidas por el impacto de un cuerpo sobre el lecho de la corriente que lo transporta. En esencia son series regulares de "bounce casts" o "brush casts". *Marcas de saltación regular.*

Slide:

See slump; slides are considered by some to have large lateral displacement in contrast to slumps which are local or restricted displacements.

Siehe Rutschung. Gleitungen werden zum Teil als mit großen seitlichen Verschiebungen verbunden angesehen; im Gegensatz zu Rutschungen, die lokale oder geringe Verschiebungen darstellen. *Gleitung, subaquatische —.*

Cf. slump; certains auteurs considèrent que les «slides» montrent un déplacement latéral important alors que les «slumps» sont plus localisés et restreints. *Eboulement, coulée.*

Algunos autores consideran que es un desplazamiento lateral amplio, en contraste con "slump", desplazamiento local o restringido; véase "slump". *Deslizamiento.*

Slide casts (KUENEN and SANDERS, 1956, pl. 2; TEN HAAF, 1959, p. 38, Fig. 23):

The casts of marks made by sliding objects, such as a mass of sediment, large soft-bodied animal, plant mats, etc.

Die Eindrücke oder deren Ausfüllungen, die von gleitenden Gegenständen hervorgerufen werden, z.B. von Sedimentmassen, großen Weichkörper-Tieren, Pflanzenmassen usw. *Gleitmarke.*

Moulages d'empreintes laissées par des objets qui glissent, par exemple une masse de sédiments, un gros animal à corps mou, un radeau végétal, etc. *Empreintes de glissement.*

Calco de marcas originadas por deslizamiento en masa de sedimentos, restos de animales, tapices vegetales, etc. *Calco de deslizamiento.*

Slide mark (KUENEN, 1957, p. 251) (slide casts): See plates 65 A and B.

A parallel system of scratches or grooves left on sediments by subaqueous gliding or slumping. The "slide casts" of TEN HAAF (1959, p. 38) are one to several decimeters wide and commonly smoothly curved, less than a meter in length and ascribed to larger clots of mud, sand, or perhaps colonial organisms that slipped or slid.

Parallele Reihen von Striemen oder Rillen, die von subaquatischen Gleitungen oder Rutschungen im Sediment hinterlassen werden. TEN HAAF'S (1959, S. 38) Gleitmarken sind ein bis mehrere Dezimeter breit, leicht gekrümmt, weniger als 1 m lang und werden als Gleiterscheinung von größeren Klumpen aus Schlamm, Sand oder Organismen-Kolonien angesehen. *Gleitmarke, Schleifmarke.*

Groupe de rayures ou de rainures parallèles gravées sur les sédiments par un éboulement ou glissement sous-aquatique. Les «slide casts» de TEN HAAF ont d'un à plusieurs décimètres de largeur, moins d'un mètre de longueur, et en général ont une courbure régulière; il les attribue à de grosses mottes de boue ou de sable, ou peut-être à des colonies animales qui ont glissé. *Marques de glissement.*

Sistema paralelos de surcos originados por deslizamiento subácueo. Los "slide casts" de TEN HAAF (1959, p. 38) son marcas cuyas dimensiones varían entre uno y varios decímetros de ancho, una longitud no mayor de un metro y de perfil comunmente poco curvado. Su formación es atribuida al deslizamiento de grandes "coágulos" de fango, arena o tal vez colonias de organismos. *Calco de deslizamiento.*

Slip bedding (HILLS, 1941, p. 15):
The contortion of stratification planes caused by gliding. See convolute bedding.

Durch Gleitung entstandene Verfältelung von Einzel-Schichten. Siehe Wulstschichtung. *Gleitfaltung, Gleitstauchung.*

Contournement des plans de stratification à la suite d'un glissement. *Lit.: Surfaces contournées par glissement; structures fluidales convulsées.*

Estratificación deforma por deslizamiento. Véase "convolute bedding". *(Estratificación intraplegada).*

Slip block (KUENEN, 1948, p. 371):
" ... a separate mass ... which has slid away from its original position and come to rest down the slope without being much deformed."

„ ... eine losgetrennte Masse ... die von ihrer ursprünglichen Lage abgeglitten und am Fuße eines Hanges zu liegen gekommen ist, ohne nennenswerte Deformation erlitten zu haben." *Gleitpacken.*

«Une masse indépendante ... qui a quitté sa position d'origine par glissement, et qui est arrivée au pied de la pente sans avoir subi de déformation importante.» *Lit.: Bloc glissé, translaté.*

" ...masa independiente ... deslizada de su posición original por una pendiente hasta su pie, sin haber sufrido mayor deformación". *Bloque deslizado.*

Slip face:
The lee surface of a sand wave. The foreset laminations of a cross-bedded layer are slip-face accumulations. See foreset. Also referred to as slip slope.

Die Leeseite einer Sandwelle. Die Leeblätter einer schräggeschichteten Lage sind An-häufungen an Rutschflächen. Siehe Leeblatt. Auch als Rutsch-Hang bezeichnet. *Rutschfläche, Gleitfläche.*

La face aval d'une vague de sable. Les couches frontales d'une unité à stratification croisée sont des accumulations sur une «slip face». Appelée aussi «slip slope». *Talus croulant.*

Superficie a sotavento de una onda de arena. Las capas frontales son superficies de acumulación por deslizamiento; véase "foreset". Sinónimo de "slip slope". *Cara de deslizamiento.*

Slip slope: See slip face.

Sludge cast (WOOD and SMITH, 1959, p. 169): See furrow cast.

Slump:

The deposit produced by a downslope, *en masse* movement of material; may be subaqueous or subaerial; see also slide.

Die von hangabwärts gerichteten Massen-Bewegungen abgelagerten Sedimente, sei es unter Wasser, sei es auf Land. Siehe auch Gleitung. *Rutschung, subaquatische (submarine) Rutschung, Gleitung.*

Dépôt produit par la descente en masse de matériaux le long d'une pente. Peut être sous-aquatique ou terrestre. *Lit.: paquet de terrain glissé, glissement.*

Depósito formado en una pendiente por movimiento en masa de material, ya sea subácueo o subaereal. Véase "slide". *Depósito por deslizamiento.*

Slump balls (KUENEN, 1948, p. 369, pl. 26):

More or less flattened balls, 2—3 cm to 2—3 m across; commonly laminated, with internal contortions and a smooth external form. Probably the same as pillow-structure.

Mehr oder weniger abgeflachte Ballen, 2—3 cm bis 2—3 m im Durchmesser. Im allgemeinen feingeschichtet, mit inneren Verfältelungen und einer glatten Oberfläche. Wahrscheinlich dasselbe wie Ballenstruktur. *Gleitblöcke, Sedimentrolle, Rutschkörper, -wülste, Aufrollung.*

Boules plus ou moins aplaties, allant de 2—3 cm à 2—3 mètres. En général feuilletées, avec des contournements internes et un contour externe lisse. Probablement l'équivalent de «pillow structure». *Pseudo-nodules arrondis.*

Esferoides más o menos achatados, de 2—3 cm a 2—3 m de diámetro, comunmente con laminación deformada y superficies lisas. Probablemente similar a "pillow-structure". *Esferoides lisos.*

Slump bedding:

Term loosely applied to any disturbed bedding; strictly applicable only to structures produced by lateral movement or slump.

Eine Bezeichnung, die lose auf jegliche gestörte Schichtung angewandt wird. Strenggenommen nur auf Strukturen anwendbar, die durch seitliche Bewegungen oder Rutschungen erzeugt sind. *Gleitfaltung, Rutschfaltung, Gleitstauchung.*

Terme assez vague que l'on applique à toute stratification dérangée. Ne s'applique littéralement qu'aux structures produites par des mouvements latéraux ou des glissements.

Término general aplicado a toda estratificación deformada; en sentido estricto se emplea para definir estructuras originadas por movimientos laterales o deslizamientos. *Estratificación deslizado.*

Slump mark (MCKEE, 1945, p. 320—323):

Marks and structures made by dry or wet sand avalanching down the lee side of a sand wave or dune.

Marken und Strukturen, die entstehen, wenn trockener oder feuchter Sand die Leeseite einer Sandwelle oder Düne lawinenartig herunterrollt. *Rutschmarke (w).*

Empreintes et structures formées par du sable sec ou mouillé qui dévale sur le talus aval d'une vague de sable ou d'une dune.

Marcas y estructuras originadas por desmoronamiento de arena seca o húmeda, a sotavento de una duna u onda de arena. *Marca a sotavento.*

Slump overfold (CROWELL, 1957, p. 998, Fig. 5):

Hook-shaped masses of sandstone produced during slump (KSIAZKIEWICZ, 1958, Figs. 18 and 19).

Hakenförmige Massen aus Sandstein, die bei Rutschung entstehen (KSIAZKIEWICZ, 1958, Abb. 18 und 19). *Rutschfalten* (n).

Masses de grès en forme de crochets qui se forment pendant un glissement. *Lit.: couches déversées par glissement.*

Masas arenosas plegadas a modo de ganchos originadas por deslizamientos (KSIAZKIE-WICZ, 1958, figs. 18, 19). *Pliegues volcados por deslizamiento.*

Slump sheet (KUENEN, 1948, p. 373):

A bed of limited thickness and wide extent consisting of slumped materials.

Eine weit ausgedehnte Schicht beschränkter Dicke, die aus gerutschtem Material besteht. *Rutschungs-Schicht* (w).

Couche d'épaisseur limitée et de grande extension formée de matériaux glissés. *Lit.: nappe de glissement.*

Capa de espesor limitado y amplia extensión, constituida por materiales deslizados. *Manto de deslizamiento.*

Slurry slump (DZULYNSKI and SLACZKA, 1959, p. 217):

A slump in which the moving mass disintegrates into a quasi-liquid slurry.

Eine Rutschung, in der die sich bewegende Masse in einen flüssigkeitsähnlichen Zustand übergeht. *Trübestrom.*

Glissement dans lequel la masse en mouvement se désintègre pour former une boue fluente quasi-liquide. *Lit.: glissement de boue fluente.*

Deslizamiento por el cual la masa en movimiento se desintegra formando una suspensión viscosa. *Lodo por deslizamiento.*

Snowball structure (HADDING, 1931, p. 390):

Same as slump ball according to KUENEN (1948, p. 369). See ball-and-pillow structure.

Nach KUENEN (1948, S. 369) dasselbe wie Rutschungsballen. Siehe Ballenstruktur. *Schneeball-Struktur, Sandstein-Rolle.*

Synonyme de «slump ball», d'après KUENEN.

Estructura similar a "slump balls" según KUENEN (1948, p. 369). Véase "ball-and-pillow structure". *(Esferoides lisos).*

Sole mark (KUENEN, 1957): See plates 54A to 66.

A term which has come to be generally applied to the various hieroglyphs found on the undersides or soles of sandstone (and in some cases limestone) beds. These marks are largely casts of structures formed on the surface of the underlying mud by currents, organisms, or other agents. After consolidation and exposure, the underlying shale weathers away leaving the cast as a raised positive feature on the sole of the overlying sandstone.

Eine jetzt allgemein gebrauchte Bezeichnung für die vielfältigen Hieroglyphen, die sich auf den Unterseiten oder Sohlen von Sandstein- und in manchen Fällen Kalkstein-Schichten finden. Diese Marken sind im allgemeinen Ausgüsse von Strukturen, die sich auf der Oberfläche des unterliegenden Schlammes durch die Tätigkeit von Strömungen, Organismen usw. formen. Nach Verfestigung wittert im Aufschluß der liegende Schiefer-ton aus, und so zeigen sich die verschiedenen Marken als erhabene Strukturen an der Sohlfläche des hangenden Sandsteines. *Schichtflächen-Marke, Schichtflächen-Erscheinung, Sohlmarke, Strömungsmarke.*

Terme devenu d'utilisation générale pour désigner divers hiéroglyphes qui se rencontrent à la face inférieure («sole»: lit. semelle, plante) de bancs de grès (et quelquefois calcaires).

Ces empreintes sont le plus souvent des moulages de structures formées à la surface de la boue sous-jacente par les courants, des organismes vivants, ou autres agents. Après leur consolidation et affleurement, l'argile schisteuse sous-jacente se désintègre et laisse le moulage comme une figure en relief à la face inférieure du grès sus-jacent. *Lit.: empreintes plantaires. Hiéroglyphes de mur; empreintes de mur.*

Término aplicado a diferentes hieroglifos hallados en la superficie inferior de ciertas areniscas y calizas. Estas marcas son generalmente calcos de estructuras originadas en la superficie de capas fangosas por corrientes, organismos u otros agentes. Cuando los sedimentos han sido consolidados y expuestos a la intemperie, los estratos pelíticos son destruidos con mayor facilidad, dejando al descubierto en la base de las areniscas, formas en realce o calcos. *Marca de base.*

Sorted bedding (BOUMA, 1961, p. 145):

A type of graded bedding in which only one grain size is present at each horizon within the bed and the size decreases upwards.

Eine Art gradierter Schichtung, in der nur eine Korngröße in jeder Lage einer Schicht vorkommt, aber die Korngröße nach oben zu abnimmt. *Sortierte Schichtung* (w).

Type de granoclassement dans lequel il n'y a qu'une taille de grain dans chaque niveau de la couche, cette taille allant en diminuant vers le haut de la couche.

Tipo de estratificación gradada formada por un sólo tamaño de grano en cada nivel del estrato, el cual disminuye hacia arriba. *Estratificación seleccionada.*

Spiral balls (HADDING, 1931, p. 389):

Sandstone bodies with rolled-up spiral structure due to lateral mass flowage of thin interbedded sands and shales.

Sandstein-Körper mit aufgerollten Wickelstrukturen, die durch seitliche Fließbewegungen dünner wechsellagernder Sande und Schiefertone bedingt sind. *Wickelstrukturen (-falten), Sedimentrolle.*

Masses de grès à structure enroulée en spirale, et due à l'écoulement latéral en masse de lits fins intercalés de sable et d'argile schisteuse.

Masas areniscosas espiraladas hacia arriba, formadas por desplazamiento lateral en masa de sedimentos arenosos y pelíticos interestratificados. *Esferoides espiralados.*

Spiral structure (FAIRBRIDGE, 1946, p. 87):

Probably same as snow-ball structure, which see; term also applied to organic burrows (CAROZZI, 1960; KUENEN, 1961).

Wahrscheinlich das gleiche wie Schneeball-Strukturen; s. dort. Bezeichnung ist auch auf organische Bauten angewandt worden (CAROZZI, 1960; KUENEN, 1961). *Spiralstruktur.*

Probablement équivalent de «snowball structure»; terme appliqué aussi à des terriers d'origine animale.

Probablemente similar a "snowball structure"; término también empleado para orificios originados por especies animales excavadoras (CAROZZI, 1960; KUENEN, 1961). *Estructura espiralada.*

Spring pits (QUIRKE, 1930, p. 88):

Small craterlets found on sand beaches produced by ascending waters. Range in diameter from 30 to 60 cm with a depth of about 15 cm.

Von aufsteigendem Wasser geformte kleine Trichter, die sich am Strand finden. Ihr Durchmesser reicht von 30 bis 60 cm mit einer durchschnittlichen Tiefe von 15 cm. *Quelltrichter.*

Petits cratères que l'on rencontre sur les plages de sable et qui sont formés par la montée de l'eau. Leur diamètre va de 30 à 60 cm et leur profondeur est d'environ 15 cm.

Cractercillos originados por aguas ascendentes en arenas de playa. Diámetro entre 30 y 60 cm; profundidad 15 cm aproximadamente. *Cractercillos de playa.*

Squamiform load cast (TEN HAAF, 1959, p. 46) (squamiform cast): See plate 59B.

". . . a type of load casting that covers some soles with crowded lobate casts overlapping down-current"; resembles sagged flute casts that have an opposite orientation with respect to current direction.

„ . . . eine Art Belastungsmarken, die einige Schicht-Unterseiten als enggescharte zungenförmige Marken bedecken, die sich nach stromabwärts überlappen". Sie ähneln gesackten Strömungswülsten, die eine der Strömung entgegengesetzte Orientierung aufweisen. *Schuppenförmige Belastungsmarke* (w), *schuppenförmige Gefließ-Marke* (?).

«Type de figures de charge qui couvrent certaines faces inférieures de bancs de moulages lobés, tassés, se recouvrant vers l'aval»; Ressemble à des «flute casts» aplatis mais qui auraient une direction inverse par rapport à celle du courant. *Empreinte de charge en écailles.*

Tipo de calco de carga con formas lobuladas, encimadas en sentido de la corriente; se asemejan a turboglifos ahondados, pero con orientación opuesta respecto al sentido de la corriente. *Calco de carga escamiforme.*

Starved ripples: See plates 90A and 90B.

Only isolated crests of ripple marks are present. See incomplete ripple marks.

Es sind nur isolierte Kämme von Rippelmarken vorhanden. Siehe Einzelrücken.

Seules les crêtes de rides isolées sont présentes.

Crestas aisladas de una óndula. Véase "incomplete ripple marks". *(Ondula incompleta).*

Storm roller (CHADWICK, 1931): See pillow-structure.

Straticulate (CHADWICK, 1940, p. 1923):

A term applied to polygonal columnar structure in limestone. See columnar structure.

Eine auf polygonale Säulenstruktur in Kalksteinen angewandte Bezeichnung. Siehe Säulenstruktur.

Terme qui désigne une structure columnaire polygonale dans un calcaire.

Término aplicado a aquella estructura columnar de sección poligonal desarrollada en ciertas calizas. Véase "columnar structure". *(Estructura columnar).*

Stratification:

The term describing a layered or bedded sequence. See bedding.

Ein Ausdruck für geschichtete Gesteinsabfolgen. Siehe Schichtung.

Terme qui désigne une séquence stratifiée ou litée. *Stratification.*

Término que indica una secuencia estratificada. Véase "bedding". *Estratificación.*

Stratum:

Commonly applied to a single sedimentary bed or layer. Also defined as a layer greater than 1 cm in thickness (PAYNE, 1942, p. 1724; McKEE and WEIR, 1953, p. 382).

Im allgemeinen auf eine einzelne Schicht oder Lage aus Sediment angewandt. Auch als Lage definiert, die dicker ist als 1 cm (PAYNE, 1942, S. 1724; McKEE und WEIR, 1953, S. 382). *Schicht.*

Employé généralement pour une couche sédimentaire ou un lit; parfois défini comme une couche de plus d'un cm d'épaisseur. *Strate.*

Término comunmente aplicado a una unidad de sedimentación. También definido como todo depósito cuyo espesor sobrepasa el centímetro (PAYNE, 1942, p. 1724; MCKEE and WEIR, 1953, p. 382). *Estrato.*

Streaked-out ripples: See flame structures.

Striation cast (PETTIJOHN, 1957, p. 181) (micro-groove cast):
Casts of striations or very small grooves.
Ausgüsse von Riefen oder sehr kleinen Rillen. *Riefenmarke* (n), *Driftstreifung.*
Moulages de stries, ou de très petites rainures.
Calco de estrías o surcos muy delgados. *Calco de microsurco.*

Stromatolite (KALKOWSKI, 1908, p. 68): See plates 48 and 49.
The term stromatolite has been generally applied to laminated structures attributed to the work of blue-green algae. Commonly called "algal structures". Characteristically laminated with varied gross forms, from near-horizontal, to markedly convex, columnar and sub-spherical (CLOUD, 1942).

Der Name Stromatolith (KALKOWSKI, 1908, S. 68) wird im allgemeinen auf feingeschichtete Strukturen angewandt, die man auf die Tätigkeit von blaugrünen Algen zurückführt. Oft „Algenstruktur" genannt. Charakteristisch sind feinschichtige Strukturen, die von nahezu horizontalen bis zu ausgesprochen konvexen, säulenförmigen und subsphärischen Formen variieren (CLOUD, 1942). *Stromatolith.*

Ce terme a été généralement appliqué à des structures qu'on attribue au travail des algues bleues. Appelées souvent «algal structures» (structures d'algues). En général elles sont finement litées, et leur forme générale varie beaucoup, de presque plate à nettement convexe, columnaire, ou presque sphérique. *Stromatolithe.*

El término "stromatolite" se ha aplicado a aquella estructura laminada atribuida a la actividad biológica de algas verdeazuladas. También denominada "algal structure". Estructura en forma columnar y subesférica, de dimensiones variadas, constituida por láminas horizontales hasta marcadamente convexas (CLOUD, 1942). *Estromatolita.*

Subsolifluction (HEIM, 1908, p. 142):
Subaqueous solifluction.
Subaquatische Solifluktion. *Subsolifluktion.*
Solifluxion sous-aquatique. *Subsolifluxion.*
Solifluxion subácuea. *Subsolifluxión.*

Substratal lineation (CROWELL, 1955, p. 1358):
Defined to include groove casts and load-cast lineation but commonly extended to include all substratal lineations.
Schließt der Definition nach Rillenmarken und Belastungs-Lineation ein, wird jedoch im allgemeinen auf alle Lineationen auf Schicht-Unterseiten ausgedehnt. *Lineation auf Schicht-Unterseite* (w).
Terme défini à l'origine de façon à englober les alignements de cannelures et d'empreintes de charge, mais étendu à toutes les linéations de faces inférieures de bancs. *Lit.: Alignements de faces inférieures.*
Término referido a la lineación de calcos de surcos y calcos de carga, incluyendo generalmente a toda aquella lineación de marcas o calcos, localizados en la superficie inferior de un estrato. *Lineación subestratal.*

Sun cracks: See mud cracks.

Surf ripples (KUENEN, 1950, p. 292):

A general name for ripples formed by wave-generated currents in the surf zone.

Ein allgemeiner Name für Rippeln, die von den Wellen geformt werden, die sich in der Brandungszone brechen. *Brandungsrippeln* (w).

Terme général qui désigne les rides formées par les vagues qui déferlent dans la zone de déferlement. *Lit.: rides de déferlement.*

Término general para óndulas formadas por olas en la zona de rompiente. *Ondula de rompiente.*

Swash mark:

A thin wavy line on a beach marking the upper limit of the swash of a wave. Also called wave line (HALL, 1843, p. 54) (SHROCK, 1948, Fig. 89).

Eine dünne, wellenartige Linie am Strand, die die obere Reichweite der Wellen markiert. Auch „Wellenlinie" genannt (HALL, 1843, S. 54) (SHROCK, 1948, Abb. 89). *Spülbogen, Spülmarke.*

Line fine et ondulée qui marque la limite supérieure du jet de rive d'une vague sur la plage. Appelée aussi «wave line». *Tracé du jet de rive.*

Línea delgada de trazado ondulante que marca el límite del oleaje en una playa. También denominada "wave line" (HALL, 1843, p. 54) (SHROCK, 1948, fig. 89). *Marca de resaca.*

Symmetrical ripple marks: See plates 86 B and 89 A.

Ripple marks with symmetrical cross-section. Straight crests predominate.

Rippelmarken mit symmetrischem Querschnitt. Gerade Kämme herrschen vor. *Symmetrische Rippelmarken* (w).

Rides de plage à profil symétrique. Les crête rectilignes sont prédominantes. *Rides symétriques.*

Ondula con crestas de perfil simétrico; predominan las crestas de trazado recto (en planta). *Ondula simétrica.*

Syndromous load cast (TEN HAAF, 1958, p. 48):

A peculiar form of load casting characterized by elongate shallow casts with sharp creases which combine to form a dendritic pattern. Junctures occur without exception in the down-current sense. See also rill cast, rill mark, and furrow flute cast.

Eine eigenartige Form von Belastungsmarken, die durch lange, flache Ausgüsse mit scharfen Falten gekennzeichnet sind, die ohne Ausnahme nach stromabwärts zusammenlaufen und so ein dendritisches Muster formen. Siehe auch Rinnenmarke, Rieselmarke und gefurchte Strömungswülste. *Zusammengesetzte Belastungsmarke* (w).

Type particulier de figure de charge, caractérisé par des moulages allongés, peu élevés, avec des replis nets qui s'anastomosent en réseau dendritique.

Calco de carga con formas elongadas y suave realce, limitadas por surcos agudos de diseño dendrítico que confluyen invariablemente en sentido de la corriente. Véase "rill cast", "rill mark" y "furrow flute cast". *Calco de carga dendriseptado.*

Syneresis cracks:

Cracks developed by shrinkage related to dehydration; usually applied only to cracks formed upon aging of gels.

Risse, die bei von Entwässerung verursachter Schrumpfung entstehen. Gewöhnlich nur auf Risse angewandt, die sich beim Altern von Gelen bilden. *Synäresisrisse* (w).

Fissures formées par retrait lié à la déshydratation; en général ne s'utilise que pour les fissures dues au vieillissement d'un gel. *Fissures de dessiccation.*

Desarrollo de grietas por contracción, debido a la deshidratación de geles en proceso de envejecimiento. *Grietas de deshidratación.*

Synglyph (VASSOEVICH, 1953, p. 33):
A hieroglyph formed during the sedimentation process.
Eine synsedimentäre Hieroglyphe. *Synglyphe.*
Hiéroglyphe formé pendant le processus de sédimentation. *Synglyphe.*
Hieroglifo formado durante procesos sedimentarios. *Singlifo.*

Tabular cross-bedding (LAHEE, 1952, Fig. 54):
A cross-bedded unit with a flat base and flat top forming a tabular body.
Eine schräggeschichtete Einheit, die wegen ihrer flachen Sohl- und Dachfläche einen tafelförmigen Körper bildet. *Tafelige Schrägschichtung* (w), *Diagonalschichtung, ebenflächig begrenzte Schrägschichtung.*
Ensemble à stratification oblique, à base et sommet plans, formant un bloc tabulaire. *Stratification oblique de type tabulaire.*
Unidad sedimentaria con estratificación entrecruzada, limitada por una base y techo planos, de forma tabular. *Estratificación entrecruzada tabular.*

Tadpole nests (HITCHCOCK, 1858, p. 121—123):
An early name used for interference ripple mark.
Eine alte, früher auf Interferenzrippeln angewandte Bezeichnung. *Kaulquappen-Nester.*
Terme ancien qui désigne des rides interférentes. *Lit.: nids de têtards; rides en fossettes.*
Término en desuso sinónimo de "interference ripple mark". *(Ondula de interferencia).*

Tangential cross-bedding: See plate 36.
Foreset beds tangential to underlying surface.
Leeblätter, die sich tangential an die Sohlfläche schmiegen. *Bogige Schrägschichtung, schaufelförmige Schrägschichtung.*
Couche frontales tangentielles à la surface sous-jacente.
Capas frontales tangentes a la superficie de depositación. *Estratificación entrecruzada tangencial.*

Taphoglyph (VASSOEVICH, 1953, p. 72):
Imprints of dead animal bodies.
Eindrücke von Kadavern. *Taphoglyphe.*
Empreintes de cadavres d'animaux. *Taphoglyphe.*
Impresiones de animales muertos. *Tafoglifo.*

Teggoglyph (VASSOEVICH, 1953, p. 59):
A load structure; see load casts.
Eine Belastungs-Struktur. Siehe Belastungsmarke. *Teggoglyphe.*
Empreinte de charge. *Teggoglyphe.*
Estructura de carga; véase "load casts". *Teggoglifo.*

Terraced flute cast (TEN HAAF, 1959, p. 28, Fig. 14; KUENEN, 1957, p. 241):
Flute casts with external sculpturing resembling differentially-weathered bedding laminations. In reality, a cast of differentially eroded laminations in the underlying shale and unrelated to internal structure of the cast.

Strömungswülste mit äußerem Relief, das unterschiedliche Verwitterung von Feinschichten vortäuscht. Hat aber in Wirklichkeit nichts mit der inneren Struktur der Strömungswülste zu tun, sondern entsteht beim Ausguß von unterschiedlich erodierten Feinschichten. *Terrassierter Strömungswulst* (w).

Empreintes de sillons d'érosion dont le relief superficiel ferait penser que le moulage est constitué de lits fins érodés différemment. En réalité c'est le schiste sous-jacent dont les laminations avaient subi une érosion differentielle; leurs moulages apparaissent sur l'empreinte mais n'ont rien à voir avec sa structure interne. *Lit.: Empreintes de sillons d'érosion en gradins.*

Turboglifo de perfil transversal escalonado. Calco de marca originada por desbaste escalonado de una pelita laminada e independiente de la estructura interna del mismo. *Turboglifo aterrazado.*

Toeset (JOPLING, 1960, p. 126—129):
The forward, lower part of a tangential foreset bed. Name derived from deposition at the "toe" of a tangential foreset.

Der vordere, untere Teil eines tangentialen Leeblattes. Name rührt von der Ablagerung am Fuße eines tangentialen Leeblattes her. *Sohlblatt.*

La partie avancée, la plus basse, d'une couche frontale tangentielle. Ce nom est dérivé du fait que le dépôt se fait au bout («toe»: orteil, bout du pied) de la couche frontale.

Parte avanzada e inferior de una capa frontal tangente a la superficie de depositación. *Pie de capa frontal.*

Tool marks (DZULYNSKI and SANDERS, 1959):
Sole marks produced by engraving "tools" such as shells, sand grains, pebbles, and others swept over firm lutite bottoms by currents, generally preserved as casts on base of overlying bed.

Schichtflächen-Marken, die von sich eindrückenden Gegenständen wie Schalen, Sandkörnern, Geröllen und dergleichen verursacht werden, wenn diese von Strömungen über widerstandsfähigen tonigen Boden getrieben werden. *Gegenstands-Marken* (w).

Empreintes produites par des «outils» graveurs, tels que coquilles, grains de sable, cailloux et autres, entraînés par les courants sur des fonds assez fermes de lutite, et généralement conservées comme des moulages à la base de la couche sus-jacente. *Lit.: Empreintes «d'outils».*

Marcas originadas por conchillas, granos de arena, rodados, etc. al ser desplazados por corrientes sobre fondos pelíticos semiconsolidados; preservadas como calcos en la base de estratos suprayacentes. *Marcas labradas.*

Top set (beds):
Originally used for essentially horizontal beds that overlie the inclined foreset beds of a delta.

Ursprünglich auf im wesentlichen horizontale Schichten angewandt, die die geneigten Leeschichten eines Deltas überlagern. *Dachschicht.*

A l'origine, terme utilisé pour désigner les couches presque horizontales qui recouvrent les couches frontales inclinées d'un delta. *Couches supérieures planes, couches sommitales.*

Capas horizontales que yacen sobre las capas frontales de un delta. *Capas dorsales.*

Torose load cast (CROWELL, 1955, p. 1360, pl. 2, Fig. 4): See plate 54B.
A sole mark formed by elongate ridges which pinch and swell along their trends; unlike flute casts, the down-current terminations are bulbous or tear-drop shaped.

Eine Schichtflächen-Marke, die von länglichen Rücken gebildet wird, welche in der Richtung ihrer Längsausdehnung an- und abschwellen. Unähnlich Strömungswülsten sind die stromabwärts gerichteten Enden gewulstet oder tropfenförmig. *Wulstige Belastungsmarke* (w).

Empreintes formées de bourrelets allongés, qui s'étranglent et s'élargissent sur leur longueur; à l'encontre des sillons d'érosion, elles ont des terminaisons avales bulbeuses ou en forme de larme. *Lit.: Empreintes de charge torsadées, tordues, cordées.*

Marcas de base en forma de lomos elongados con estrechamientos periódicos a lo largo de su longitud. Sus extremos bulbosos están orientados en sentido de la corriente. *Calco de carga polilobulado.*

Torrential cross-bedding (HOBBS, 1906, Fig. 2):

A term used to define, by implication, cross-bedding resulting from rapid, or "torrential," deposition; essentially tabular cross-bedding.

Alte Bezeichnung für Schrägschichtung, die bei rascher, sturzflutartiger Ablagerung entstanden sein soll. Im wesentlichen tafelige Schrägschichtung. *Diagonalschichtung.*

Terme qui implique une stratification croisée due à un dépôt rapide, «torrentiel». En essence, une stratification croisée tabulaire.

Define a aquella estratificación entrecruzada formada por depositación rápida o torrencial. Similar a "tabular cross-bedding". *Estratificación entrecruzada torrencial.*

Torrential cross-lamination: See torrential cross-bedding.

Transverse ripple marks:

Ripple mark formed with ridges transverse to current. Cross-sections may be symmetrical or asymmetrical.

Quer zur Strömung verlaufende Rippelmarken. Symmetrische oder unsymmetrische Profile. *Transversalrippeln.*

Rides dont les crêtes sont perpendiculaires à la direction du courant. Peuvent avoir un profil symétrique ou asymétrique. *Rides transversales.*

Ondula con crestas orientadas transversalmente a la dirección de la corriente; pueden ser de perfil simétrico o asimétrico. *Ondula transversa.*

Transverse scour marks (DZULYNSKI and SANDERS, 1963, p. 68):

Scour marks with long axes transverse to main current direction; regular spacing leads to confusion with ordinary transverse ripples; have been called "current ripple casts" (KUENEN, 1957, Fig. 6; DZULYNSKI and SLACZKA, 1958, p. 230, pl. 28, Fig. 2).

Kolkmarken, deren Längsachsen quer zur Strömungsrichtung verlaufen. Die regelmäßigen Abstände geben zu Verwechslungen mit gewöhnlichen Transversalrippeln Anlaß. Auch Ausgüsse von Strömungsrippeln genannt (KUENEN, 1957, Abb. 6; DZULYNSKI und SLACZKA, 1958, S. 230, Tafel 28, Abb. 2). *Transversale Kolkmarken* (w).

Traces de creusement dont les grands axes sont perpendiculaires à la direction principale du courant; leur espacement régulier tend à les faire confondre avec les rides transversales ordinaires; appelées «current ripple casts» (moulages de rides de courant) par KUENEN et autres. *Lit.: Empreintes transversales de creusement, d'affouillement.*

Marcas de desbaste con sus ejes longitudinales transversos a la dirección de la corriente. Se encuentran regularmente espaciadas y ello hace que sean facilmente confundidas con óndulas transversales. También denominada "current ripple casts" (KUENEN, 1957, Fig. 6; DZULYNSKI and SLACZKA, 1958, p. 230, pl. 28, fig. 2). *Marcas de desbaste transversales.*

Trash line:

A line on the beach, defined by debris and surface texture, that marks the farthermost advance of high tide.

Durch Treibgut und Struktur gekennzeichnete Bänder, die an Stränden die äußerste Reichweite der Flut markieren. *Spülsaum, Wellenlinie.*

Ligne marquée sur la plage par des débris et par la texture de la surface, et qui indique la limite extrême de la haute mer. *Laisse de haute mer, frange de marée.*

Línea y textura de superficie formadas por escombros que marcan el máximo avance de pleamar en una playa. *Línea de escombros.*

Trough cross-lamination: See trough cross-stratification.

Trough cross-stratification (McKee and Weir, 1953, Fig. 2):

Lower bounding surface of cross-stratified unit is a curved surface of erosion. See festoon cross-bedding.

Schräggeschichtete Einheit, die im Liegenden von einer gekrümmten Erosionsfläche begrenzt ist. Siehe bogige Schrägschichtung. *Trogförmige Schrägschichtung* (w), *bogige, synklinale, gekrümmte, schaufelförmige Schrägschichtung, Muldenschichtung.*

La surface limite inférieure d'une unité à stratification croisée est une surface d'érosion concave, en creux. *Lit.: Stratification croisée en auge.*

Unidad sedimentaria con estratificación entrecruzada depositada sobre una superficie curva de erosión. Véase "festoon cross-bedding". *(Entrecruzamiento festoneado).*

Turboglyph (Vassoevich, 1953, p. 36, 65): See flute cast.

Undertow mark (Clarke, 1917, p. 204): See groove cast.

Unilateral rolling strata (Andersen, 1931, Fig. 38):

Asymmetrical ripple or wavy bedding the steeper side of which indicates the direction of flow.

Unsymmetrische Rippeln oder wellige Schichtung, deren steilere Seite die Strömung anzeigt.

Rides asymétriques, ou couches onduleuses dont la pente est plus raide vers l'amont. *Couches à ondulations asymétriques.*

Estratificación ondulada u óndulas asimétricas, cuyos flancos más abruptos indican el sentido de la corriente. *Estratificación corrugada unilateral.*

Upslope ripple (McKee, 1939, p. 72):

A ripple that climbs a sloping surface.

Rippeln, die hangaufwärts wandern.

Une ride qui remonte le long d'une pente.

Ondula que asciende por una superficie inclinada. *Ondula ascendente.*

Vibration mark (Dzulynski and Slaczka, 1959, p. 234) ("chatter mark," herringbone mark): See plates 62 and 63 A.

A rare modification of a groove consisting of crescentic depressions, concave upcurrent. Presumed to result from unsteady action of inscribing tool. See also chevron mark and ruffled groove casts.

Eine seltene Abart von Rillenmarken, die aus bogenförmigen Vertiefungen besteht, deren konkave Seite stromaufwärts weist. Man nimmt als Ursache die unstetige Tätigkeit des sich eingrabenden Gegenstandes an. Siehe auch Fiedermarke und gefiederte Rillenmarke. *Schwingungsmarke* (w).

Type particulier (et rare) de cannelure, caractérisé par des dépressions en croissant, concaves vers l'amont. On les croit dues à l'action inégale d'un outil graveur. *Lit.: Empreinte de vibration.*

Rara modificación de un calco de surco, que consiste en depresiones en forma de media luna; sus extremos están orientados en sentido contrario a la corriente. Se suponen originadas por la acción irregular de un cuerpo. Véase "chevron mark" y "ruffled groove cast". *Marcas de vibración.*

Vortex cast (WOOD and SMITH, 1957, p. 192): See flute cast.

Wash-over crescents (TANNER, 1960, Fig. 2h):

Crescentic barchan-like *depressions* whose plane of symmetry parallels current direction.

Bogenförmige, barchanenähnliche Vertiefungen, deren Symmetrieebene der Strömungs-richtung parallel läuft. *Aufspül-Bögen* (n).

Dépressions en croissant, entre de petites crêtes comparables à des barkhanes, et dont le plan de symétrie est parallèle à la direction du courant. Petites dépressions en croissant.

Depresiones en forma de barjanes cuyos planos de simetría están orientados paralelamente a la dirección de la corriente. *Desbaste en media luna.*

Washout (Cut-out): See plate 20B.

Channel-like features which cut or transgress the stratification of the underlying beds; may be small scour-and-fill structures or large erosional channels.

Rinnenartige Formen, die in die Schichtung des unterliegenden Gesteins einschneiden oder sie transgredieren. Sie können kleine Auskolk-Rinnen oder große Erosionsrinnen sein. *Priel, Rinne, Erosionsrinne, -furche, Auswaschungs-Rinne, Wasser-Rinne.*

Accidents ressemblant à des chenaux, et qui recoupent la stratification des couches sous-jacentes, ou leur sont transgressives; peuvent être de petits accidents de creusement et remplissage, ou de grands chenaux d'érosion. *Chenaux d'érosion, érosions de base, dicho-tomie, dédoublement de couches.*

Desbaste por agua que corta a sedimentos estratificados; puede dar lugar a pequeños o amplios cauces posteriormente rellenados por sedimentos clásticos (relleno de cauce). *Cauce.*

Water-level mark (HÄNTZSCHEL, 1938, p. 8):

Small horizontal, wave-cut "terraces" on an inclined surface of unconsolidated sediment that mark a former water level.

Kleine, horizontale, von Wellen eingeschnittene „Terrassen" auf geneigten Flächen, in unverfestigtem Sediment, die einen früheren Wasserstand anzeigen. *Wasserstandsmarke.*

Petites «terrasses» (ressauts, gradins) horizontales, formées par les vagues sur une pente de sédiments meubles, et qui marquent les niveaux précédents de l'eau.

Terraza pequeña formada sobre una superficie inclinada de sedimentos no consolidados, que marca el nivel de un cuerpo de agua. *Marca de nivel.*

Wave lines: See swash mark.

Wave ripple mark:

A synonym of oscillation ripple mark (KINDLE, 1917b, p. 913) which commonly has symmetrical cross-sections.

Synonym für Oszillations-Rippeln (KINDLE, 1917b, S. 913), mit häufig symmetrischen Querschnitten. *Seegangsrippeln.*

Synonyme de «oscillation ripple mark»; profil généralement symétrique. *Ride de vague d'oscillation.*

Sinónimo de "oscillation ripple marks" (KINDLE, 1917b, p. 913). Ondula formada por crestas de perfil generalmente simétrico. *Ondula de oscilación.*

Waved stratum:

An obsolete term for a ripple-marked bed (cf. LOCKE, 1838, p. 246—247).

Ein veralteter Name für gerippelte Schichtflächen (s. LOCKE, 1838, S. 246—247).

Terme ancien pour une couche à rides de plage.

Término en desuso empleado para aquellos estratos con óndulas (cf. LOCKE, 1838, p.246—247). *Estrato ondulítico.*

Wavy bedding (wavy lamination): See plate 11.

Bedding characterized by undulatory bounding surfaces. May be related to ripple bedding if regular and to nodular bedding if less regular.

Schichtung, deren Begrenzungsflächen gewellt sind. Bei regelmäßiger Ausbildung mögen sie zur Rippelschichtung gehören; zur knolligen Schichtung, sind sie unregelmäßig. *Wellige Schichtung.*

Stratification caractérisée par des couches limites ondulées; peut-être apparentée à une stratification de rides si elle est régulière, et à des couches à nodules si elle est moins régulière. *Stratification ondulée.*

Estratos limitados por superficies ondulantes. Si la ondulación es regular, su origen está vinculado a corrientes de agua o aire (óndula); si es menos regular puede estar relacionada con la formación de nódulos (estratificación nodular). *Estratificación ondulada.*

Wedge-shaped cross-bedding: See plate 32 A.

Cross-bedding with wedge-shaped outline in vertical sections (LAHEE, 1952, Fig. 60).

Schrägschichtung, deren Vertikalschnitt keilförmig ist (LAHEE, 1952, Abb. 60). *Keilförmige Schrägschichtung* (w).

Stratification oblique où les couches se terminent en coin en coupe verticale. *Couches à stratification oblique en coin.*

Estratificación entrecruzada acuñada. *Estratificación entrecruzada cuneiforme.*

Wedge torrential cross-bedding:

Torrential cross-bedding with wedge-shaped cross-bedded units. See torrential cross-bedding.

„Sturzflutartige" Schrägschichtung mit keilförmig schräggeschichteten Einheiten. Siehe Diagonalschichtung. *Kreuzschichtung.*

Stratification «torrentielle» à paquets de couches obliques terminés en coin.

Unidades acuñadas con estratificación entrecruzada torrencial. Véase "torrential cross-bedding". *Estratificación entrecruzada torrencial cuneiforme.*

Whirl-balls (DZULYNSKI et al., 1957, p. 107):

Spindle, tubular, ellipsoidal, or spherical balls of fine sandstone embedded in silt with their long axes vertical or steeply inclined; attributed to vortices in mudflows.

Spindel-, röhrenförmige, ellipsoidale oder kugelige Ballen aus feinkörnigem Sandstein, eingebettet in Silt, deren Längsachsen senkrecht oder steil geneigt sind. Man bringt sie mit Wirbeln in Schlammströmen in Zusammenhang. *Wirbelbälle* (w).

Masses de grès fin en fuseau, cylindriques, ovoïdes ou sphériques, incluses dans une argile, avec leurs grands axes verticaux ou très inclinés; on les attribue à des tourbillons formés dans une coulée de boue.

Masas de arenisca fina con formas de husos, tubulares y elipsoidales, englobadas por limo; sus dimensiones mayores están dispuestas verticalmente o con marcada inclinación. Su desarrollo es atribuido a vórtices formados en torrentes de barro. *Arenisca vorticelada.*

Whirl-zone (BEETS, 1946, p. 233; LIPPERT, 1937, p. 360):

Transition zone between slump sheet and overlying strata.

Bezeichnet den allmählichen Übergang der von subaquatischer Gleitung verfalteten Schichten zum Hangenden. *Aufwirbelungs-Zone.*

Zone de transition entre un paquet de terrain glissé et les couches sus-jacentes. *Stratification tourbillonnaire.*

Zona deformada por deslizamiento. *Zona de vórtices.*

Windrow ridges (TANNER, 1960, p. 482):

A variety of ripple mark consisting of straight, tapered ridges parallel to current.

Eine Art Rippelmarken, die aus geraden, spitz zulaufenden Rücken bestehen, parallel zur Strömungsrichtung. *Windreihen-Kämme* (w).

Type de rides de plage formées de crêtes rectilignes, parallèles au courant, et qui s'atténuent vers l'aval.

Variedad de óndula que consiste en lomos rectos y ahusados paralelos a la dirección del viento. *Lomos eólicos.*

Alphabetisches Verzeichnis der deutschen Fachausdrücke

In das nachfolgende Verzeichnis wurden alle deutschen Ausdrücke aufgenommen, die im vorangehenden „glossary" erscheinen, einschließlich wörtlicher Übersetzungen. Meist aus dem Griechischen abgeleitete, neue Bezeichnungen dagegen, z. B. Abioglyphe, finden sich nur im „glossary", da ihre alphabetische Reihenfolge in den einzelnen Sprachen übereinstimmt. Eine Ausnahme bilden im Deutschen eingebürgerte Namen wie Fukoide und Hieroglyphe.

Abscherungs-Struktur: décollement structure
Antidüne: antidune
Aufprallmarke: bounce cast
Aufrollung: slump ball
Aufspül-Bögen: wash-over crescents
Aufstoßmarke: bounce cast
Aufwirbelungs-Zone: whirl-zone
Auskolkung: flute, scour mark; current crescent
Auswaschungs-Rinne: washout, channel
Ballenstruktur: ball-and-pillow structure
Bankinnere Verfältelung: convolute bedding
Bankung: massive bedding
Bau: burrow
Belastete Strömungsmarken: load-casted current markings
Belastungsfahne: load wave
Belastungsfalte: load fold
Belastungslineation: load-cast lineation
Belastungsmarke: load cast
Belastungsmulde: load mold
Belastungsriefen: load-cast lineation
Belastungs-Setzung: sag structure
Belastungstasche: load pocket
Bodenschichten: bottomset beds
Bogige Schrägschichtung: festoon cross-lamination, trough cross-stratification, concave cross-bedding, concave inclined bedding, tangential cross-bedding
Bogiger Bruch: crescentic fracture
Boudinage: boudinage
Brandungs-Kiesrücken: beach cusps
Brandungsrippeln: surf ripples
Breccienstruktur: brecciated structure
Ctenoiden-Marke: ctenoid cast
Dachschicht: top set
Dachziegel-Lagerung (-Stellung; dachziegelartige —): imbricate structure
Deckschicht: topset
Dehnungsmarke: parting cast
Deltaschichtung: delta bedding
Diagonale Kolkmarken: diagonal scour marks

Diagonalschichtung: torrential cross-bedding; planar cross-bedding; inclined bedding; diagonal bedding; tabular cross-bedding
Driftmarke (-spur): drag mark; groove cast
Driftstreifung: striation cast
Düne: dune
Ebene Schrägschichtung: planar cross-bedding
Ebenflächig begrenzte Schrägschichtung: planar cross-stratification; tabular cross-bedding
Einfache Schrägschichtung: simple cross-stratification
Eingerollte Schlickgerölle: clay galls
Einschlagmarke: impact cast
Einwälzstruktur: convolute bedding
Einzelrücken: incomplete ripples
Eiskristall-Marken (-Spuren, -Abdrücke, -Pseudomorphosen): ice crystal marks
Endostratische Sedifluktion: convolute bedding; intraformational folds; intrastratal flow structure; intraformational corrugation
Entgasungs-Kanal: air heave structure
Entgasungs (-Marken, -Trichter): gas pits
Erosionsfurche: washout; flute
Erosionsmarken: scour marks
Erosionsrillen: erosion grooves
Erosionsrinne: channel; washout; flute
Erschütterungs-Schicht: quake sheet
Fächermarke (fächerförmige Fließmarke): frondescent cast; cabbage leaf marking
Fältelung: convolute bedding
Fältelungs-Rutschung: convolute bedding
Falsche Trockenrisse: false mud cracks
Feinblättrige Schrägschichtung: cross-lamination
Feine Parallelstriemung: parting lineation
Feinschichtung: lamination
Fiedermarke: chevron mark
Fiederschichtung: chevron cross-bedding; herringbone cross-bedding
Flachkämmige Rippelmarken: flat-topped ripple marks

Flachzapten: depressed flute cast
Flammenstruktur: flame structure
Flaserschichtung: flaser structure
Fließfältelung (synsedimentäre —): convolute
    bedding
Fließmarke: current mark; flute cast
Fließrille: scour mark, flute cast
Fließrinne: flute
Fließrippel: current ripple mark
Fließwulst: flute cast; load cast; ball-and-
    pillow structure
Fossitextur: mottled structure
Fukoide: fucoid
Fulgurit: fulgurite
Furche: groove
Furchenmarke: furrow flute cast
Gas-Trichter: gas pits, bubble impressions
Gefältelte Netzleisten: crumpled mud-crack
    casts
Gefiederte Rillenmarke: ruffled groove cast
Gefiederte Schleifrille: chevron mark
Gefleckte Schichtung: mottled structure
Gefließ-Marke: flowage cast, flow cast
Gegenstands-Marken: tool marks
Gekneteter Sandstein: kneaded sandstone
Gekrümmte Rippelmarke: curved ripple mark
Gekrümmte Schrägschichtung: trough cross-
    stratification
Geneigte Schichtung: inclined bedding
Gerade gestreckte Schrägschichtung: planar
    cross-bedding
Geradlinige Riefung: parting lineation
Gerichtete Belastungsmarke: directional load
    cast
Gerippelte Erosionsmulde: ripple scour
Gescharte große Fließmarken: shooting flow
    cast
Geschwänzte Buckelstruktur: knob-and-trail
Gespickte Tongerölle: armored mud balls
Girlandenschichtung: loop-bedding
Gleitblöcke: slump balls
Gleitfältelung: convolute bedding
Gleitfaltung: slip bedding, slump bedding;
    convolute bedding
Gleitfläche: slip face
Gleitmarke: slide mark, slide cast
Gleitpacken: slip block
Gleitstauchung: convolute bedding; slip bed-
    ding; slump bedding
Gleitung: slide, slump
Gradierte Schichtung (— Bettung): graded
    bedding
Großrippel: megaripple, metaripple, sand wave
Haftrippeln: anti-ripplets
Hagel-Eindrücke: hail imprints
Hauptwülste: elongate irregular marks
Hexagonale Interferenzrippeln: hexagonal in-
    terference ripples

Hieroglyphe: hieroglyph
Horizontal-Rippeln: level-surface ripples
Hüpfmarke: skip cast, skip mark
Hufeisen-Wulst: current crescent; horse-shoe
    flute cast
Ichnofossil: ichnofossil
Innere Wulst-Textur: convolute bedding
Interferenzrippeln: interference ripples
Kammförmige Aufstoßmarke: ctenoid cast
Kantengestelltes Konglomerat: edgewise
    structure
Kaulquappen-Nester: tadpole nests
Kegelwulst: flute cast
Keilförmige Schrägschichtung: wedge-shaped
    cross-bedding
Kissenstruktur: pillow structure
Kleindimensionale Erosionsrinne: scour and fill
Kleindimensionale Schrägschichtung: micro-
    cross-lamination
Kleine Rillenmarke: microgroove cast
Kleinrippel-Schrägschichtungs-Flaserung:
    ripple cross-lamination
Kletternde Rippeln: climbing ripples
Knauer: nodule
Knittermarken: crinkle marks; creep wrinkles
Knolle: nodule
Knollige Schichtung: nodular bedding
Kolk: flute
Kolkausguß: flute cast
Kolkförmige Erosion: scour and fill
Kolkmarke: scour mark, flute; flute cast;
    current crescent
Kolkriefen: scour lineation
Kolkrinne: channel cast
Konvexe Schrägschichtung: convex inclined
    bedding
Koprolith: coprolite
Korkzieher-Zapfen: corkscrew flute cast
Korrosionsfläche: corrosion surface
Kreuzrillen: cross-grooves
Kreuzschichtung: cross-bedding; wedge torren-
    tial cross-bedding; chevron cross-bedding;
    festoon cross-bedding
Kristall-Pseudomorphosen (-Ausgüsse): crystal
    casts
Küstenspitzen: beach cusps
Lakustrische Pflanzenknäuel: lake balls
Lamellargefüge: lamination
Lamination: lamination
Lebensspuren: lebensspuren
Leeblatt: foreset bed
Leeschicht: foreset bed
Lineation: lineation, substratal-
Lineation auf Schicht-Unterseite: substratal
    lineation
Linguoidrippeln: linguoid ripple marks
Linsenschichtung, linsige Schichtung: lenti-
    cular bedding

Linsige Schrägschichtung: lenticular cross-bedding
Longitudinale Rippelmarken: longitudinal ripple marks
Luftriß: mud crack
Luvblatt-Schichtung: backset bedding
Luvgraben: current crescent
Marine Pflanzenknäuel: Sea balls
Massige (massive) Schichtung: massive bedding
Mittelschicht: foreset bed
Mudlump: mud lump
Muldenschichtung: festoon cross-lamination, trough cross-stratification
Nagelkopf-Schrammen: nail head scratch
Neptunische Gänge: neptunian dikes
Netzleiste: mud crack
Normale Rippelmarken: normal ripple marks
Oszillationsrippeln: oscillation ripple marks
Parallel-Rippelmarke: parallel ripple mark
Pararippeln: para-ripples
Perikline Rippelmarke: pericline ripple mark
Pflanzenballen: sea balls
Priel: washout
Prielfüllung: channel cast
Primärbreccie: edgewise structure
Progressive Sandwellen: progressive sand waves
Pseudo-Schrägschichtung: pseudo-cross-stratification
Quastenmarke: brush cast
Quelltrichter: spring pits
Radiale Trockenrisse: radial mud cracks
Rautenförmige Rippeln: rhomboid ripple marks
Rechteckige Interferenzrippeln: rectangular interference ripples
Regentropfen-Marken (-Eindrücke): rain prints
Regressive Rippeln: regressive ripples
Regressive Sandwelle: regressive sand wave
Rhomboidrippel: rhomboid ripple mark
Riefenmarke: striation cast
Rieselmarke: rill mark
Rille: groove
Rillenmarke: groove cast
Ringmarke: ring mark
Rinne: groove; channel, washout
Rinnenausfüllung (-Ausguß): groove cast
Rinnenfüllung: channel cast
Rinnenmarke: channel cast
Rippelflaserung: ripple cross-lamination
Rippelmarke: ripple mark
Rippelschichtung: ripple bedding, ripple cross-lamination, lee-side concentration, rolling strata, ripple drift
Rollmarke: roll-mark
Rückprallmarke: bounce cast

Rückstrom-Rippelmarken: backwash ripple marks
Runzelmarken: creep wrinkles
Runzelschichtung: crinkled bedding
Rutschfalten: slump overfold
Rutschfaltung: slump bedding; convolute bedding
Rutschfläche: slip face
Rutschkörper (-wülste): slump balls
Rutschmarke: slump mark
Rutschung: slump
Rutschungsballen (-körper): ball-and-pillow structure, flow roll
Rutschungs-Schicht: slump sheet
Sackungs-Struktur: sag structure
Saigerungs-Schichtung: graded bedding
Sandkegel: sand volcanoes; pit-and-mound structure
Sandrücken: sand wave
Sandstein-Gang: sandstone dike
Sandstein-Rolle: snowball structure
Sandstein-Wülste: load casts; flute casts
Sandvulkane: sand volcanoes
Sandwelle: sand wave
Säulenstruktur: columnar structure
Schaufelförmige Schrägschichtung: trough cross-stratification; tangential cross-bedding
Scheinrippeln: pseudoripples
Schicht: bed, stratum
Schichtflächen-Erscheinung: sole mark
Schichtflächen-Marke: sole mark
Schichtinterne Verfältelung: convolute bedding
Schichtung: bedding
Schieferton-Schmitzen: shale crescents
Schlammbedeckte Rippelmarken: mud-buried ripple marks
Schlammvulkan: mud volcano
Schleifmarke: drag mark; groove cast; slide mark
Schleifriefen: drag striae
Schneeball-Struktur: snowball structure
Schräg-Feinschichtung: cross-lamination
Schrägschicht: foreset bed
Schrägschichtung: cross-bedding, -lamination
Schrägschichtungs-Bögen: rib-and-furrow
Schrumpfungsriß: mud crack, desiccation crack
Schürfmarken: gouge marks
Schuppenförmige Belastungsmarke: squamiform load cast
Schuppenförmige Gefließmarke: squamiform load cast
Schwingungsmarke; vibration mark
Schwundriß: mud crack, desiccation crack
Sedimentations-Einheit: sedimentation unit
Sedimentfahne: sand shadow
Sedimentgang: clastic dike, sandstone dike

Sedimentrolle: slump ball, flow roll, ball-and-pillow structure; spiral ball
Sedimentwalze: flow roll
Seegangsrippeln: wave ripple marks, oscillation ripple marks
Sichelförmige Rippelmarke: cuspate ripple mark
Sohlblatt: toe set
Sohlmarke: sole mark
Sortierte Schichtung: sorted bedding
Spiralstruktur: spiral structure
Splittermarke: chatter mark
Springmarken: saltation marks
Spülbogen: swash mark
Spülmarke: swash mark
Spülsaum: trash line
Stauchfältelung: convolute bedding
Stechmarke: prod cast
Steigmarke: air heave structure
Stirn-Absatz: foreset bed
Stoßmarke (-eindruck): prod cast, impact cast
Strandhörner: beach cusps
Striemung: lineation
Strömungskamm: current crescent
Strömungsmarke: current mark; flute cast; sole mark
Strömungsriefen: sand streak; lineation; parting lineation
Strömungsrinne: channel, scour mark
Strömungsrippel: current ripple mark
Strömungs-Streifung: parting lineation
Strömungswulst: flute cast
Stromatolith: stromatolite
Strombank: sand wave
Stromstreifung: parting lineation
Subaquatische Gleitung: slide
Subaquatische Rutschung (submarine —): slump; convolute bedding
Subsolifluktion: subsolifluction
Symmetrische Rippelmarken: symmetrical ripple marks
Synäresisrisse: syneresis cracks
Synklinale Schrägschichtung: festoon cross-bedding, trough cross-stratification
Synsedimentäre Falten: intraformational folds
Synsedimentäre Fließfältung: convolute bedding
Synsedimentäre Runzelung: intraformational corrugation
Tafelige Schrägschichtung: tabular cross-bedding
Terrassierter Strömungswulst: terraced flute cast
Tongallen: clay galls
Tonrollen: clay galls
Transversale Kolkmarken: transverse scour marks
Transversalrippeln: transverse ripple marks

Trichter- und Kegelstruktur: pit and mound structure
Trockenriß: mud crack, desiccation crack
Trogförmige Schrägschichtung: festoon cross-bedding, trough cross-stratification
Trübestrom: slurry slump
Unsymmetrische Rippelmarken: asymmetrical ripple marks
Unterseiten-Wulst: load cast
Verfältelte Strömungsrippel-Schichtung: convolute current-ripple lamination
Verfältelung: convolute bedding
Verformte Schrägschichtung: deformed cross-bedding
Verknäuelung: crumpled ball
Verknäuelter Rutschungsballen: crumpled ball
Vorschüttungsblatt: foreset bed
Wasser-Rinne: washout
Wasserstandsmarke: water-level mark
Wellenfurche: ripple mark
Wellenlinie: trash line
Wellige Schichtung: wavy bedding
Wellung: scalloping
Wickelfalten (-strukturen): convolutional ball, flow roll, spiral ball, ball-and-pillow structure, roll-up ball; convolute bedding
Wickelung (-sstruktur): convolute bedding
Windreihen-Kämme: windrow-ridges
Winkelige Schrägschichtung: angular cross-bedding
Winkelschichtung: cross-bedding
Wirbelbälle: whirl-balls
Wirbelmarken: eddy markings
Wirre, kleindimensionale Schrägschichtung: choppy cross-lamination
Wühlgefüge: mottled structure
Wulstbank: convolute bedding; ball-and-pillow structure
Wulstige Belastungsmarke: torose load cast
Wulstschichtung: convolute bedding
Wulststruktur: ball-and-pillow structure
Wulst-Textur: convolute bedding
Wulstung: convolute bedding
Wurzelstruktur: root cast
Zapfenwülste: conical flute casts
Zerrungs-Struktur: pull-apart structure
Zusammengefaltete Schichtung: prolapsed bedding
Zusammengesetzte Belastungsmarke: syndromous load cast
Zusammengesetzte Leeblatt-Schichtung: compound foreset bedding
Zusammengesetzte Rippelmarken: compound ripples, compound ripple marks
Zusammengesetzte Schrägschichtung: furious cross-lamination
Zylindrische Struktur: cylindrical structure

Index des termes français

La liste alphabétique qui suit contient les expressions françaises citées dans le glossaire, avec un renvoi à un ou plusieurs termes anglais. Que le lecteur veuille bien garder à l'esprit que cette liste n'est pas un lexique, que les termes anglais ne sont que des renvois aux en-tête du glossaire, auxquels il faudra se reporter pour trouver les définitions et références, et pour comprendre l'applicabilité des termes français.

Empreintes de grêle: hail imprints
Empreintes de grêlons: hailstone imprints
Empreintes de pluie: rain prints
Empreintes de roulement: roll-marks
Empreintes de vibration: chattermarks; vibration marks
Empreintes en chevrons: chevron mark
Empreintes en tourbillons: eddy markings
Empreintes froncées: crinkle marks
Empreintes transversales d'affouillement: transverse scour marks
Empreintes transversales de creusement: transverse scour marks
Ensemble sédimentaire: sedimentation unit
Eraflure: brush mark
Erosion de base: wash-out
Fausse stratification: false bedding, false stratification
Fausse stratification angulaire: angular cross-bedding
Fausses rides: pseudo-ripple marks
Fentes de dessiccation radiaires: radiate mud-cracks
Fentes de retrait: desiccation cracks
Festons: scalloping
Figures d'enfoncement: sag structure
Figures de bondissement: saltation marks
Figures de décollement: décollement structure
Figures de fluxion: flowage cast
Figures de saltation: saltation marks
Filets de sable: sand streak
Filon clastique: clastic dike, sandstone dike
Filon de grès: sandstone dike
Filon neptunien: neptunian dike
Fissures de dessiccation: syneresis cracks
Frange de marée: trash line
Fronces d'écoulement: flowage cast
Fronces transversales: creep wrinkles
Fucoïde: fucoid
Galet mou: clay gall
Glissement: slump
Glissement de boue fluente: slurry slump
Grande ride: meta-ripple
Granoclassement: graded bedding
Granocroissance: graded bedding (type inverse)
Granodécroissance: graded bedding (type normal)
Granuloclassement: graded bedding
Grès pétris: kneaded sandstone
Groupe planaire de laminae: planar cross-stratification
Hiéroglyphe de mur: sole mark
Laisse de haute mer: trash line
Laminae: lamination
Laminae en volutes: convolute bedding
Linéation de délit: parting lineation
Litage: bedding
Marques de glissement: slide marks

Marques de percussion: chattermarks
Mégaride: megaripple
Métaride: meta-ripple
Microstratification croisée: micro-cross-lamination
Moulage d'échancrure: flute cast
Moulage de cannelure: groove cast
Moulage de chenal: channel cast
Moulage de rainure: groove cast
Moulage de stries: microgroove cast
Moulage de surcharge: load cast
Moulages de cristaux: crystal casts
Moulages de cristaux de glace: ice crystal marks
Moulages de fentes de retrait: mud crack (casts)
Moulages de fissures chiffonnés: crumpled mud-crack casts
Moulages de fissures froissés: crumpled mud-crack casts
Moulages de racines: root casts
Moule de charge: load mold
Nappe de glissement: slump sheet
Nodule: nodule
Olisthostrome: olistostrome
Paquet de terrain glissé: slump
Peau de dinosaure: dinosaur leather
Pelotes de mer: sea balls
Pelotes végétales: lake balls
Phénomène d'arrachage: pull-apart structure
Plissement de charge: load fold
Plissements intraformationnels: intraformational folds
Plissotements dûs à un dégagement d'air: Air heave structure
Plissotements intraformationnels: intraformational corrugation
Poche de surcharge: load pocket
Poches d'enfoncement: sag structure
Pseudo-fentes de dessiccation: false mud cracks
Pseudo-nodules: slump balls
Pseudo-nodules irréguliers: crumpled balls
Pseudo-rides: pseudo-ripple marks
Pseudo-stratification oblique: pseudo-cross-stratification
Rainures: grooves, groove cast
Ride d'oscillation: wave ripple mark
Ride de vague: wave ripple mark
Ride géante: sand wave
Rides arquées en croissant: cuspate ripple marks; linguoid ripple marks
Rides asymétriques: asymmetrical ripple marks
Rides chevauchantes: climbing ripples
Rides complexes: compound ripples, compound ripple marks
Rides d'interférence: compound ripples; interference ripple marks

Rides d'oscillation: oscillation ripple marks
Rides de courant: asymmetrical ripple marks; current ripple marks
Rides de déferlement: surf ripples
Rides de grande taille: para-ripples
Rides de plage: ripple marks
Rides de plage arquées: curved ripple marks
Rides de plage en croissant: curved ripple marks
Rides de plage géantes: megaripple
Rides de retrait: backwash ripple marks
Rides en fossettes: dimpled current marks, tadpole nests, interference ripple marks
Rides en losange: rhomboid ripple marks
Rides hexagonales: hexagonal interference ripples
Rides interférentes: interference ripple marks
Rides longitudinales: longitudinal ripple-marks
Rides losangiques: rhomboid ripple marks
Rides normales: normal ripple marks
Rides parallèles: parallel ripple marks
Rides périclinales: pericline ripple marks
Rides rhomboïdales: rhomboid ripple marks
Rides superposées: compound ripple marks
Rides symétriques: symmetrical ripple marks
Rides transversales: transverse ripple marks
Rigole: channel
Rigoles de plage: rill marks
Sédimentation dégradée (vers le haut): graded bedding
Série massive: massive bedding
Sillons d'érosion allongés en faisceaux parallèles: furrow flute cast
Strate: bed; stratum
Straticule: lamination
Stratification: bedding; stratification
Stratification croisée: cross-bedding
Stratification croisée arquée: concave cross-bedding
Stratification croisée convexe vers le bas: concave cross-bedding
Stratification croisée de rides: ripple-cross-lamination
Stratification croisée en auge: trough cross-stratification
Stratification croisée en festons: festoon cross-lamination
Stratification croisée en lentille: lenticular cross-bedding
Stratification croisée incurvée: concave cross-bedding
Stratification de rides: lee-side concentration
Stratification deltaïque: delta bedding
Stratification en diagonale: diagonal bedding
Stratification entrecroisée en arêtes de poisson: chevron cross-bedding
Stratification entrecroisée en chevrons: chevron cross-bedding

Stratification inclinée: inclined bedding
Stratification lenticulaire: lenticular bedding
Stratification oblique: cross-bedding
Stratification oblique arquée: concave inclined-bedding
Stratification oblique couchée: prolapsed bedding
Stratification oblique dérangée: deformed cross-bedding
Stratification oblique fine: cross-lamination
Stratification oblique incurvée: concave inclined-bedding
Stratification oblique tabulaire: tabular cross-bedding
Stratification ondulée: convolute bedding
Stratification ondulée: wavy bedding
Stratification onduleuse asymétrique: rolling incline bedding
Stratification tourbillonnaire: whirl-zone
Strie en tête de clou: nailhead striation
Stries d'entraînement: drag striae
Stries de traînage: drag striae
Stromatolithe: stromatolite
Structure à chenaux: channel cast
Structure brèchique: brecciated structure
Structure de champ: edgewise structure
Structure en boules: ball structure
Structure en boules-et-coussins: ball-and-pillow structure
Structure en brèche: brecciated structure
Structure en côtes et sillons: rib-and-furrow
Structure en coussins: pillow-structure
Structure en écailles: imbricate structure
Structure en flamme: flame structure
Structure fluidale convulsée: slip bedding
Structure imbriquée: imbricate structure
Structure prismée: columnar structure
Structures cylindriques: cylindrical structures
Subsolifluxion: subsolifluction
Surface durcie: corrosion surface
Surface limite: corrosion surface
Surfaces contournées par glissement: slip bedding
Talus croulant: slip face
Terrier: burrow
Texture mouchetée: mottled structure
Texture tachetée: mottled structure
Trace d'organisme: bioglyph; ichnofossil
Tracé du jet de rive: swash mark
Trace vivante: bioglyph
Trou: burrow
Tube: burrow
Vague de charge: load wave
Vague de sable: meta-ripple; sand wave
Vestiges fossiles de vie: ichnofossil
Volcan de boue: mud volcano
Volcans de sable: sand volcanoes

Indice Castellano-Inglés

Abioglifo: Abioglyph
Antiduna: antidune, regressive sand wave
Anti-óndula: anti-ripplets
Arenisca malaxada: kneaded sandstone
Arenisca vorticelada: whirl-balls, sandstone
Bioglifo: bioglyph
Bloque deslizado: slip block
Bolsillo de carga: load pocket
Calco caulifoliado: cabbage leaf marking, deltoidal cast, frondescent cast
Calco ctenoide: ctenoid cast
Calco de carga: load cast
Calco de carga dendriseptado: syndromous load cast
Calco de carga escamiforme: squamiform load cast
Calco de carga polilobulado: torose load cast
Calco de cristales: crystal casts
Calco de cristales de hielo: ice crystal marks (casts)
Calco de deslizamiento: slide cast
Calco de empuje: brush mark (brush cast)
Calco de estriaciones de carga: load-cast striations
Calco de flujo: flow cast, flowage cast
Calco de grietas de desecación, deformado: crumpled mud-crack casts
Calco de impacto: impact cast
Calco de microsurco: drag striae, microgroove cast, striation cast
Calco de punzamiento: prod mark (prod cast)
Calco de raíces: root casts
Calco de roce: bounce cast
Calco de surco: groove cast, mud furrow, undertow mark
Calco de surco espigado: groove ruffling, ruffled groove cast
Calco espigado: chevron mark, herringbone marking, chevron cast
Calco en herradura: crescent cast, current crescent, horse-shoe flute cast
Calco junco-radiciforme: reed casts
Capa: bed
Capa basal: bottomset bed
Capa dorsal: top set (bed)
Capa frontal: foreset bed
Cataglifo: kataglyph

Cauce: channel, scour channel, washout (cut-out)
Cobertura de óndula: mud-buried ripple mark
Coprolito: coprolite
Corrugación intraformacional: intraformational corrugation, laminar corrugation
Corrugaciones por escape de aire: air-heave structure
Cractercillos de playa: spring pits
Criosombra de obstáculo: knob-and-trail
"Cuero de dinosaurio": "dinosaur leather"
Cuspilitos: beach cusps
Depósito por desmoronamiento: slump
Desbaste en media luna: wash-over crescents
Deslizamiento: slide
Diaglifo: diaglyph
Dique areniscoso: sandstone dike
Dique clástico: clastic dike, neptunian dike
Duna: dune
Endoglifo: endoglyph
Entrecruzamiento festoneado: crescent-type cross-bedding, festoon cross-bedding, festoon cross-lamination, trough cross-lamination, trough cross-stratification
Entrecruzamiento de crestas: lee-side concentration
Entrecruzamiento de óndulas superpuestas: climbing ripples
Epeiroglifo: epeiroglyph
Epiglifo: epiglyph
Erpoglifo: Erpoglyph
Esferoides algáceos: algal balls
Esferoides corrugados: crumpled balls
Esferoides espiralados: spiral balls
Esferoides intraformacionales: convolute balls, roll-up ball
Esferoides lacustres: lake balls
Esferoides lisos: slump balls, snowball structure
Esferoides marinos: sea balls
Estratificación: bedding, stratification
Estratificación corrugada: ordinary rolling strata
Estratificación corrugada unilateral: unilateral rolling strata
Estratificación deformada por desmoronamiento: slump bedding
Estratificación deltaica: delta bedding

361

Estratificación entrecruzada: cross-bedding, cross-stratification, current bedding, diagonal lamination, false bedding, false stratification, foreset bedding, oblique lamination

Estratificación entrecruzada angular: angular cross-bedding

Estratificación entrecruzada basiplana: planar cross-stratification

Estratificación entrecruzada compuesta: compound foreset bedding

Estratificación entrecruzada cóncava: concave cross-bedding

Estratificación entrecruzada cuneiforme: wedge-shaped cross-bedding

Estratificación entrecruzada deformada: deformed cross-bedding

Estratificación entrecruzada doble: "furious" cross-lamination

Estratificación entrecruzada espigada: chevron cross-bedding, herringbone cross-bedding

Estratificación entrecruzada lenticular: lenticular cross-bedding

Estratificación entrecruzada plana: planar cross-bedding

Estratificación entrecruzada simple: simple cross-stratification

Estratificación entrecruzada tabular: tabular cross-bedding

Estratificación entrecruzada tangencial: tangential cross-bedding

Estratificación entrecruzada torrencial: torrential cross-bedding

Estratificación entrecruzada torrencial cuneiforme: wedge torrential cross-bedding

Estratificación eslabonada: loop-bedding

Estratificación flaser: flaser structure

Estratificación gradada: graded bedding

Estratificación inclinada: diagonal bedding, drift bedding, inclined bedding, oblique bedding, oblique stratification

Estratificación inclinada cóncava: concave inclined-bedding

Estratificación inclinada convexa: convex incline-bedding

Estratificación inclinada inversa: backset bedding

Estratificación intraplegada: convolute bedding, curly bedding, gnarly bedding, hassock bedding, hassock structure, intrastratal contortions, "sealing-wax flow", slip bedding

Estratificación lenticular: lenticular bedding

Estratificación masiva: massive bedding

Estratificación nodular: nodular bedding

Estratificación ondulada: wavy bedding

Estratificación ondulítica: ripple bedding

Estratificación pseudo-inclinada: rolling incline-bedding

Estratificación rizada: crinkled bedding

Estratificación seleccionada: sorted bedding

Estrato: bed, stratum

Estrato ondulítico: waved stratum

Estría roma: nailhead striation

Estromatolita: stromatolite

Estructura almohadillada: ball-and-pillow structure, ball structure, flow roll, flow structure, force-aparts, intra-stratal flowage, mammillary structure, pillow-structure, pseudo-nodules, sandstone balls, storm roller

Estructura brechosa: brecciated structure

Estructura cilíndrica: cylindrical structures, sandstone pipes

Estructura columnar: columnar structure, staticulate

Estructura de carga: sag structure

Estructura de costillas y surcos: rib-and-furrow

Estructura de despegue: décollement structure

Estructura de flujo intraestratal: intrastratal flow structure

Estructura de hoyuelo y montículo: pit-and-mound structure

Estructura espiralada: spiral structure

Estructura flamiforme: flame structure, load wave, streaked-out ripples

Estructura imbricada: imbricate structure

Estructura moniliforme: boudinage, pull-apart structure

Estructura moteada: mottled structure

Estructura tubular: burrow

Exoglifo: Exoglyph

Flanco de deslizamiento: slip face, slip slope

Fractura falciforme: crescentic fracture

Fucoide: fucoid

Fulgurita: fulgurite

Grietas de desecación: desiccation cracks, desiccation marks, mud cracks, shrinkage cracks, sun cracks

Grietas de deshidratación: syneresis cracks

Grietas radiadas de desecación: radiate mud cracks

Grietas tensionales: parting cast, pseudo mud cracks

Hieroglifo: hieroglyph

Hiperglifo: hyperglyph

Hipoglifo: hypoglyph

Hoyos por escape gaseoso: gas pits

Hoyuelos de burbujeo: bubble impressions

Icnofósil: ichnofossil

Keazoglifo: keazoglyph

Klizoglifo: klizoglyph

Ksimoglifo: ksimoglyph

Laminación: lamination

Laminación entrecruzada: cross-lamination, flow-and-plunge structure

Laminación entrecruzada torrencial: torrential cross-lamination
Laminación intraplegada: convolute lamination
Laminación micro-entrecruzada: micro-cross-lamination
Lentecillos arcillosos: shale crescents
Lineación: lineation
Línea de escombros: trash line
Lineación de calcos de carga: load-cast lineation
Lineación de cauces: scour lineation
Lineación por corriente: current lineation, parting lineation
Lineación substratal: substratal lineation
Lodo por deslizamiento: slurry slump
Lomos eólicos: windrow ridges
Manto de deslizamiento: slump sheet
Marca a sotavento: slump mark
Marca de arrastre: drag mark
Marca de base: sole mark
Marca de nivel: water-level mark
Marca de resaca: swash mark, wave line
Marca en herradura: crescenting scour mark
Marcas anulares: ring-mark
Marcas de corriente agravadas: load casted current markings, load casted sole marks
Marcas de desbaste: scour marks
Marcas de desbaste diagonales: diagonal scour marks
Marcas de desbaste transversales: transverse scour marks
Marcas de escurrimiento: rill mark
Marcas de granizo: hail (hailstone) imprints
Marcas de lluvia: rain prints (casts), rain-drop impressions
Marcas de rechinamiento: chattermarks
Marcas de remolino: eddy markings
Marcas de rolido: roll-mark, roll-spuren
Marcas de saltación: saltation marks
Marcas de saltación regular: skip marks
Marcas de vibración: vibration mark
Marcas labradas: tool marks
Marcas lunadas: crescentic gouge, gouge marks
Marcas plumiformes: feather-like flow markings
Mecanoglifo: mechanoglyph
Mega calcos de flujo: shooting flow cast
Mega-óndula: megaripple
Metaglifo: metaglyph
Meta-óndula: meta-ripple
Micro-cauces elongados: elongate irregular marks
Micro-entrecruzamiento festoneado: "choppy" cross-lamination
Migro-óndulas entrecruzadas: ripple cross-lamination, ripple drift, rolling strata
Migro-óndulas entrecruzadas, deformadas: convolute current-ripple lamination

Moldeado de carga: load mold
Nódulo: nodule
Olistoglifo: olistoglyph
Olistrostroma: olistrostrome
Onda de arena: sand wave
Onda de arena progresiva: progressive sand waves
Ondula: ripple mark
Ondula ascendente: upslope ripple
Ondula asimétrica: asymmetrical ripple marks
Ondula compuesta: compound ripples
Ondula curva: curved ripple marks
Ondula de cauce: ripple scour
Ondula de crestas planas: flat-topped ripple mark
Ondula de oscilación: oscillation ripple marks, wave ripple marks
Ondula de resaca: backwash ripple marks
Ondula de rompiente: surf ripples
Ondula descendente: downslope ripple
Ondula horizontal: level surface ripple
Ondula incompleta: incomplete ripples, starved ripples
Ondula linguoide: current mark, cuspate ripple mark, linguoid ripple marks
Ondula longitudinal: longitudinal ripple marks
Ondula normal: normal ripple mark
Ondula paralela: parallel ripple mark
Ondula periclinal: pericline ripple mark
Ondula rectangular: rectangular interference ripples
Ondula regresiva: regressive ripple
Ondula romboide: rhomboid ripple mark
Ondula simétrica: symmetrical ripple marks
Ondula transversa: current ripple mark, transverse ripple marks
Ondulas conjugadas: compound ripple marks
Ondulas hexagonales: hexagonal interference ripples
Ondulas de interferencia: cross ripples, dimpled current mark, interference ripple marks, tadpole nests
"Ondulita" en pizarra: scalloping
Para-óndula: para-ripples
Peiroglifo: peiroglyph
Pie de capa frontal: toe-set
Pliegues de carga: load fold
Pliegues intraformacionales: intraformational folds
Pliegues volcados por deslizamiento: prolapsed bedding
Pliegues volcados por desmoronamiento: slump overfold
Proglifo: proglyph
Promontorios fangosos: mud lumps
Pseudo-estratificación entrecruzada: pseudo cross-stratification

Pseudo-grietas de desecación: false mud cracks

Pseudo-óndulas: creep wrinkles, crinkle marks, pseudo-ripples

Relleno de cauce: cut-and-fill, channel cast, gouge channel, scour and fill

Reoglifo: rheoglyph

Rodado de arcilla: clay ball

Rodado de arcilla acorazado: armored mud ball

Sesgoconglomerado: edgewise structure

Singlifo: synglyph

Sombra arenosa: sand shadow

Subsolifluxión: subsolifluction

Superficie de corrosión: corrosion surface

Surco: groove

Surcos de erosión: erosion grooves

Surcos intersectos: cross-grooves

Tafoglifo: taphoglyph

Teggoglifo: teggoglyph

Trazas arenosas: sand streak

Turboglifo: directional load cast, flow mark, flute, flute cast, lobate plunge structure, lobate rill mark, scour cast, scour finger, turboglyph, vortex cast

Turboglifo aterrazado: terraced flute cast

Turboglifos abulbosos interseptos: furrow cast, sludge cast

Turboglifos cónicos: conical flute casts

Turboglifos en sacacorcho: corkscrew flute casts

Turboglifos interseptos: delicate flute casts, furrow flute cast

Turboglifos platiformes: depressed flute casts

Unidad sedimentaria: sedimentation unit

Vesículas arcillosas: clay galls

Volcanes de arena: sand volcanoes

Volcanes de fango: mud volcanoes

Zona de vórtices: whirl-zone

References

*ABEL, OTHENIO, 1935: Vorzeitige Lebensspuren. Jena: Gustav Fischer, 644 p.

ANDERSEN, S. A., 1931: Om aase og terrasser inden for Susaa's vandomraade og deres vidnesbyrd om isafsmeltningens forløb* (The eskers and terraces in the basin of the River Susaa and their evidences of the process of the ice waning.) Danmarks Geol. Undersøgelse, Raekke II, No. 45, 201 p. [Danish; Engl. Summ.].

ANDRÉE, K., 1915: Ursachen und Arten der Schichtung. Geol. Rundschau 6, 351—397.

ARAI, JUZO, 1957: Some load cast structures in the Akahira formation (Oligocene), Chichibu basin, Saitama Prefecture, Japan. Bull. Chichibu Mus. Nat. History, No. 7, 85—88.

— 1959: Cylindrical structures in the Tertiary sediments of the Chichibu basin, Saitama Prefecture, Japan. Bull. Chichibu Mus. Nat. History, No. 9, 61—68.

— 1960: The Tertiary system of the Chichibu basin, Saitama Prefecture, central Japan, Part I, Sedimentology. Tokyo, Japan Soc. Promotion of Sci., 122 p.

ATKINSON, D. J., 1962: Tectonic control of sedimentation and the interpretation of sediment alternation in the Tertiary of Prince Charles foreland, Spitzbergen. Bull. Geol. Soc. Am. 73, 343—364.

BAIRD, D. M., 1962: Prince Edward Island National Park, the living sands. Geol. Survey Canada, Misc. Rept. 3, 56 p.

BAULIG, H., 1956: Vocabulaire franco-anglo-allemand de géomorphologie. Société d'édition Les Belles Lettres, Paris (Publ. Faculté Lettres Univ. Strasbourg, fasc. 130), 230 p.

BEASLEY, H. C., 1914: Some fossils from the Keuper sandstone of Alton Staffordshire. Proc. Liverpool Geol. Soc. 12, 35—39.

BECKER, G. F., 1893: Finite homogeneous strain, flow and rupture of rocks. Bull. Geol. Soc. Am. 4, 13—90.

BEETS, C., 1946: Miocene submarine disturbances of strata in northern Italy. J. Geol. 54, 229—245.

BELL, H. S., 1940: Armored mud balls—their origin, properties and role in sedimentation. J. Geol. 48, 1—31.

*BENEO, E., 1955: Les resultats des etudes pour la recherche petrolifere en Sicile (Italy). Proc. Fourth World Petrol. Cong., Sec. I, 108—121.

BERSIER, A., 1959a: Exemples de sédimentation cyclothémique dans l'Aquitanien de Lausanne. Eclogae Geol. Helv. 51, 842—853.

— 1959b: Séquences détritiques et divagations fluviales. Eclogae Geol. Helv. 51, 854—893.

BIRKENMAJER, K., 1958: Orientouane hieroglify splywowe we fliszu Karpackimi i ich stosunek do hieroglifów pradowych i wleczniowych (Oriented flowage casts and marks in the Carpathian flysch, and their relation to flute and groove casts). Acta Geol. Polon. 8, 118—148 [Polish; Engl. summ.].

— 1959: Systematyka warstwowań w utworach fliszowych i podobynch (Classification of bedding in flysch and similar graded deposits). Studia Geol. Polonica 3, 133 p. [Polish; Engl. Summ.].

BOKMAN, J., 1953: Lithology and petrology of the Stanley and Jackfork formation. J. Geol. 61, 152—170.

— 1956: Terminology for stratification in sedimentary rocks. Bull. Geol. Soc. Am. 67, 125—126.

BOTVINKINA, L. N., 1959: Morphological classification of bedding in sedimentary rocks. Izvestiya (Amer. Geol. Inst. English translation, 1960) Acad. Sci. U.S.S.R., geol. ser., No. 6, p. 13—30).

— 1962: Sloistost osadochnykh porod (Bedding of sedimentary rocks). Akad. Nauk. S.S.S.R., Trudy geol. Inst. no. 59, 542 p.

— YU. A. ZHEMCHUZHNIKOV, P. P. TIMOFEEV, A. P. FEOFILOVA and V. S. YABLOKOV, 1956: Atlas litogeticheskikh tipov uglenosnykh otlozhenii srednego karbona Donetskogo basseina (Atlas of lithogenic types of coal-bearing deposits of the Middle Carboniferous of the Donetz Basin). Akad. Nauk. S.S.S.R., Inst. Geol. Nauk, Moscow, 368 p.

BOUMA, A. H., 1962: Sedimentology of some flysch deposits. Amsterdam: Elsevier Publishing Co., 168 p.

BRADLEY, W. H., 1930: The behavior of certain mud-crack casts during compaction. Am. J. Sci. [5] 20, 136—144.

— 1931: Origin and microfossils of the oil shale of the Green River Formation of Colorado and Utah. U.S. Geol. Survey, Prof. Paper 168, 58 p.

*BRUNS, YE P., 1954: Nablyudeniya nad osobennostyami sloistosi otlozhenig. Methodich. Rukovodstoo po geol. S"yemki i poiskam (The peculiarities of bedding in deposits. Manual

of methods used in geological surveying and
prospecting). VSEGEI, Gosgeoltekhizdat.

BUCHER, W. H., 1919: On ripples and related
sedimentary surface forms and their paleogeographic interpretation. Am. J. Sci. **47** [4],
149—210, 241—269.

BURT, F. A., 1930: Origin and significance of
clay-galls. Pan-Am. Geol. **53**, 105—110.

CAROZZI, A. V., 1953: Pétrographie des roches
sédimentaires. Lausanne: F. Rouge & Cie.,
250 p.

— 1960: Microscopic arched flow structures and
spiral structures in sedimentary rocks. Bull.
Inst. Nat. Genevois **60**, 1—23.

CASTER, K. E., 1957: Problematica. Treatise on
Marine Ecology and Paleoecology, Memoir 67,
Geol. Soc. Am. p. 1025—1032.

CHADWICK, G. H., 1931: Storm rollers (abstract).
Bull. Geol. Soc. Am. **42**, 242.

— 1940: Columnar limestone produced by sun-
cracking (abstract). Bull. Geol. Soc. Am. **51**,
1923.

— 1948: Ordovician "dinosaur-leather" marking
(exhibit). Bull. Geol. Soc. Am. **59**, 1315.

CHAMBERLIN, T. C., 1888: The rock-scorings of
the great ice invasions. U.S. Geol. Survey,
7th Ann. Rept., 1885—1886, p. 155—248.

Chambre syndicale de la recherche et de la production du pétrole et du gaz naturel, 1961:
Essai de nomenclature des roches sédimentaires. Editions Technip (Paris?), 78 p.
(Prepared by Comité des Techniciens, Com-
mission d'Exploration, Sous-commission La-
boratoires de Stratigraphie.)

CLARKE, J. M., 1917: Strand and undertow
markings of Upper Devonian times as indication of the prevailing climate. New York
State Mus., Bull. **196**, 199—210.

CLOOS, HANS, 1938: Primäre Richtungen in
Sedimenten der rheinischen Geosynkline:
Geol. Rundschau **29**, 357—367.

CLOUD, P. E., 1942: Notes on stromatolites. Am.
J. Sci. **240**, 363—379.

COOPER, J. R., 1943: Flow structures in Berea
sandstone and Bedford shale of central Ohio.
J. Geol. **51**, 190—203.

CRONEIS, CAREY, and D. M. GRUBBS, 1939:
Silurian sea balls. J. Geol. **47**, 598—612.

CROWELL, J. C., 1955: Directional-current structures from the Pre-alpine Flysch, Switzerland.
Bull. Geol. Soc. Am. **66**, 1351—1384.

— 1957: Origin of pebbly mudstones. Bull.
Geol. Soc. Am. **68**, 993—1010.

CUMMINS, W. A., 1958: Some sedimentary structures from the Lower Keuper sandstones.
Liverpool Manchester Geol. J. **2**, 32—43.

DANGEARD, L., FRANCIS DORÉ and PIERRE
JUIGNET, 1961: Le Briovérien supérieur de
basse Normandie (étage de la Laize), série à
turbidites, a tous les caractères d'un flysch.
Revue de Géogr. Physique et de Géol.
Dynamique (2) **4**, No. 4, 251—259.

DAVIS, W. M., 1890: Structure and origin of
glacial sand plains. Bull. Geol. Soc. Am. **1**,
195—202.

DILLER, J. S., 1890: Sandstone dikes. Bull. Geol.
Soc. Am. **1**, 441—442.

DOTT jr., R. H., and J. K. HOWARD, 1962: Convolute lamination in nongraded sequences.
J. Geol. **70**, 114—120.

DUNBAR, C. O., and JOHN RODGERS, 1957: Principles of Stratigraphy. New York: John
Wiley & Sons, 356 p.

DZULYNSKI, ST., M. KSIAZKIEWICZ and PH. H.
KUENEN, 1959: Turbidites in flysch of the
Polish Carpathian Mountains. Bull. Geol.
Soc. Am. **70**, 1089—1118.

—, and J. E. SANDERS, 1959: Bottom marks on
firm lutite substratum underlying turbidite
beds (abstract). Bull. Geol. Soc. Am. **70**,
1594.

— —, 1962: Current marks on firm mud bottoms. Trans. Conn. Acad. Arts. Sci. **42**,
57—96.

—, and A. SLACZKA, 1958: Sedymentacja i
wskaźniki kierunkowe transporta w warstwach
Krośnieńch (Directional structures and sedi-
mentation of the Krosno beds, Carpathian
Flysch). Ann. soc. géol. Pologne **28**, 205—260
[English; Polish summ.].

EINSELE, G. (in press): „Convolute bedding" und
ähnliche Sedimentstrukturen im rheinischen
Oberdevon und anderen Ablagerungen. Neues
Jahrb. Geol. u. Paläont., Abhandl.

FAHRIG, W. F., 1961: The geology of the Athabaska formation. Geol. Survey Canada, Bull.
68, 41 p.

FAIRBRIDGE, R. W., 1946: Submarine slumping
and location of oil bodies. Bull. Am. Assoc.
Petrol. Geologists **30**, 84—92.

FEARNSIDES, W. G., 1910: Tremadoc Slates and
associated rocks of the south-east Carnarvon-
shire. Quart. J. Geol. Soc. London **66**, 142—
187.

FOX, C. S., 1931: The natural history of Indian
coal. Mem. Geol. Survey India **57**, 283 p.

*FUCHS, TH., 1895: Studien über Fucoiden und
He iroglyphen. Akad. Wiss. Wien, Denkschr.,
Math.-nat. Kl. **62**, 369—448.

FUGGER, E., 1899: Das Salzburger Vorland.
Jahrb. Kais. Kön. Geol. Reichsanst. **49**,
287—428.

GABELMAN, J. W., 1955: Cylindrical structures
in Permian (?) siltstone, Eagle County,
Colorado. J. Geol. **63**, 214—277.

GATES, OLCOTT, 1962: The geology of the Cutler
and Moose River quadrangles, Washington
County, Maine. Maine Geol. Survey, Quadrangle Mapping Ser. No. 1, 67 p.

GILBERT, G. K. 1899: Ripple-marks and cross-
bedding. Bull. Geol. Soc. Am. **10**, 135—140.

— 1905: Crescentic gouges on glaciated surfaces.
Bull. Geol. Soc. Am. **15**, 303—316.

— 1914: The transportation of debris by run-
ning water. U.S. Geol. Survey Prof. Paper
86, 263 p.

GILL, W. D., and PH. H. KUENEN, 1958: Sand
volcanoes on slumps in the Carboniferous of
County Clare, Ireland. Quart. J. Geol. Soc.
London **113**, 441—460.

GRANGEON, M., 1961: Observations sur la sédimentation houillère. Bull. soc. géol. France [7], **2**, 630—650.

GRAY, H. H., 1955: Thickness of bedding and parting in sedimentary rocks. Bull. Geol. Soc. Am. **66**, 147—148.

GÜRICH, G., 1943: Schrägschichtungsbögen und zapfenförmige Fließwülste im „Flagstone" von Pretoria und ähnliche Vorkommnisse im Quarzit von Kuibis, SWA, dem Schilfsandstein von Maulbronn u. a. Z. dtsch. geol. Ges. **85**, 652—663.

GUILCHER, A., 1954: Morphologie littorale et sous-marine. Paris: Presses Universitaires de France, 216 p.

HAAF, E. TEN, 1956: Significance of convolute lamination. Geol. en Mijnbouw **18**, 188—194.

— 1959: Graded beds of the northern Apennines. Ph. D. thesis, Rijks University of Groningen, 102 p.

HADDING, A., 1931: On subaqueous glides. Geol. Fören. i Stockholm Förh. **53**, 378—393.

HÄNTZSCHEL, W., 1938: Bau und Bildung von Groß-Rippeln im Wattenmeer. Senckenbergiana, **20**, 1—42.

HAGN, H., 1955: Fazies und Mikrofauna der Gesteine der Bayrischen Alpen. Leiden, Brill, 174 p.

HALL, JAMES, 1843: Geology of New York, Pt. 4, Survey of the fourth district. Albany: Carroll and Cook, 683 p.

HAMBLIN, W. K., 1961: Micro-cross-lamination in upper Keweenawan sediments of northern Michigan. J. Sediment. Petrol. **31**, 380—401.

— 1962: X-ray radiography in the study of structures in homogeneous sediments. J. Sediment. Petrol. **32**, 201—210.

HANSEN, EDWARD, S. C. PORTER, B. A. HALL and ALLAN HILLS, 1961: Décollement structures in glacial-lake sediments. Bull. Geol. Soc. Am. **72**, 1415—1418.

HARDY, C. T., and J. S. WILLIAMS, 1959: Columnar contemporaneous deformation. J. Sediment. Petrol. **29**, 281—282.

HARRIS jr., S. E., 1943: Friction cracks and the direction of glacial movement. J. Geol. **51**, 244—258.

HAWLEY, J. E., and R. C. HART, 1934: Cylindrical structures in sandstones. Bull. Geol. Soc. Am. **45**, 1017—1034.

HEIM, A., 1908: Über rezente und fossile subaquatische Rutschungen und deren lithologische Bedeutung. Neues Jahrb. Mineral. Geol. Paläon. **2**, 136—157.

HESSLAND, IVAR, 1955: Studies in the lithogenesis of the Cambrian and basal Ordovician of the Böda Hamn sequence of strata. Bull. Geol. Inst. Univ. Upsala. **35**, 35—109.

HILLS, E. S., 1941: Outlines of structural geology. New York: Nordeman Publ. Co., 172 p.

HITCHCOCK, EDWARD, 1858: Ichnology of New England, a report on the sandstone of the Connecticut Valley. Boston: William White, State Printer, 220 p.

HOBBS, W. H., 1906: Guadix formation of Granada, Spain. Bull. Geol. Soc. Am. **17**, 285—294.

HOPPE, W., 1930: Beiträge zur Geologie und Petrographie des Buntsandsteins im Odenwald. III. Schichtung und Bankung. Notizbl. Vereins f. Erdkunde [V] **12**, 149—170.

ILLIES, H., 1949: Die Schrägschichtung in fluviatilen und litoralen Sedimenten, ihre Ursachen, Messung und Auswertung. Mitt. Geol. Staatsinst. Hamburg, **19**, 89—109.

INGERSON, EARL, 1940: Fabric criteria for distinguishing pseudo ripple marks from ripple marks. Bull. Geol. Soc. Am. **51**, 557—570.

INGRAM, R. L., 1954: Fissility of mudrocks. Bull. Geol. Soc. Am. **64**, 869—878.

JAMIESON, T. F., 1860: On the drift and rolled gravel of the north of Scotland. Quart. J. Geol. Soc. London **16**, 347—371.

JOPLING, A. V., 1960: An experimental study of the mechanics of bedding. Ph. D. thesis, Harvard University, 358 p.

— 1961: Origin of regressive ripples explained in terms of fluid mechanic processes *in* Short papers in the geologic and hydrologic sciences, articles 293—435. U.S. Geol. Survey. Prof. Paper **424-D**, D-15—D-17.

*JUKES, J. B., 1872: The Students Manual of Geology, 3rd ed. Edinburgh: A. and C. Black, 778 p.

KALKOWSKI, ERNST, 1908: Oolith und Stromatolith im Norddeutschen Buntsandstein. Z. dtsch. geol. Ges. **60**, 68—125.

KAUFMAN, D. W., and C. B. SLAWSON, 1950: Ripple mark in rock salt of the Salina formation. J. Geol. **58**, 24—29.

KEMP, J. F., 1922: Handbook of rocks, 5th ed. New York: D. van Nostrand Co., 282 p.

KHVOROVA, I. V., 1958: Atlas karbonatnych porod srednego i verhnego karbon Russkoy platformy (Atlas of the Middle and Upper Carboniferous carbonate rocks of the Russian Platform). Izdatelstvo, Akademii Nauk SSSR, Moscow, 1958, 167 p.

KINDLE, E. M., 1914: An inquiry into the origin of *Batrachioides* the *antiquor* of the Lockport dolomite of New York. Geol. Mag. [6] **1**, 158—161.

— 1916: Small pit and mound structure. Geol. Mag. [6] **3**, 542—547.

— 1917a: Recent and fossil ripple-mark. Geol. Survey Canada, Museum Bull. No. 25, Geol. Series, No. 34, 121 p.

— 1917b: Diagnostic characteristics of marine clastics. Bull. Geol. Soc. Am. **28**, 905—916.

— 1926: Contrasted types of mud cracks. Proc. and Trans. Roy. Soc. Canada [3] **20**, pt. 4, 71—75.

KINGMA, J. T., 1958: The Tongaporutuan sedimentation in central Hawkes Bay. New Zealand J. Geol. Geophys. **1**, 1—30.

KNIGHT, S. H., 1929: The Fountain and Casper formations of the Laramie Basin. Univ. Wyoming Publ. in Science, Geology **1**, 82 p.

KREJCI-GRAF, K., 1932: Definition der Begriffe Marken, Spuren, Fährten, Bauten, Hieroglyfen und Fucoiden. Senckenbergiana **14**, 19—39.

KSIAZKEWICZ, M., 1958: Osuwiska podmorskie we fliszu karpackim (Submarine slumping in the Carpathian flysch). Ann. soc. géol. Pologne **28**, 123—150. [English with Polish summ.].

KUENEN, PH. H., 1948a: The formation of beach cusps. J. Geol. **56**, 34—40.

— 1948b: Slumping in the Carboniferous rocks of Pembrokeshire. Quart. J. Geol. Soc. London **104**, 365—368.

— 1950: Marine geology. New York: John Wiley & Sons, 568 p.

— 1952: Paleogeographic significance of graded bedding and associated features. Koninkl. Ned. Akad. Wetenschap. Proc., Ser. B **55**, 28—36.

— 1953: Significant features of graded bedding. Bull. Am. Assoc. Petrol. Geologists **37**, 1044—1066.

— 1957: Sole markings of graded graywacke beds. J. Geol. **65**, 231—258.

— 1958: Experiments in geology. Trans. Geol. Soc. Glasgow **23**, 1—28.

— 1961: Some arched and spiral structures in sediments. Geol. en Mijnbouw **40**, 71—74.

— A. FAURE-MURET, M. LANTEAUME and P. FALLOT, 1957: Observations sur les flyschs des Alpes Maritimes françaises et italiennes. Bull. soc. géol. France. [6] **7**, 11—26.

—, and J. E. SANDERS, 1956: Sedimentation phenomena in Kulm and Flözleeres graywackes, Sauerland and Oberharz, Germany. Am. J. Sci. **254**, 649—671.

LAHEE, F. H., 1952: Field geology, 5th ed. New York: McGraw-Hill Book Co., 883 p.

LAMONT, ARCHIE, 1957: Slow anti-dunes and flow marks. Geol. Mag. **94**, 472—480.

LANE, E. W., 1947: Report of the subcommittee on sediment terminology. Trans. Am. Geophys. Union **28**, 938.

LIPPERT, H., 1937: Gleit-Faltung in subaquatischem und subaerischem Gestein. Senckenbergiana **19**, 355—375.

LLARENA, JOAQUIN GOMEZ DE, 1954: Observaciones geologicas en el flysch Cretacico-Numulitico de Guipuzcoa. Mono. Instit. "Lucas Mallada" Invest. Geol., No. 13, pt. 1, 98 p.

LOCKE, JOHN, 1838: Professor Locke's geological report *in* Second Annual Report. Ohio Geol. Survey, Columbus, p. 203—286.

LOMBARD, A., 1956: Géologie sédimentaire. Les séries marines. Paris: Masson & Cie.; Liége: H. Vaillaut-Caimanne, 722 p.

LYELL, CHARLES, 1839: Elements of geology, 1st ed. Philadelphia: James Kay, Jun. and Brother, 316 p.

MACAR, P., 1948: Les pseudo-nodules du Famennien et leur origine. Ann. soc. géol. Belg. **72**, 47—74.

— 1952: Les critères indiquant l'ordre de succession des couches et leur emploi en Belgique. Ann. soc. géol. Belg. **75**, B259—B278.

MAXON, J. H., 1940: Gas pits in non-marine sediments. J. Sediment. Petrol. **10**, 142—145.

— and IAN CAMPBELL, 1935: Stream fluting and stream erosion. J. Geol. **43**, 729—744.

McBRIDE, E. F., 1960: Martinsburg flysch of the central Appalachians. Ph.D. thesis, The Johns Hopkins University [Baltimore], 375 p.

— 1962: Flysch and associated beds of the Martinsburg formation (Ordovician), central Appalachians. J. Sediment. Petrol. **32**, 39—91.

—, and L. S. YEAKEL, 1963: Relationship between lineation and rock fabric. J. Sediment. Petrol. 33, 179—182.

McCROSSAN, R. G., 1958: Sedimentary boudinage structures in the Upper Devonian Ireton formation of Alberta. J. Sediment. Petrol. **28**, 316—320.

McDOWELL, J. P., 1957: The sedimentary petrology of the Mississagi Quartzite in the Blind River area. Ontario Dept. Mines, Geol. Circular **6**, 31 p.

McIVER, N. L., 1961: Upper Devonian marine sedimentation in the central Appalachians. Ph.D. thesis, The Johns Hopkins University, 347 p.

McKEE, EDWIN D., 1938: Original structures in Colorado flood deposits of Grand Canyon. J. Sediment. Petrol. **8**, 77—83.

— 1939: Some types of bedding in the Colorado River delta. J. Geol. **47**, 64—81.

— 1945: Small-scale structures in the Coconino sandstone of northern Arizona. J. Geol. **53**, 313—325.

— 1954: Stratigraphy and history of the Moenkopi formation of Triassic age. Geol. Soc. Am. Mem. **61**, 133 p.

—, and G. W. WEIR, 1953: Terminology of stratification and cross-stratification. Bull. Geol. Soc. Am. **64**, 381—390.

*MIGLIORINI, C. I., 1950: Data a conferma della risedimentazione della arenarie macigno. Atti Soc. toscana sci. nat. Pisa Mem. **57**, Ser. A, 3—15.

MOBERLY jr., RALPH, 1960: Morrison, Cloverly and Sykes Mountain formations, northern Bighorn Basin, Wyoming and Montana. Bull. Geol. Soc. Am. **71**, 1137—1176.

MOORE, D. G., and P. C. SCRUTON, 1957: Minor internal structures of some Recent unconsolidated sediments. Bull. Am. Assoc. Petrol. Geologists **41**, 2723—2751.

MORGAN, J. P., 1961: Genesis and paleontology of Mississippi mud lumps, Part 1. Louisiana Geol. Survey, Bull. **35**, 116 p.

MÜLLER, G., 1961: Das Sand-Silt-Ton-Verhältnis in rezenten marinen Sedimenten. Neues Jahrb. Mineral., Monatsh. **7**, 148—163.

NATLAND, M. L., and PH. H. KUENEN, 1951: Sedimentary history of the Ventura Basin, California, and the action of turbidity currents. Tulsa, Soc. Econ. Paleon. Min. Spec. Publ., No. 2, 76—107.

NIEHOFF, W., 1958: Die primär gerichteten Sedimentstrukturen, insbesondere die Schrägschichtung im Koblenzquarzit am Mittelrhein. Geol. Rundschau **47**, 252—321.

368

NORMAN, E., 1959: Petrography of some Pennsylvanian underclay carbonate beds in Illinois. Master's thesis, University of Illinois, 66p.

OHLSON, BIRGER, 1962: Observations on Recent lake balls and ancient *Corycium* inclusions in Finland. Bull. comm. géol. Finlande, No. 196, 377—390.

OSBORNE, G. D., 1948: A review of some aspects of the stratigraphy structure and physiography of the Sydney Basin. Proc. Linnean. Soc. N.S. Wales 73, vii.

OTTO, GEORGE H., 1938: The sedimentation unit and its use in field sampling. J. Geol. 46, 509—582.

PAVONI, N., 1959: Rollmarken von Fischwirbeln aus den oligozänen Fischschiefern von Engi-Matt (Kt. Glarus). Ecologae Geol. Helv. 52, 941—949.

PAYNE, T. G., 1942: Stratigraphical analysis and environmental reconstruction. Bull. Am. Assoc. Petrol. Geologists 26, 1697—1770.

PEABODY, F. E., 1947: Current crescents in the Triassic Moenkopi formation. J. Sediment. Petrol. 17, 73—76.

PEPPER, J. F., R. J. DE WITT and D. F. DESMAREST, 1954: Geology of the Bedford shale and Berea sandstone in the Appalachian Basin. U.S. Geol. Survey, Prof. Paper 259, 111 p.

PETTIJOHN, F. J.: Sedimentary rocks, 2nd ed. New York: Harper and Brothers, 718 p.

PHILLIPS, JOHN, 1836: Illustrations of the geology of Yorkshire, Part II, The Mountain Limestone District. London: John Murray, 253 p.

PLESSMANN, W., 1961: Strömungsmarken in klastischen Sedimenten und ihre geologische Auswertung. Geol. Jahrb. 78, 503—566.

POTTER, P. E., and H. D. GLASS, 1958: Petrology and sedimentation of the Pennsylvanian sediments in southern Illinois: A vertical profile. Illinois Geol. Survey, Rept. Invest. 204, 60 p.

—, and F. J. PETTIJOHN, 1963: Paleocurrents and Basin Analysis. Berlin-Göttingen-Heidelberg: Springer, 296 b.

PRENTICE, J. E., 1956: The interpretation of flow-markings and load-casts. Geol. Mag. 93, 393—401.

— 1960: Flow structure in sedimentary rocks. J. Geol. 68, 217—225.

QUIRKE, T. T., 1930: Spring pits: Sedimentation phenomena (?). J. Geol. 38, 88—91.

RABIEN, A., 1956: Zur Stratigraphie und Fazies des Ober-Devons in der Waldecker Hauptmulde. Abhandl. hess. Landesamtes Bodenforsch. 16, 83 p.

RADOMSKI, A., 1958: Charakterystyka sedymentologiczna fliszu podbalańskiego (The sedimentological character of the Podhale flysch). Acta Geol. Polon. 8, 335—409 [Polish; English summ.].

REICHE, PERRY, 1938: An analysis of cross-lamination; the Coconino sandstone. J. Geol. 46, 905—932.

REZAK, RICHARD, 1957: Stromatolites of the Belt series in Glacier National Park and vicinity, Montana. U.S. Geol. Survey, Prof. Paper 294D, 127—154.

RICH, J. L., 1950: Flow markings, groovings, and intrastratalcrumplings as criteria for recognition of slope deposits with illustrations from Silurian rocks of Wales. Bull. Am. Assoc. Petrol. 34, 717—741.

RICHTER, R., 1937: Zerlegung von „Buchstabendemonstrationszügen". Natur u. Volk 67, 625—626.

RIGBY, J. K., 1959: Possible eddy markings in the Shinarump conglomerate of northeastern Utah. J. Sediment. Petrol. 29, 283—284.

RÜCKLIN, H., 1938: Strömungsmarken im unteren Muschelkalk des Saarlandes. Senckenbergiana 20, 94—114.

SARDESON, F. W., 1914: Characteristics of a corrosion conglomerate. Bull. Geol. Soc. Am. 25, 269—276.

SCHIEFERDECKER, A. A. G., ed., 1959: Geological nomenclature. Royal Geological and Mining Society of the Netherlands, 523 pp.

SCHMIDT, WO., 1959: Das Setzen von Erläuterungs-Bindestrichen in wissenschaftlichen Arbeiten. Neues Jahrb. Geol. u. Paläont., Monatsh. p. 139—142.

SEILACHER, A., 1953: Studien zur Palichnologie. I: Über die Methoden der Palichnologie. Neues Jahrb. Geol. u. Paläont., Abhandl. 96, 421—452.

— 1960: Strömungsanzeichen im Hunsrückschiefer. Notizbl. hess. Landesamtes Bodenforsch. 88, 88—106.

SHACKLETON. J. S., 1962: Cross-strata of the Rough Rock (Millstone Grit Series) in the Pennines. Liverpool Manchester Geol. J. 3, 109—120.

SHAW, E. W., 1914: The mud lumps at the mouths of the Mississippi. U.S. Geol. Survey, Prof. Paper 85, 11—28.

SHROCK, R. R., 1948: Sequence in layered rocks. New York: McGraw-Hill Book Co., 507 p.

SIMONEN, A., and O. KUOVO, 1951: Archean varved schists north of Tampere in Finland. Bull. comm. géol. Finlande, No. 154, 93—114.

SMITH, B., 1961: Ball or pillow-form structures in sandstones. Geol. Mag. 3, 146—156.

SNAVELY, P. D., and H. C. WAGNER, 1963: Tertiary geologic history of western Oregon and Washington. Rept. Invest. No. 22 Div. Mines and Geol. Washington.

SORBY, H. C., 1856: On the physical geography of the Old Red Sandstone sea of the central district of Scotland. Edinburgh New Philos. J., n.s. 3, 112—122.

— 1857: On the physical geography of the Tertiary estuary of the Isle of Wright. Edinburgh New Philos. J., n.s. 5, 275—298.

SPURR, J. E., 1894: False bedding in stratified drift deposits. Am. Geologist 13, 43—47.

STEWART jr., H. B., 1956: Contorted sediments in modern coastal lagoon explained by laboratory experiments. Bull. Am. Assoc. Petrol. Geologists **40**, 153—160.

STEWART, J. B., 1961: Origin of cross-strata in fluvial sandstone layers in the Chinle formation (Upper Triassic) on the Colorado Plateau *in* Geological Survey Research 1961. U.S. Geol. Survey, Prof. Paper **424**B, B127—B129.

STOKES, W. L., 1947: Primary lineation in fluvial sandstones: a criterion of current direction. J. Geol. **55**, 52—54.

— 1953: Primary sedimentary trend indicators as applied to ore finding in the Carrizo Mountains, Arizona and New Mexico. U.S. Atomy Energy Comm. RME-3043, 48 p.

STRAATEN, L. M. J. U. VAN, 1950: Giant ripples in tidal channels. Tijdschr. van Het Kan. Ned. Aardrijks. Genootschop **67**, 76—81.

— 1951: Longitudinal ripple marks in mud and sand. J. Sediment. Petrol. **21**, 47—54.

— 1953a: Megaripples in the Dutch Wadden sea and in the basin of Arcachon (France). Geol. en Mijnbouw, n.s. **15**, 1—11.

— 1953b: Rhythmic patterns in Dutch North Sea beaches. Geol. en Mijnbouw, n.s. **15**, 31—43.

STRAKHOV, N. M., ed., 1958: Méthodes d'étude des roches sédimentaires. Ann. Service d'Inform. Géol. **2**, No. 35, 353 p. (translated by J. P. DE SAINT-AUBIN, A. VATAN, *et al.*, from 1957 Russian work).

SULLWOLD jr., H. H., 1959: Nomenclature of load deformation in turbidites. Bull. Geol. Soc. Am. **70**, 1247—1248.

— 1960: Load cast terminology and origin of convolute bedding: Further comments. Bull. Geol. Soc. Am. **71**, 635—636.

TANNER, W. F., 1958: An occurrence of flat-topped ripple marks. J. Sediment. Petrol. **28**, 95—96.

— 1960: Shallow ripple mark varieties. J. Sediment. Petrol. **30**, 481—485.

TERMIER, H., and G. TERMIER, 1960: Erosion et sédimentation. Paris: Masson & Cie., 412 p.

*TRUSHKOVA, N. N., and A. A. KUKHARENKO, 1961: Atlas of placer minerals. M. Gosgeoltekhizdat, 437 p. [Russian].

TWENHOFEL, W. H., 1921: Impressions made by bubbles, rain drops and other agencies. Bull. Geol. Soc. Am. **32**, 359—371.

— 1939: Principles of sedimentation. New York: McGraw-Hill Book Co., 610 p.

UDDEN, J. A., 1918: Fossil ice crystals. Univ. Texas Bull., No. 1821, 8 p.

VASSOEVICH, N. B., 1953: O nekotorykh flishe-vykh teksturakh (znakakh) (On some flysch textures). Trudy Lvovs. Geol. Obsh., Univ. Ivan Franko, Geol. ser. No. 3, 17—85 [Russian].

WALTON, E. K., 1956: Limitations of graded bedding and alternate criteria of upward sequence in the rocks of the Southern Upland. Trans. Edinburgh Geol. Soc. **16**, 267—271.

WEGNER, TH., 1932: Unter Gezeiteneinwirkung entstandene Wellenfurchen. Zentr. Mineral. Geol. u. Paläont., Abt. B, 31—34.

WEISS, M. P., 1954: Corrosion zones in carbonate rocks. Ohio J. Sci. **54**, 289—293.

WHITE, I. C., 1881: The geology of Susquehanna County and Wayne County. Second Geological Survey of Pennsylvania, Rept. Prog. 6s, 243 p.

WILLIAMS, B. J., and J. E. PRENTICE, 1957: Slump structures in the Ludlovian rocks of North Herefordshire. Proc. Geol. Assoc. **68**, 268—293.

WILLIAMS, E., 1960: Intrastratal flow and convolute folding. Geol. Mag. **97**, 208—214.

WOOD, ALAN, and A. J. SMITH, 1959: The sedimentation and sedimentary history of the Aberystwyth grits (Upper Llandoverian). Quart. J. Geol. Soc. London **114**, 163—189.

*ZHEMCHUZHNIKOV, YU. A., 1940: Opyt morfologicheskoy Klassifikatsii sloistosti osadochnykh porod (An attempt at a morphological classification of bedding in sedimentary rocks). Gornyatskaya Pravda, Nauchno-tekhnicheskiy listok.